Materials Chemistry

Materials Chemistry

Edited by
Sean Fraser

Willford Press
www.willfordpress.com

Published by Willford Press,
118-35 Queens Blvd., Suite 400,
Forest Hills, NY 11375, USA

ISBN: 978-1-68285-489-1

Cataloging-in-Publication Data

Materials chemistry / edited by Sean Fraser.
p. cm.
Includes bibliographical references and index.
ISBN 978-1-68285-489-1
1. Materials science. 2. Materials--Analysis. 3. Chemistry, Technical.
4. Chemistry, Physical and theoretical. I. Fraser, Sean.
TA403.6 .M38 2018
620.11--dc23

For information on all Willford Press publications
visit our website at www.willfordpress.com

WILLFORD PRESS

Printed in China.

Contents

Preface

The process of designing and discovering new materials is known as materials chemistry. It is a sub-field of materials sciences and is very important in the present age. The important aspects covered under this field are structure, which includes nanostructure, atomic structure, macro structure, crystallography, bonding, etc. Some of the main materials studied under this field are polymers, ceramic materials, composites, metal alloys, etc. This book is compiled in such a manner, that it will provide in-depth knowledge about the theory and practice of materials chemistry. The topics covered herein deal with the core subjects of the area. This textbook is a complete source of knowledge on the present status of this important field.

A foreword of all chapters of the book is provided below:

Chapter 1- Solid-state chemistry is devoted to the study of solid phase materials in relation to their structure, properties, and synthesis. It is also focused on studying molecular solids. This is an introductory chapter which will introduce briefly all the significant aspects of solid-state chemistry; **Chapter 2 -** Graphene is an allotrope of carbon that forms the fundamental structure of all other allotropes. It is considered to be 200 times stronger than steel and is also a good conductor of heat and electricity. This section will provide an integrated understanding of graphene and its applications; **Chapter 3 -** Nanotechnology is the science that deals with modification of elements and particles at their molecular or atomic state. It also includes manufacturing products that are very small in size. One nanometre can be defined as one billionth of a metre. This section elucidates the crucial theories and principles of nanotechnology and nanoscience; **Chapter 4 -** The allotrope of carbon having a cylindrical structure is called a carbon nanotube. Their properties are of extreme significance in nanotechnology. They can be normally classified into single-walled and multi-walled nanotubes. Carbon nanotubes are best understood in confluence with the major topics listed in the following chapter; **Chapter 5 -** Quantum dots are semiconductors that are measurable in nanoparticles. They emit light of a particular frequency when brought in contact with electricity or magnetic force. These frequencies can be manipulated to change the dot's properties. The topics discussed in the chapter are of great importance to broaden the existing knowledge on quantum dots; **Chapter 6 -** A light emitting diode (LED) is a type of semiconductor that is used to produce light. A widely used type of light emitting diode is organic light emitting diodes (OLEDs). The light-creating part of an OLED is manufactured from a thin layer of organic compound. The aspects elucidated in this section are of vital importance, and provide a better understanding of organic light emitting diodes.

At the end, I would like to thank all the people associated with this book devoting their precious time and providing their valuable contributions to this book. I would also like to express my gratitude to my fellow colleagues who encouraged me throughout the process.

Editor

1

Basics of Solid State Chemistry

Solid-state chemistry is devoted to the study of solid phase materials in relation to their structure, properties, and synthesis. It is also focused on studying molecular solids. This is an introductory chapter which will introduce briefly all the significant aspects of solid-state chemistry.

Solid State Chemistry

Solid-state chemistry, also sometimes referred to as materials chemistry, is the study of the synthesis, structure, and properties of solid phase materials, particularly, but not necessarily exclusively of, non-molecular solids. It therefore has a strong overlap with solid-state physics, mineralogy, crystallography, ceramics, metallurgy, thermodynamics, materials science and electronics with a focus on the synthesis of novel materials and their characterization.

History

Because of its direct relevance to products of commerce, solid state inorganic chemistry has been strongly driven by technology. Progress in the field has often been fueled by the demands of industry, well ahead of purely academic curiosity. Applications discovered in the 20th century include zeolite and platinum-based catalysts for petroleum processing in the 1950s, high-purity silicon as a core component of microelectronic devices in the 1960s, and "high temperature" superconductivity in the 1980s. The invention of X-ray crystallography in the early 1900s by William Lawrence Bragg enabled further innovation. Our understanding of how reactions proceed at the atomic level in the solid state was advanced considerably by Carl Wagner's work on oxidation rate theory, counter diffusion of ions, and defect chemistry. Because of this, he has sometimes been referred to as the *father of solid state chemistry.*

Synthetic Methods

Given the diversity of solid state compounds, an equally diverse array of methods are used for their preparation. For organic materials, such as charge transfer salts, the methods operate near room temperature and are often similar to the techniques of organic synthesis. Redox reactions are sometimes conducted by electrocrystallisation, as illustrated by the preparation of the Bechgaard salts from tetrathiafulvalene.

Oven Techniques

For thermally robust materials, high temperature methods are often employed. For example, bulk solids are prepared using tube furnaces, which allow reactions to be conducted up to ca. 1100° C.

Special equipment e.g. ovens consisting of a tantalum tube through which an electric current is passed can be used for even higher temperatures up to 2000° C. Such high temperatures are at times required to induce diffusion of the reactants, but this depends strongly on the system stud-ied. Some solid state reactions already proceed at temperatures as low as 100°C.

Melt Methods

One method often employed is to melt the reactants together and then later anneal the solidified melt. If volatile reactants are involved the reactants are often put in an ampoule that is evacuated -often while keeping the reactant mixture cold e.g. by keeping the bottom of the ampoule in liquid nitrogen- and then sealed. The sealed ampoule is then put in an oven and given a certain heat treatment.

Solution Methods

It is possible to use solvents to prepare solids by precipitation or by evaporation. At times the solvent is used hydrothermally, i.e. under pressure at temperatures higher than the normal boiling point. A variation on this theme is the use of flux methods, where a salt of relatively low melting point is added to the mixture to act as a high temperature solvent in which the desired reaction can take place.

Gas Reactions

Many solids react vigorously with reactive gas species like chlorine, iodine, oxygen etc. Others form adducts with other gases, e.g. CO or ethylene. Such reactions are often carried out in a tube that is open ended on both sides and through which the gas is passed. A variation of this is to let the reaction take place inside a measuring device such as a TGA. In that case stoichiometric information can be obtained during the reaction, which helps identify the products.

A special case of a gas reaction is a chemical transport reaction. These are often carried out in a sealed ampoule to which a small amount of a transport agent, e.g. iodine is added. The ampoule is then placed in a zone oven. This is essentially two tube ovens attached to each other which allows a temperature gradient to be imposed. Such a method can be used to obtain the product in the form of single crystals suitable for structure determination by X-ray diffraction.

Chemical vapour deposition is a high temperature method that is widely employed for the preparation of coatings and semiconductors from molecular precursors

Air and Moisture Sensitive Materials

Many solids are hygroscopic and/or oxygen sensitive. Many halides e.g. are very 'thirsty' and can only be studied in their anhydrous form if they are handled in a glove box filled with dry (and/or oxygen-free) gas, usually nitrogen.

Characterization

New Phases, Phase Diagrams, Structures

The synthetic methodology and the characterization of the product often go hand in hand in the sense that not one but a series of reaction mixtures are prepared and subjected to heat treatment. The stoichiometry is typically *varied* in a systematic way to find which stoichiometries will lead to new solid compounds or to solid solutions between known ones. A prime method to characterize the reaction products is powder diffraction, because many solid state reactions will produce polycristalline ingots or powders. Powder diffraction will facilitate the identification of known phases in the mixture. If a pattern is found that is not known in the diffraction data libraries an attempt can be made to index the pattern, i.e. to identify the symmetry and the size of the unit cell. (If the product is not crystalline the characterization is typically much more difficult.)

Once the unit cell of a new phase is known, the next step is to establish the stoichiometry of the phase. This can be done in a number of ways. Sometimes the composition of the original mixture will give a clue, if one finds only one product -a single powder pattern- or if one was trying to make a phase of a certain composition by analogy to known materials but this is rare. Often considerable effort in refining the synthetic methodology is required to obtain a pure sample of the new material. If it is possible to separate the product from the rest of the reaction mixture elemental analysis can be used. Another way involves SEM and the generation of characteristic X-rays in the electron beam. The easiest way to solve the structure is by using single crystal X-ray diffraction.

The latter often requires *revisiting* and refining the preparative procedures and that is linked to the question which phases are stable at what composition and what stoichiometry. In other words, what does the phase diagram looks like. An important tool in establishing this is thermal analysis techniques like DSC or DTA and increasingly also, thanks to the advent of synchrotrons temperature-dependent powder diffraction. Increased knowledge of the phase relations often leads to further refinement in synthetic procedures in an iterative way. New phases are thus characterized by their melting points and their stoichiometric domains. The latter is important for the many solids that are non-stoichiometric compounds. The cell parameters obtained from XRD are particularly helpful to characterize the homogeneity ranges of the latter.

Further Characterization

In many -but certainly not all- cases new solid compounds are further characterized by a variety of techniques that straddle the fine line that (hardly) separates solid-state chemistry from solid-state physics.

Optical Properties

For non-metallic materials it is often possible to obtain UV/VIS spectra. In the case of semiconductors that will give an idea of the band gap.

Crystal

A crystal or crystalline solid is a solid material whose constituents (such as atoms, molecules, or ions) are arranged in a highly ordered microscopic structure, forming a crystal lattice that extends in all directions. In addition, macroscopic single crystals are usually identifiable by their geometrical shape, consisting of flat faces with specific, characteristic orientations. The scientific study of crystals and crystal formation is known as crystallography. The process of crystal formation via mechanisms of crystal growth is called crystallization or solidification.

A crystal of amethyst quartz

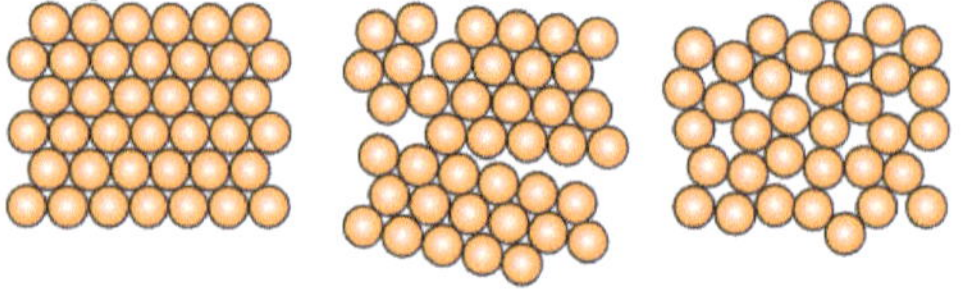

Microscopically, a single crystal has atoms in a near-perfect periodic arrangement; a polycrystal is composed of many microscopic crystals (called "crystallites" or "grains"); and an amorphous solid (such as glass) has no periodic arrangement even microscopically.

The word *crystal* derives from the Ancient Greek word (*krustallos*), meaning both "ice" and "rock crystal", from (*kruos*), "icy cold, frost".

Examples of large crystals include snowflakes, diamonds, and table salt. Most inorganic solids are not crystals but polycrystals, i.e. many microscopic crystals fused together into a single solid. Examples of polycrystals include most metals, rocks, ceramics, and ice. A third category of solids is amorphous solids, where the atoms have no periodic structure whatsoever. Examples of amorphous solids include glass, wax, and many plastics.

Crystals are often used in pseudoscientific practices such as crystal therapy, and, along with gemstones, are sometimes associated with spellwork in Wiccan beliefs and related religious movements.

Crystal Structure (Microscopic)

Halite (table salt, NaCl): Microscopic and croscopic

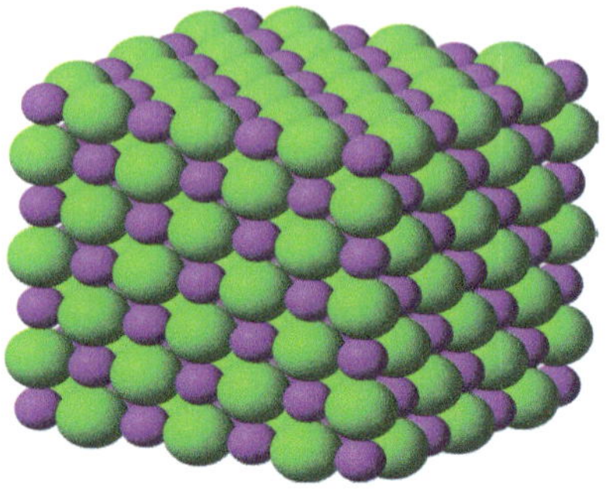

Microscopic structure of a halite crystal. (Purple is sodium ion, green is chlorine ion.) There is cubic symmetry in the atoms' arrangement.

Macroscopic (~16cm) halite crystal. The right-angles between crystal faces are due to the cubic symmetry of the atoms' arrangement.

The scientific definition of a "crystal" is based on the microscopic arrangement of atoms inside it, called the crystal structure. A crystal is a solid where the atoms form a periodic arrangement.

Not all solids are crystals. For example, when liquid water starts freezing, the phase change begins with small ice crystals that grow until they fuse, forming a *polycrystalline* structure. In the final block of ice, each of the small crystals (called "crystallites" or "grains") is a true crystal with a periodic arrangement of atoms, but the whole polycrystal does *not* have a periodic arrangement of atoms, because the periodic pattern is broken at the grain boundaries. Most macroscopic inorganic solids are polycrystalline, including almost all metals, ceramics, ice, rocks, etc. Solids that are neither crystalline nor polycrystalline, such as glass, are called *amorphous solids*, also called glassy, vitreous, or noncrystalline. These have no periodic order, even microscopically. There are distinct differences between crystalline solids and amorphous solids: most notably, the process of forming a glass does not release the latent heat of fusion, but forming a crystal does.

A crystal structure (an arrangement of atoms in a crystal) is characterized by its *unit cell*, a small imaginary box containing one or more atoms in a specific spatial arrangement. The unit cells are stacked in three-dimensional space to form the crystal.

The symmetry of a crystal is constrained by the requirement that the unit cells stack perfectly with no gaps. There are 219 possible crystal symmetries, called crystallographic space groups. These are grouped into 7 crystal systems, such as cubic crystal system (where the crystals may form cubes or rectangular boxes) or hexagonal crystal system (where the crystals may form hexagons, such as ordinary water ice).

Crystal Faces and Shapes

Crystals are commonly recognized by their shape, consisting of flat faces with sharp angles. These shape characteristics are not *necessary* for a crystal—a crystal is scientifically defined by its microscopic atomic arrangement, not its macroscopic shape—but the characteristic macroscopic shape is often present and easy to see.

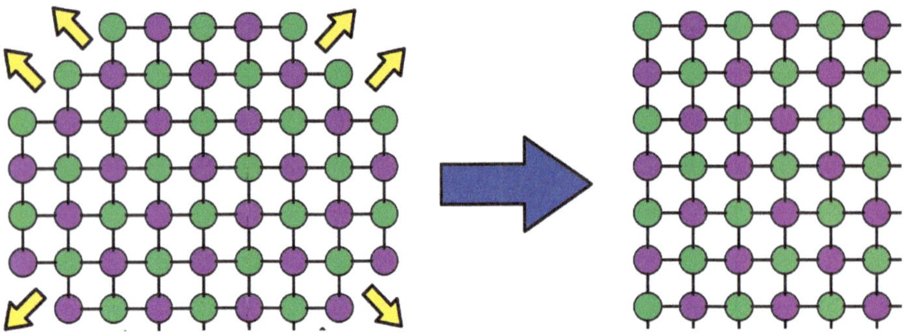

As a halite crystal is growing, new atoms can very easily attach to the parts of the surface with rough atomic-scale structure and many dangling bonds. Therefore, these parts of the crystal grow out very quickly (yellow arrows). Eventually, the whole surface consists of smooth, stable faces, where new atoms cannot as easily attach themselves.

Euhedral crystals are those with obvious, well-formed flat faces. Anhedral crystals do not, usually because the crystal is one grain in a polycrystalline solid.

The flat faces (also called facets) of a euhedral crystal are oriented in a specific way relative to the underlying atomic arrangement of the crystal: they are planes of relatively low Miller index. This occurs because some surface orientations are more stable than others (lower surface energy). As a crystal grows, new atoms attach easily to the rougher and less stable parts of the surface, but less easily to the flat, stable surfaces. Therefore, the flat surfaces tend to grow larger and smoother, until the whole crystal surface consists of these plane surfaces.

One of the oldest techniques in the science of crystallography consists of measuring the three-dimensional orientations of the faces of a crystal, and using them to infer the underlying crystal symmetry.

A crystal's habit is its visible external shape. This is determined by the crystal structure (which restricts the possible facet orientations), the specific crystal chemistry and bonding (which may favor some facet types over others), and the conditions under which the crystal formed.

Occurrence in Nature

Ice crystals

Fossil shell with calcite crystals

Rocks

By volume and weight, the largest concentrations of crystals in the Earth are part of its solid bedrock. Crystals found in rocks typically range in size from a fraction of a millimetre to several centimetres across, although exceptionally large crystals are occasionally found. As of 1999, the world's largest known naturally occurring crystal is a crystal of beryl from Malakialina, Madagascar, 18 m (59 ft) long and 3.5 m (11 ft) in diameter, and weighing 380,000 kg (840,000 lb).

Some crystals have formed by magmatic and metamorphic processes, giving origin to large masses of crystalline rock. The vast majority of igneous rocks are formed from molten magma and the degree of crystallization depends primarily on the conditions under which they solidified. Such rocks as granite, which have cooled very slowly and under great pressures, have completely crystallized; but many kinds of lava were poured out at the surface and cooled very rapidly, and in this latter group a small amount of amorphous or glassy matter is common. Other crystalline rocks, the metamorphic rocks such as marbles, mica-schists and quartzites, are recrystallized. This means that they were at first fragmental rocks like limestone, shale and sandstone and have never been in a molten condition nor entirely in solution, but the high temperature and pressure conditions of metamorphism have acted on them by erasing their original structures and inducing recrystallization in the solid state.

Other rock crystals have formed out of precipitation from fluids, commonly water, to form druses or quartz veins. The evaporites such as halite, gypsum and some limestones have been deposited from aqueous solution, mostly owing to evaporation in arid climates.

Ice

Water-based ice in the form of snow, sea ice and glaciers is a very common manifestation of crystalline or polycrystalline matter on Earth. A single snowflake is typically a single crystal, while an ice cube is a polycrystal.

Organigenic Crystals

Many living organisms are able to produce crystals, for example calcite and aragonite in the case of most molluscs or hydroxylapatite in the case of vertebrates.

Polymorphism and Allotropy

The same group of atoms can often solidify in many different ways. Polymorphism is the ability of a solid to exist in more than one crystal form. For example, water ice is ordinarily found in the hexagonal form Ice I_h, but can also exist as the cubic Ice I_c, the rhombohedral ice II, and many other forms. The different polymorphs are usually called different *phases*.

In addition, the same atoms may be able to form noncrystalline phases. For example, water can also form amorphous ice, while SiO_2 can form both fused silica (an amorphous glass) and quartz (a crystal). Likewise, if a substance can form crystals, it can also form polycrystals.

For pure chemical elements, polymorphism is known as allotropy. For example, diamond and graphite are two crystalline forms of carbon, while amorphous carbon is a noncrystalline form. Polymorphs, despite having the same atoms, may have wildly different properties. For example, diamond is among the hardest substances known, while graphite is so soft that it is used as a lubricant.

Polyamorphism is a similar phenomenon where the same atoms can exist in more than one amorphous solid form.

Crystallization

Vertical cooling crystallizer in a beet sugar factory.

Crystallization is the process of forming a crystalline structure from a fluid or from materials dissolved in a fluid.

Crystallization is a complex and extensively-studied field, because depending on the conditions, a single fluid can solidify into many different possible forms. It can form a single crystal, perhaps with various possible phases, stoichiometries, impurities, defects, and habits. Or, it can form a polycrystal, with various possibilities for the size, arrangement, orientation, and phase of its grains. The final form of the solid is determined by the conditions under which the fluid is being solidified, such as the chemistry of the fluid, the ambient pressure, the temperature, and the speed with which all these parameters are changing.

Specific industrial techniques to produce large single crystals (called *boules*) include the Czochralski process and the Bridgman technique. Other less exotic methods of crystallization may be used, depending on the physical properties of the substance, including hydrothermal synthesis, sublimation, or simply solvent-based crystallization.

Large single crystals can be created by geological processes. For example, selenite crystals in excess of 10 meters are found in the Cave of the Crystals in Naica, Mexico.

Crystals can also be formed by biological processes. Conversely, some organisms have special techniques to *prevent* crystallization from occurring, such as antifreeze proteins.

Defects, Impurities, and Twinning

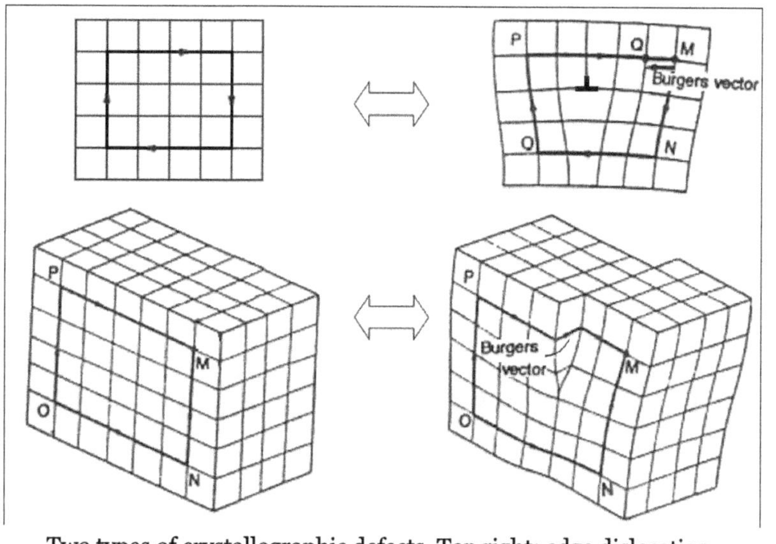

Two types of crystallographic defects. Top right: edge dislocation.
Bottom right: screw dislocation.

An *ideal* crystal has every atom in a perfect, exactly repeating pattern. However, in reality, most crystalline materials have a variety of crystallographic defects, places where the crystal's pattern is interrupted. The types and structures of these defects may have a profound effect on the properties of the materials.

A few examples of crystallographic defects include vacancy defects (an empty space where an atom should fit), interstitial defects (an extra atom squeezed in where it does not fit), and dislocations. Dislocations are especially important in materials science, because they help determine the mechanical strength of materials.

Another common type of crystallographic defect is an impurity, meaning that the "wrong" type of atom is present in a crystal. For example, a perfect crystal of diamond would only contain carbon atoms, but a real crystal might perhaps contain a few boron atoms as well. These boron impurities change the diamond's color to slightly blue. Likewise, the only difference between ruby and sapphire is the type of impurities present in a corundum crystal.

Twinned pyrite crystal group.

In semiconductors, a special type of impurity, called a dopant, drastically changes the crystal's electrical properties. Semiconductor devices, such as transistors, are made possible largely by putting different semiconductor dopants into different places, in specific patterns.

Twinning is a phenomenon somewhere between a crystallographic defect and a grain boundary. Like a grain boundary, a twin boundary has different crystal orientations on its two sides. But unlike a grain boundary, the orientations are not random, but related in a specific, mirror-image way.

Mosaicity is a spread of crystal plane orientations. A mosaic crystal is supposed to consist of smaller crystalline units that are somewhat misaligned with respect to each other.

Chemical Bonds

In general, solids can be held together by various types of chemical bonds, such as metallic bonds, ionic bonds, covalent bonds, van der Waals bonds, and others. None of these are necessarily crystalline or non-crystalline. However, there are some general trends as follows.

Metals are almost always polycrystalline, though there are exceptions like amorphous metal and single-crystal metals. The latter are grown synthetically. (A microscopically-small piece of metal may naturally form into a single crystal, but larger pieces generally do not.) Ionic compound materials are usually crystalline or polycrystalline. In practice, large salt crystals can be created by solidification of a molten fluid, or by crystallization out of a solution. Covalently bonded solids (sometimes called covalent network solids) are also very common, notable examples being diamond and quartz. Weak van der Waals forces also help hold together certain crystals, such as crystalline molecular solids, as well as the interlayer bonding in graphite. Polymer materials generally will form crystalline regions, but the lengths of the molecules usually prevent complete crystallization—and sometimes polymers are completely amorphous.

Quasicrystals

The material holmium–magnesium–zinc (Ho–Mg–Zn) forms quasicrystals, which can take on the macroscopic shape of a dodecahedron. (Only a quasicrystal, not a normal crystal, can take this shape.) The edges are 2 mm long.

A quasicrystal consists of arrays of atoms that are ordered but not strictly periodic. They have many attributes in common with ordinary crystals, such as displaying a discrete pattern in x-ray diffraction, and the ability to form shapes with smooth, flat faces.

Quasicrystals are most famous for their ability to show five-fold symmetry, which is impossible for an ordinary periodic crystal.

The International Union of Crystallography has redefined the term "crystal" to include both ordinary periodic crystals and quasicrystals ("any solid having an essentially discrete diffraction diagram").

Quasicrystals, first discovered in 1982, are quite rare in practice. Only about 100 solids are known to form quasicrystals, compared to about 400,000 periodic crystals known in 2004. The 2011 Nobel Prize in Chemistry was awarded to Dan Shechtman for the discovery of quasicrystals.

Special Properties from Anisotropy

Crystals can have certain special electrical, optical, and mechanical properties that glass and polycrystals normally cannot. These properties are related to the anisotropy of the crystal, i.e. the lack of rotational symmetry in its atomic arrangement. One such property is the piezoelectric effect, where a voltage across the crystal can shrink or stretch it. Another is birefringence, where a double image appears when looking through a crystal. Moreover, various properties of a crystal, including electrical conductivity, electrical permittivity, and Young's modulus, may be different in different directions in a crystal. For example, graphite crystals consist of a stack of sheets, and although each individual sheet is mechanically very strong, the sheets are rather loosely bound to each other. Therefore, the mechanical strength of the material is quite different depending on the direction of stress.

Not all crystals have all of these properties. Conversely, these properties are not quite exclusive to crystals. They can appear in glasses or polycrystals that have been made anisotropic by working or stress—for example, stress-induced birefringence.

Crystallography

Crystallography is the science of measuring the crystal structure (in other words, the atomic arrangement) of a crystal. One widely used crystallography technique is X-ray diffraction. Large numbers of known crystal structures are stored in crystallographic databases.

Insulin crystals grown in earth orbit.

Hoar frost: A type of ice crystal (picture taken from a distance of about 5 cm).

Gallium, a metal that easily forms large crystals.

Crystal Structure

In crystallography, crystal structure is a description of the ordered arrangement of atoms, ions or molecules in a crystalline material. Ordered structures occur from the intrinsic nature of the constituent particles to form symmetric patterns that repeat along the principal directions of three-dimensional space in matter.

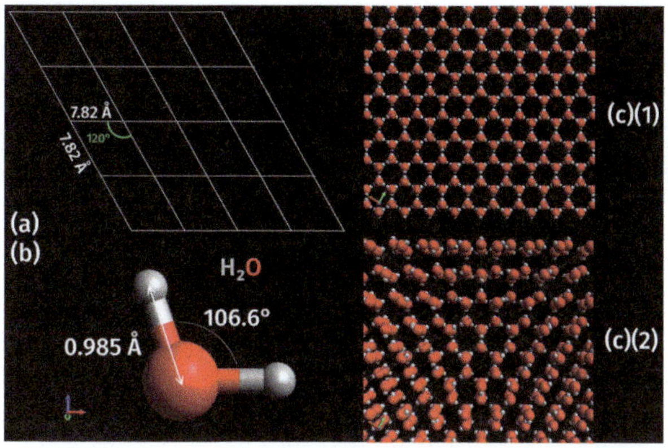

The (3-D) crystal structure of H_2O ice Ih (c) consists of bases of H_2O ice molecules (b) located on lattice points within the (2-D) hexagonal space lattice (a). The values for the H–O–H angle and O–H distance have come from *Physics of Ice* with uncertainties of ±1.5° and ±0.005 Å, respectively. The white box in (c) is the unit cell defined by Bernal and Fowler

The smallest group of particles in the material that constitutes the repeating pattern is the unit cell of the structure. The unit cell completely defines the symmetry and structure of the entire crystal lattice, which is built up by repetitive translation of the unit cell along its principal axes. The repeating patterns are said to be located at the points of the Bravais lattice.

The lengths of the principal axes, or edges, of the unit cell and the angles between them are the lattice constants, also called *lattice parameters*. The symmetry properties of the crystal are described by the concept of space groups. All possible symmetric arrangements of particles in three-dimensional space may be described by the 230 space groups.

The crystal structure and symmetry play a critical role in determining many physical properties, such as cleavage, electronic band structure, and optical transparency.

Unit Cell

The crystal structure of a material (the arrangement of atoms within a given type of crystal) can be described in terms of its unit cell. The unit cell is a box containing one or more atoms arranged in three dimensions. The unit cells stacked in three-dimensional space describe the bulk arrangement of atoms of the crystal. The unit cell is represented in terms of its lattice parameters, which are the lengths of the cell edges (a, b and c) and the angles between them (alpha, beta and gamma), while the positions of the atoms inside the unit cell are described by the set of atomic positions (x_i, y_i, z_i) measured from a reference lattice point. Commonly, atomic positions are represented in terms of fractional coordinates, relative to the unit cell lengths.

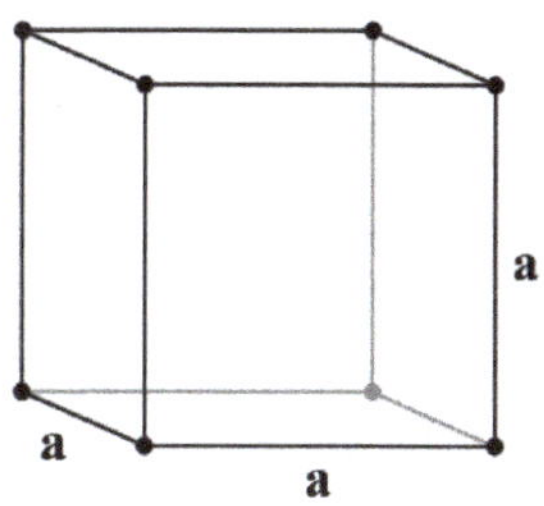

Simple cubic (P)

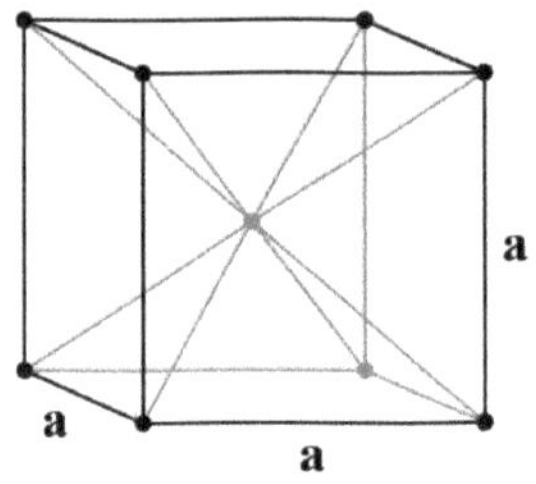

Body-centered cubic (I)

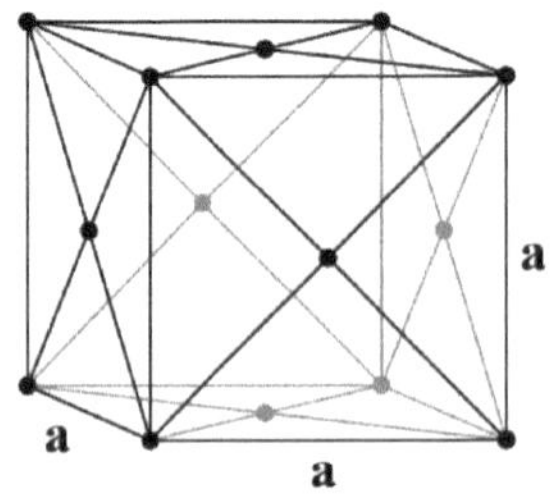

Face-centered cubic (F)

The atom positions within the unit cell can be calculated through application of symmetry operations to the asymmetric unit. The asymmetric unit refers to the smallest possible occupation of space within the unit cell. This does not, however imply that the entirety of the asymmetric unit must lie within the boundaries of the unit cell. Symmetric transformations of atom positions are calculated from the space group of the crystal structure, and this is usually a black box operation performed by computer programs. However, manual calculation of the atomic positions within the unit cell can be performed from the asymmetric unit, through the application of the symmetry operators described within the *International Tables for Crystallography: Volume A.*

Miller Indices

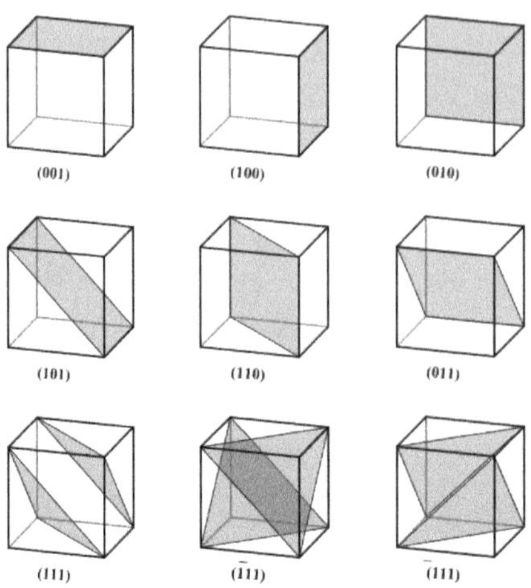

Planes with different Miller indices in cubic crystals

Vectors and planes in a crystal lattice are described by the three-value Miller index notation. It uses the indices ℓ, m, and n as directional parameters, which are separated by 90°, and are thus orthogonal.

By definition, the syntax (ℓ*mn*) denotes a plane that intercepts the three points a_1/ℓ, a_2/m, and a_3/n, or some multiple thereof. That is, the Miller indices are proportional to the inverses of the intercepts of the plane with the unit cell (in the basis of the lattice vectors). If one or more of the indices is zero, it means that the planes do not intersect that axis (i.e., the intercept is "at infinity"). A plane containing a coordinate axis is translated so that it no longer contains that axis before its Miller indices are determined. The Miller indices for a plane are integers with no common factors. Negative indices are indicated with horizontal bars, as in $(1\bar{2}3)$. In an orthogonal coordinate system for a cubic cell, the Miller indices of a plane are the Cartesian components of a vector normal to the plane.

Considering only (ℓ*mn*) planes intersecting one or more lattice points (the *lattice planes*), the distance *d* between adjacent lattice planes is related to the (shortest) reciprocal lattice vector orthogonal to the planes by the formula

$$d = \frac{2\pi}{|g_{\ell mn}|}$$

Planes and Directions

The crystallographic directions are geometric lines linking nodes (atoms, ions or molecules) of a crystal. Likewise, the crystallographic planes are geometric *planes* linking nodes. Some directions and planes have a higher density of nodes. These high density planes have an influence on the behavior of the crystal as follows:

- Optical properties: Refractive index is directly related to density (or periodic density fluctuations).
- Adsorption and reactivity: Physical adsorption and chemical reactions occur at or near surface atoms or molecules. These phenomena are thus sensitive to the density of nodes.
- Surface tension: The condensation of a material means that the atoms, ions or molecules are more stable if they are surrounded by other similar species. The surface tension of an interface thus varies according to the density on the surface.

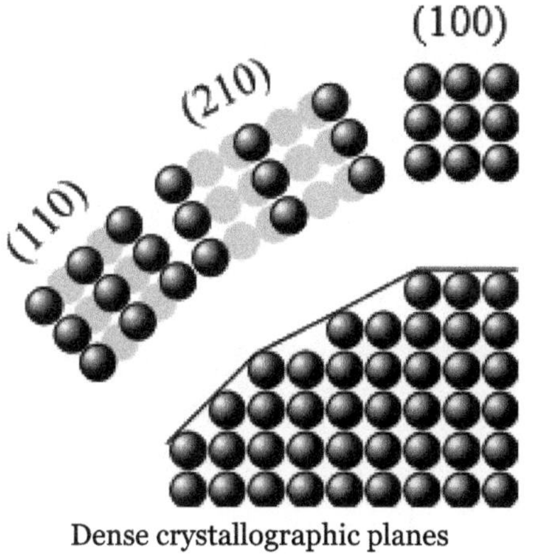

Dense crystallographic planes

- Microstructural defects: Pores and crystallites tend to have straight grain boundaries following higher density planes.

- Cleavage: This typically occurs preferentially parallel to higher density planes.
- Plastic deformation: Dislocation glide occurs preferentially parallel to higher density planes. The perturbation carried by the dislocation (Burgers vector) is along a dense direction. The shift of one node in a more dense direction requires a lesser distortion of the crystal lattice.

Some directions and planes are defined by symmetry of the crystal system. In monoclinic, rhombohedral, tetragonal, and trigonal/hexagonal systems there is one unique axis (sometimes called the principal axis) which has higher rotational symmetry than the other two axes. The basal plane is the plane perpendicular to the principal axis in these crystal systems. For triclinic, orthorhombic, and cubic crystal systems the axis designation is arbitrary and there is no principal axis.

Cubic Structures

For the special case of simple cubic crystals, the lattice vectors are orthogonal and of equal length (usually denoted *a*); similarly for the reciprocal lattice. So, in this common case, the Miller indices (ℓmn) and [ℓmn] both simply denote normals/directions in Cartesian coordinates. For cubic crystals with lattice constant *a*, the spacing *d* between adjacent (ℓmn) lattice planes is (from above):

$$d_{\ell mn} = \frac{a}{\sqrt{\ell^2 + m^2 + n^2}}$$

Because of the symmetry of cubic crystals, it is possible to change the place and sign of the integers and have equivalent directions and planes:

- Coordinates in *angle brackets* such as ⟨100⟩ denote a *family* of directions that are equivalent due to symmetry operations, such as ,[100], [010], [001], or the negative of any of those directions.
- Coordinates in *curly brackets* or *braces* such as {100} denote a family of plane normals that are equivalent due to symmetry operations, much the way angle brackets denote a family of directions.

For face-centered cubic (fcc) and body-centered cubic (bcc) lattices, the primitive lattice vectors are not orthogonal. However, in these cases the Miller indices are conventionally defined relative to the lattice vectors of the cubic supercell and hence are again simply the Cartesian directions.

Interplanar Spacing

The spacing *d* between adjacent (*hkl*) lattice planes is given by:

- Cubic:

$$\frac{1}{d^2} = \frac{h^2 + k^2 + l^2}{a^2}$$

- Tetragonal:

$$\frac{1}{d^2} = \frac{h^2 + k^2}{a^2} + \frac{l^2}{c^2}$$

- Hexagonal:

$$\frac{1}{d^2}=\frac{4}{3}\left(\frac{h^2+hk+k^2}{a^2}\right)+\frac{l^2}{c^2}$$

- Rhombohedral:

$$\frac{1}{d^2}=\frac{(h^2+k^2+l^2)\sin^2\alpha+2(hk+kl+hl)(\cos^2\alpha-\cos\alpha)}{a^2(1-3\cos^2\alpha+2\cos^3\alpha)}$$

- Orthorhombic:

$$\frac{1}{d^2}=\frac{h^2}{a^2}+\frac{k^2}{b^2}+\frac{l^2}{c^2}$$

- Monoclinic:

$$\frac{1}{d^2}=\left(\frac{h^2}{a^2}+\frac{k^2\sin^2\beta}{b^2}+\frac{l^2}{c^2}-\frac{2hl\cos\beta}{ac}\right)\csc^2\beta$$

- Triclinic:

$$\frac{1}{d^2}=\frac{\frac{h^2}{a^2}\sin^2\alpha+\frac{k^2}{b^2}\sin^2\beta+\frac{l^2}{c^2}\sin^2\gamma+\frac{2kl}{bc}\cos\alpha+\frac{2hl}{ac}\cos\beta+\frac{2hk}{ab}\cos\gamma}{1-\cos^2\alpha-\cos^2\beta-\cos^2\gamma+2\cos\alpha\cos\beta\cos\gamma}$$

Classification by Symmetry

The defining property of a crystal is its inherent symmetry, by which we mean that under certain 'operations' the crystal remains unchanged. All crystals have translational symmetry in three directions, but some have other symmetry elements as well. For example, rotating the crystal 180° about a certain axis may result in an atomic configuration that is identical to the original configuration. The crystal is then said to have a twofold rotational symmetry about this axis. In addition to rotational symmetries like this, a crystal may have symmetries in the form of mirror planes and translational symmetries, and also the so-called "compound symmetries," which are a combination of translation and rotation/mirror symmetries. A full classification of a crystal is achieved when all of these inherent symmetries of the crystal are identified.

Lattice Systems

These lattice systems are a grouping of crystal structures according to the axial system used to describe their lattice. Each lattice system consists of a set of three axes in a particular geometric arrangement. There are seven lattice systems. They are similar to but not quite the same as the seven crystal systems.

Crystal family	Lattice system	Schönflies	14 Bravais Lattices			
			Primitive	Base-centered	Body-centered	Face-centered
triclinic		C_i	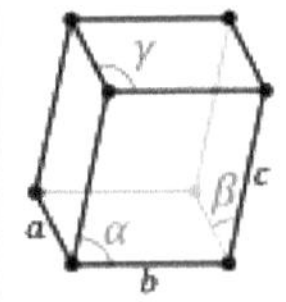			
monoclinic		C_{2h}	$\beta \neq 90°$ $a \neq c$ 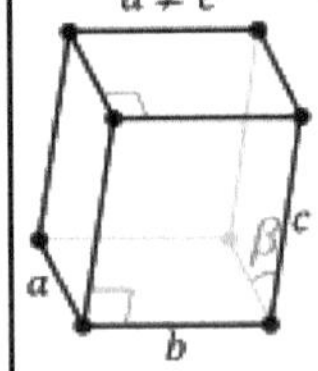	$\beta \neq 90°$ $a \neq c$ 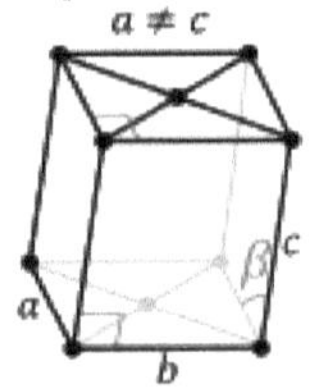		
orthorhombic		D_{2h}	$a \neq b \neq c$ 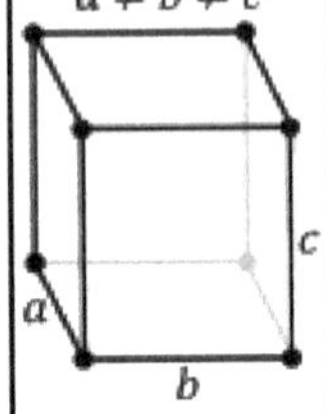	$a \neq b \neq c$ 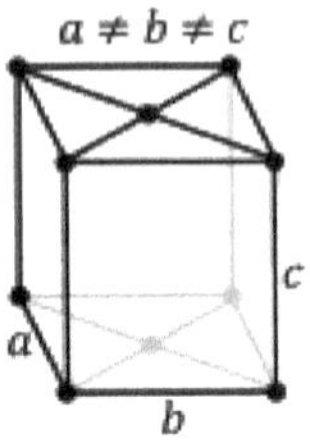	$a \neq b \neq c$ 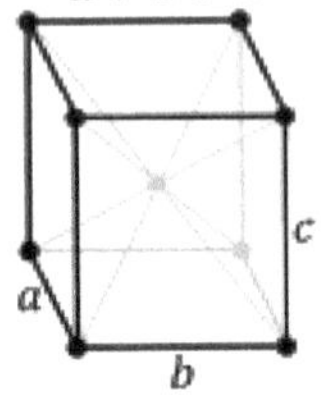	$a \neq b \neq c$ 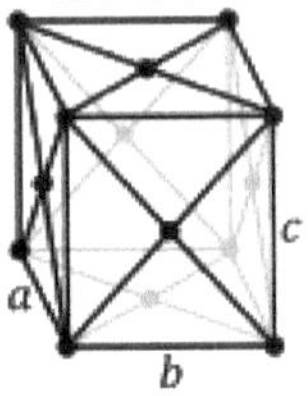
tetragonal		D_{4h}	$a \neq c$ 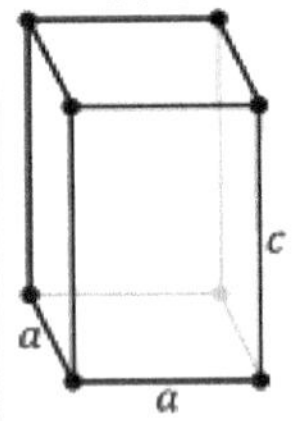		$a \neq c$ 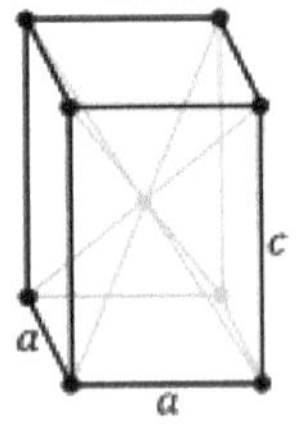	
hexagonal	rhombohedral	D_{3d}	$\alpha \neq 90°$ 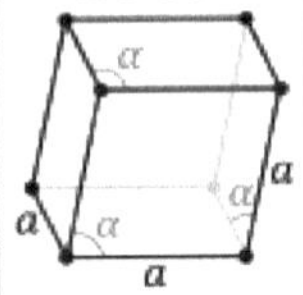			
	hexagonal	D_{6h}	$\gamma = 120°$ 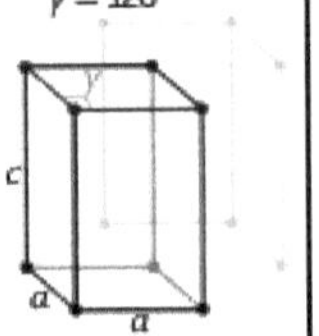			
cubic		O_h	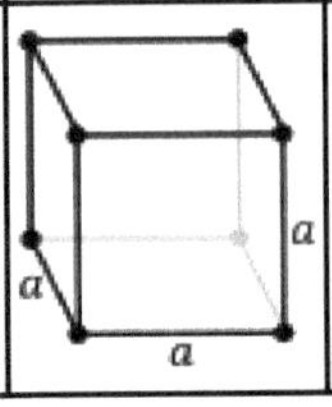		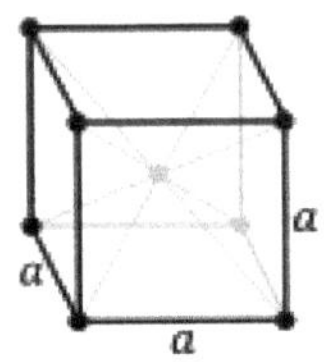	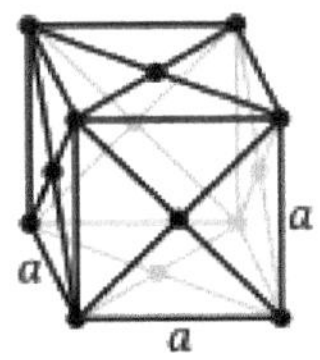

The simplest and most symmetric, the cubic (or isometric) system, has the symmetry of a cube, that is, it exhibits four threefold rotational axes oriented at 109.5° (the tetrahedral angle) with respect to each other. These threefold axes lie along the body diagonals of the cube. The other six lattice systems, are hexagonal, tetragonal, rhombohedral (often confused with the trigonal crystal system), orthorhombic, monoclinic and triclinic.

Bravais Lattices

Bravais lattices, also referred to as *space lattices,* describe the geometric arrangement of the lattice points, and therefore the translational symmetry of the crystal. The three dimensions of space afford 14 distinct Bravais lattices describing the translational symmetry. All crystalline materials recognized today, not including quasicrystals, fit in one of these arrangements. The fourteen three-dimensional lattices, classified by lattice system.

The crystal structure consists of the same group of atoms, the *basis,* positioned around each and every lattice point. This group of atoms therefore repeats indefinitely in three dimensions according to the arrangement of one of the Bravais lattices. The characteristic rotation and mirror symmetries of the unit cell is described by its crystallographic point group.

Crystal Systems

A crystal system is a set of point groups in which the point groups themselves and their corresponding space groups are assigned to a lattice system. Of the 32 point groups that exist in three dimensions, most are assigned to only one lattice system, in which case the crystal system and lattice system both have the same name. However, five point groups are assigned to two lattice systems, rhombohedral and hexagonal, because both lattice systems exhibit threefold rotational symmetry. These point groups are assigned to the trigonal crystal system.

Crystal family	Crystal system	Point group / Crystal class	Schönflies	Point symmetry	Order	Abstract group
triclinic		triclinic-pedial	C_1	enantiomorphic polar	1	trivial $\mathbb{Z}_1$
		triclinic-pinacoidal	C_i	centrosymmetric	2	cyclic $\mathbb{Z}_2$
monoclinic		monoclinic-sphenoidal	C_2	enantiomorphic polar	2	cyclic $\mathbb{Z}_2$
		monoclinic-domatic	C_s	polar	2	cyclic $\mathbb{Z}_2$
		monoclinic-prismatic	C_{2h}	centrosymmetric	4	Klein four $\mathbb{V} = \mathbb{Z}_2 \times \mathbb{Z}_2$
orthorhombic		orthorhombic-sphenoidal	D_2	enantiomorphic	4	Klein four $\mathbb{V} = \mathbb{Z}_2 \times \mathbb{Z}_2$
		orthorhombic-pyramidal	C_{2v}	polar	4	Klein four $\mathbb{V} = \mathbb{Z}_2 \times \mathbb{Z}_2$
		orthorhombic-bipyramidal	D_{2h}	centrosymmetric	8	$\mathbb{V} \times \mathbb{Z}_2$
tetragonal		tetragonal-pyramidal	C_4	enantiomorphic polar	4	cyclic $\mathbb{Z}_4$
		tetragonal-disphenoidal	S_4	non-centrosymmetric	4	cyclic $\mathbb{Z}_4$
		tetragonal-dipyramidal	C_{4h}	centrosymmetric	8	$\mathbb{Z}_4 \times \mathbb{Z}_2$
		tetragonal-trapezoidal	D_4	enantiomorphic	8	dihedral $\mathbb{D}_8 = \mathbb{Z}_4 \rtimes \mathbb{Z}_2$
		ditetragonal-pyramidal	C_{4v}	polar	8	dihedral $\mathbb{D}_8 = \mathbb{Z}_4 \rtimes \mathbb{Z}_2$
		tetragonal-scalenoidal	D_{2d}	non-centrosymmetric	8	dihedral $\mathbb{D}_8 = \mathbb{Z}_4 \rtimes \mathbb{Z}_2$
		ditetragonal-dipyramidal	D_{4h}	centrosymmetric	16	$\mathbb{D}_8 \times \mathbb{Z}_2$

hexagonal	trigonal	trigonal-pyramidal	C_3	enantiomorphic polar	3	cyclic $\mathbb{Z}_3$
		rhombohedral	S_6 (C_{3i})	centrosymmetric	6	cyclic $\mathbb{Z}_6 = \mathbb{Z}_3 \times \mathbb{Z}_2$
		trigonal-trapezoidal	D_3	enantiomorphic	6	dihedral $\mathbb{D}_6 = \mathbb{Z}_3 \rtimes \mathbb{Z}_2$
		ditrigonal-pyramidal	C_{3v}	polar	6	dihedral $\mathbb{D}_6 = \mathbb{Z}_3 \rtimes \mathbb{Z}_2$
		ditrigonal-scalahedral	D_{3d}	centrosymmetric	12	dihedral $\mathbb{D}_{12} = \mathbb{Z}_6 \rtimes \mathbb{Z}_2$
	hexagonal	hexagonal-pyramidal	C_6	enantiomorphic polar	6	cyclic $\mathbb{Z}_6 = \mathbb{Z}_3 \times \mathbb{Z}_2$
		trigonal-dipyramidal	C_{3h}	non-centrosymmetric	6	cyclic $\mathbb{Z}_6 = \mathbb{Z}_3 \times \mathbb{Z}_2$
		hexagonal-dipyramidal	C_{6h}	centrosymmetric	12	$\mathbb{Z}_6 \rtimes \mathbb{Z}_2$
		hexagonal-trapezoidal	D_6	enantiomorphic	12	dihedral $\mathbb{D}_{12} = \mathbb{Z}_6 \rtimes \mathbb{Z}_2$
		dihexagonal-pyramidal	C_{6v}	polar	12	dihedral $\mathbb{D}_{12} = \mathbb{Z}_6 \rtimes \mathbb{Z}_2$
		ditrigonal-dipyramidal	D_{3h}	non-centrosymmetric	12	dihedral $\mathbb{D}_{12} = \mathbb{Z}_6 \rtimes \mathbb{Z}_2$
		dihexagonal-dipyramidal	D_{6h}	centrosymmetric	24	$\mathbb{D}_{12} \times \mathbb{Z}_2$
cubic		tetrahedral	T	enantiomorphic	12	alternating $\mathbb{A}_4$
		hextetrahedral	T_d	non-centrosymmetric	24	symmetric $\mathbb{S}_4$
		diploidal	T_h	centrosymmetric	24	$\mathbb{A}_4 \times \mathbb{Z}_2$
		gyroidal	O	enantiomorphic	24	symmetric $\mathbb{S}_4$
		hexoctahedral	O_h	centrosymmetric	48	$\mathbb{S}_4 \times \mathbb{Z}_2$

In total there are seven crystal systems: triclinic, monoclinic, orthorhombic, tetragonal, trigonal, hexagonal, and cubic.

Point Groups

The crystallographic point group or *crystal class* is the mathematical group comprising the symmetry operations that leave at least one point unmoved and that leave the appearance of the crystal structure unchanged. These symmetry operations include

- *Reflection*, which reflects the structure across a *reflection plane*
- *Rotation*, which rotates the structure a specified portion of a circle about a *rotation axis*
- *Inversion*, which changes the sign of the coordinate of each point with respect to a *center of symmetry* or *inversion point*
- *Improper rotation*, which consists of a rotation about an axis followed by an inversion.

Rotation axes (proper and improper), reflection planes, and centers of symmetry are collectively called *symmetry elements*. There are 32 possible crystal classes. Each one can be classified into one of the seven crystal systems.

Space Groups

In addition to the operations of the point group, the space group of the crystal structure contains translational symmetry operations. These include:

- Pure *translations*, which move a point along a vector
- *Screw axes*, which rotate a point around an axis while translating parallel to the axis.
- *Glide planes*, which reflect a point through a plane while translating it parallel to the plane.

There are 230 distinct space groups.

Atomic Coordination

By considering the arrangement of atoms relative to each other, their coordination numbers (or number of nearest neighbors), interatomic distances, types of bonding, etc., it is possible to form a general view of the structures and alternative ways of visualizing them.

Close Packing

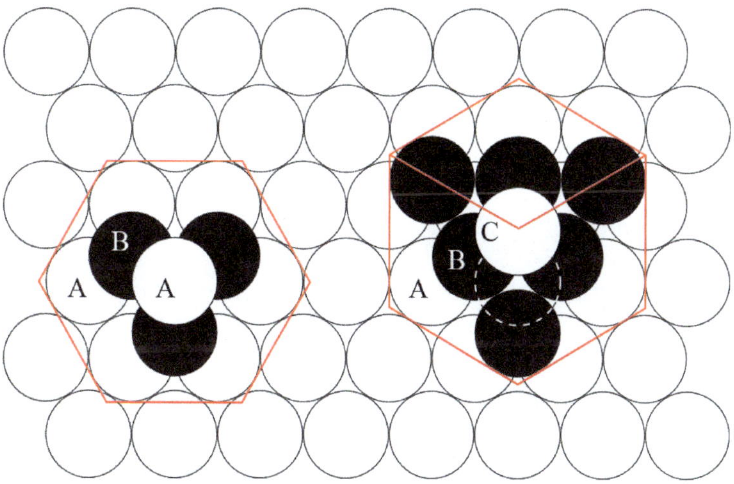

The hcp lattice (left) and the fcc lattice (right)

The principles involved can be understood by considering the most efficient way of packing together equal-sized spheres and stacking close-packed atomic planes in three dimensions. For example, if plane A lies beneath plane B, there are two possible ways of placing an additional atom on top of layer B. If an additional layer was placed directly over plane A, this would give rise to the following series:

...ABABABAB...

This arrangement of atoms in a crystal structure is known as hexagonal close packing (hcp).

If, however, all three planes are staggered relative to each other and it is not until the fourth layer is positioned directly over plane A that the sequence is repeated, then the following sequence arises:

...ABCABCABC...

This type of structural arrangement is known as cubic close packing (ccp).

The unit cell of a ccp arrangement of atoms is the face-centered cubic (fcc) unit cell. This is not immediately obvious as the closely packed layers are parallel to the {111} planes of the fcc unit cell. There are four different orientations of the close-packed layers.

The packing efficiency can be worked out by calculating the total volume of the spheres and dividing by the volume of the cell as follows:

$$\frac{4\times\frac{4}{3}\pi r^3}{16\sqrt{2}r^3}=\frac{\pi}{3\sqrt{2}}=0.7405...$$

The 74% packing efficiency is the maximum density possible in unit cells constructed of spheres of only one size. Most crystalline forms of metallic elements are hcp, fcc, or bcc (body-centered cubic). The coordination number of atoms in hcp and fcc structures is 12 and its atomic packing factor (APF) is the number mentioned above, 0.74. This can be compared to the APF of a bcc structure, which is 0.68.

Grain Boundaries

Grain boundaries are interfaces where crystals of different orientations meet. A grain boundary is a single-phase interface, with crystals on each side of the boundary being identical except in orientation. The term "crystallite boundary" is sometimes, though rarely, used. Grain boundary areas contain those atoms that have been perturbed from their original lattice sites, dislocations, and impurities that have migrated to the lower energy grain boundary.

Treating a grain boundary geometrically as an interface of a single crystal cut into two parts, one of which is rotated, we see that there are five variables required to define a grain boundary. The first two numbers come from the unit vector that specifies a rotation axis. The third number designates the angle of rotation of the grain. The final two numbers specify the plane of the grain boundary (or a unit vector that is normal to this plane).

Grain boundaries disrupt the motion of dislocations through a material, so reducing crystallite size is a common way to improve strength, as described by the Hall–Petch relationship. Since grain boundaries are defects in the crystal structure they tend to decrease the electrical and thermal conductivity of the material. The high interfacial energy and relatively weak bonding in most grain boundaries often makes them preferred sites for the onset of corrosion and for the precipitation of new phases from the solid. They are also important to many of the mechanisms of creep.

Grain boundaries are in general only a few nanometers wide. In common materials, crystallites are large enough that grain boundaries account for a small fraction of the material. However, very small grain sizes are achievable. In nanocrystalline solids, grain boundaries become a significant volume fraction of the material, with profound effects on such properties as diffusion and plasticity. In the limit of small crystallites, as the volume fraction of grain boundaries approaches 100%, the material ceases to have any crystalline character, and thus becomes an amorphous solid.

Defects and Impurities

Real crystals feature defects or irregularities in the ideal arrangements described above and it is these defects that critically determine many of the electrical and mechanical properties of real materials. When one atom substitutes for one of the principal atomic components within the crystal structure, alteration in the electrical and thermal properties of the material may ensue. Impurities may also manifest as spin impurities in certain materials. Research on magnetic impurities demonstrates that substantial alteration of certain properties such as specific heat may be affected by small concentrations of an impurity, as for example impurities in semiconducting ferromagnetic alloys may lead to different properties as first predicted in the late 1960s. Dislocations in the crystal lattice allow shear at lower stress than that needed for a perfect crystal structure.

Prediction of Structure

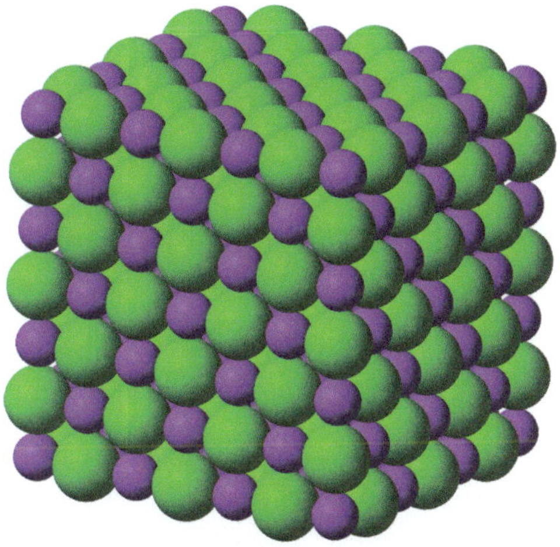

Crystal structure of sodium chloride (table salt)

The difficulty of predicting stable crystal structures based on the knowledge of only the chemical composition has long been a stumbling block on the way to fully computational materials design. Now, with more powerful algorithms and high-performance computing, structures of medium complexity can be predicted using such approaches as evolutionary algorithms, random sampling, or metadynamics.

The crystal structures of simple ionic solids (e.g., NaCl or table salt) have long been rationalized in terms of Pauling's rules, first set out in 1929 by Linus Pauling, referred to by many since as the "father of the chemical bond". Pauling also considered the nature of the interatomic forces in metals, and concluded that about half of the five d-orbitals in the transition metals are involved in bonding, with the remaining nonbonding d-orbitals being responsible for the magnetic properties. He, therefore, was able to correlate the number of d-orbitals in bond formation with the bond length as well as many of the physical properties of the substance. He subsequently introduced the metallic orbital, an extra orbital necessary to permit uninhibited resonance of valence bonds among various electronic structures.

In the resonating valence bond theory, the factors that determine the choice of one from among alternative crystal structures of a metal or intermetallic compound revolve around the energy of resonance of bonds among interatomic positions. It is clear that some modes of resonance would make larger contributions (be more mechanically stable than others), and that in particular a simple ratio of number of bonds to number of positions would be exceptional. The resulting principle is that a special stability is associated with the simplest ratios or "bond numbers": $^1/_2$, $^1/_3$, $^2/_3$, $^1/_4$, $^3/_4$, etc. The choice of structure and the value of the axial ratio (which determines the relative bond lengths) are thus a result of the effort of an atom to use its valency in the formation of stable bonds with simple fractional bond numbers.

After postulating a direct correlation between electron concentration and crystal structure in beta-phase alloys, Hume-Rothery analyzed the trends in melting points, compressibilities and bond lengths as a function of group number in the periodic table in order to establish a system of valencies of the transition elements in the metallic state. This treatment thus emphasized the increasing bond strength as a function of group number. The operation of directional forces were emphasized in one article on the relation between bond hybrids and the metallic structures. The resulting correlation between electronic and crystalline structures is summarized by a single parameter, the weight of the d-electrons per hybridized metallic orbital. The "d-weight" calculates out to 0.5, 0.7 and 0.9 for the fcc, hcp and bcc structures respectively. The relationship between d-electrons and crystal structure thus becomes apparent.

In crystal structure predictions/simulations, the periodicity is usually applied, since the system is imagined as unlimited big in all directions. Starting from a triclinic structure with no further symmetry property assumed, the system may be driven to show some additional symmetry properties by applying Newton's Second Law on particles in the unit cell and a recently developed dynamical equation for the system period vectors (lattice parameters including angles), even if the system is subject to external stress.

Polymorphism

Quartz is one of the several thermodynamically stable crystalline forms of silica, SiO_2. The most important forms of silica include: α-quartz, β-quartz, tridymite, cristobalite, coesite, and stishovite.

Polymorphism is the occurrence of multiple crystalline forms of a material. It is found in many crystalline materials including polymers, minerals, and metals. According to Gibbs' rules of phase equilibria, these unique crystalline phases are dependent on intensive variables such as pressure and temperature. Polymorphism is related to allotropy, which refers to elemental solids. The complete morphology of a material is described by polymorphism and other variables such as crystal habit, amorphous fraction or crystallographic defects. Polymorphs have different stabilities and may spontaneously convert from a metastable form (or thermodynamically unstable form) to the stable form at a particular temperature. They also exhibit different melting points, solubilities, and X-ray diffraction patterns.

One good example of this is the quartz form of silicon dioxide, or SiO_2. In the vast majority of silicates, the Si atom shows tetrahedral coordination by 4 oxygens. All but one of the crystalline forms involve tetrahedral $\{SiO_4\}$ units linked together by shared vertices in different arrangements. In different minerals the tetrahedra show different degrees of networking and polymerization. For example, they occur singly, joined together in pairs, in larger finite clusters including rings, in chains, double chains, sheets, and three-dimensional frameworks. The minerals are classified into groups based on these structures. In each of its 7 thermodynamically stable crystalline forms or polymorphs of crystalline quartz, only 2 out of 4 of each the edges of the $\{SiO_4\}$ tetrahedra are shared with others, yielding the net chemical formula for silica: SiO_2.

Another example is elemental tin (Sn), which is malleable near ambient temperatures but is brittle when cooled. This change in mechanical properties due to existence of its two major allotropes, α- and β-tin. The two allotropes that are encountered at normal pressure and temperature, α-tin and β-tin, are more commonly known as *gray tin* and *white tin* respectively. Two more allotropes, γ and σ, exist at temperatures above 161 °C and pressures above several GPa. White tin is metallic, and is the stable crystalline form at or above room temperature. Below 13.2 °C, tin exists in the gray form, which has a diamond cubic crystal structure, similar to diamond, silicon or germanium. Gray tin has no metallic properties at all, is a dull gray powdery material, and has few uses, other than a few specialized semiconductor applications. Although the α–β transformation temperature of tin is nominally 13.2 °C, impurities (e.g. Al, Zn, etc.) lower the transition temperature well below 0 °C, and upon addition of Sb or Bi the transformation may not occur at all.

Physical Properties

Twenty of the 32 crystal classes are piezoelectric, and crystals belonging to one of these classes (point groups) display piezoelectricity. All piezoelectric classes lack a center of symmetry. Any material develops a dielectric polarization when an electric field is applied, but a substance that has such a natural charge separation even in the absence of a field is called a polar material. Whether or not a material is polar is determined solely by its crystal structure. Only 10 of the 32 point groups are polar. All polar crystals are pyroelectric, so the 10 polar crystal classes are sometimes referred to as the pyroelectric classes.

There are a few crystal structures, notably the perovskite structure, which exhibit ferroelectric behavior. This is analogous to ferromagnetism, in that, in the absence of an electric field during production, the ferroelectric crystal does not exhibit a polarization. Upon the application of an electric field of sufficient magnitude, the crystal becomes permanently polarized. This polarization can be reversed by a sufficiently large counter-charge, in the same way that a ferromagnet can be reversed. However, although they are called ferroelectrics, the effect is due to the crystal structure (not the presence of a ferrous metal).

1. The unit cell

"The smallest repeat unit of a crystal structure, in 3D, which shows the full symmetry of the structure".

The unit cell is a box with:

- 3 sides - a, b, c
- 3 angles - α, β, γ

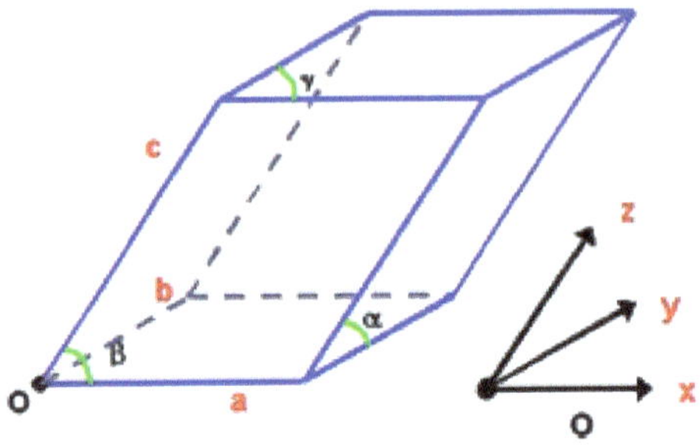

Seven unit cell shapes

• Cubic	$a=b=c$	$\alpha=\beta=\gamma=90°$
• Tetragonal	$a=b\neq c$	$\alpha=\beta=\gamma=90°$
• Orthorhombic	$a\neq b\neq c$	$\alpha=\beta=\gamma=90°$
• Monoclinic	$a\neq b\neq c$	$\alpha=\gamma=90°$, $\beta\neq 90°$
• Triclinic	$a\neq b\neq c$	$\alpha\neq\beta\neq\gamma\neq 90°$
• Hexagonal	$a=b\neq c$	$\alpha=\beta=90°$, $\gamma=120°$
• Rhombohedral	$a=b=c$	$\alpha=\beta=\gamma\neq 90°$

2D example - rocksalt (sodium chloride, NaCl)

We define lattice points; these are points with identical environments.

Choice of origin is arbitrary - lattice points need not be atoms - but unit cell size should always be the same.

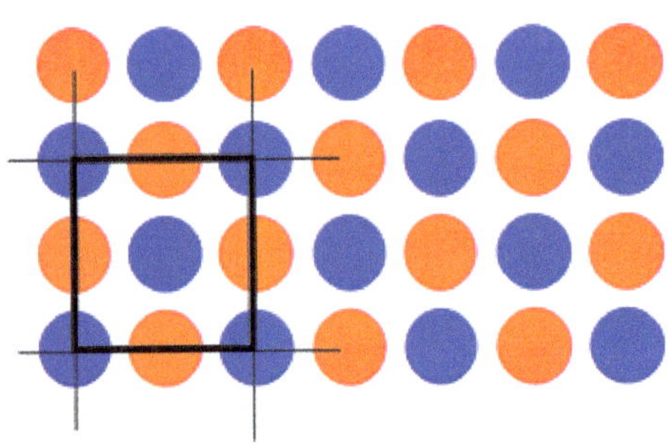

This is also a unit cell - it doesn't matter if you start from Na or Cl

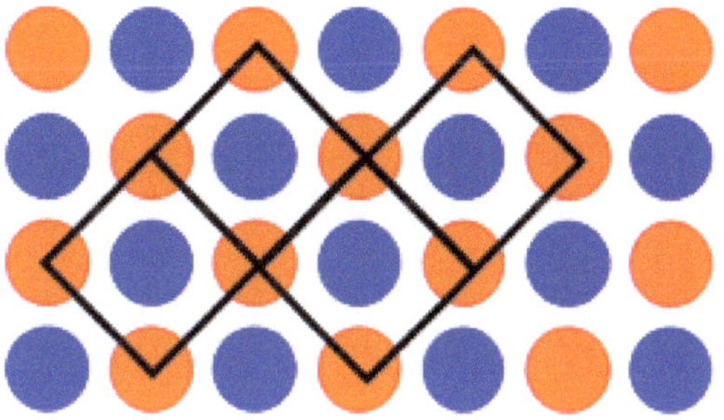

References

- Molodets, A. M.; Nabatov, S. S. (2000). "Thermodynamic Potentials, Diagram of State, and Phase Transitions of Tin on Shock Compression". High Temperature. 38 (5): 715–721. doi:10.1007/BF02755923
- Pauling, Linus (1947). "Atomic Radii and Interatomic Distances in Metals". Journal of the American Chemical Society. 69 (3): 542. doi:10.1021/ja01195a024
- Hook, J.R.; Hall, H.E. (2010). Solid State Physics. Manchester Physics Series (2nd ed.). John Wiley & Sons. ISBN 9780471928041
- Hogan, C. M. (1969). "Density of States of an Insulating Ferromagnetic Alloy". Physical Review. 188 (2): 870. Bibcode:1969PhRv..188..870H. doi:10.1103/PhysRev.188.870
- Bernal, J. D.; Fowler, R. H. (1933). "A Theory of Water and Ionic Solution, with Particular Reference to Hydrogen and Hydroxyl Ions". The Journal of Chemical Physics. 1 (8): 515. Bibcode:1933JChPh...1..515B. doi:10.1063/1.1749327
- Direct and reciprocal lattices". CSIC Dept de Cristalografia y Biologia Estructural. 6 Apr 2017. Retrieved 18 May 2017
- Zhang, X. Y.; Suhl, H (1985). "Spin-wave-related period doublings and chaos under transverse pumping". Physical Review A. 32 (4): 2530–2533. Bibcode:1985PhRvA..32.2530Z. doi:10.1103/PhysRevA.32.2530. PMID 9896377
- Encyclopaedia of Physics (2nd Edition), R.G. Lerner, G.L. Trigg, VHC publishers, 1991, ISBN (Verlagsgesellschaft) 3-527-26954-1, ISBN (VHC Inc.) 0-89573-752-3
- L. Pauling (1929). "The principles determining the structure of complex ionic crystals". J. Am. Chem. Soc. 51 (4): 1010–1026. doi:10.1021/ja01379a006
- Donald E. Sands (1994). "§4-2 Screw axes and §4-3 Glide planes". Introduction to Crystallography (Reprint of Wenjamin corrected 1975 ed.). Courier-Dover. pp. 70–71. ISBN 0486678393

- Pauling, Linus (1938). "The Nature of the Interatomic Forces in Metals". Physical Review. 54 (11): 899. Bibcode:1938PhRv...54..899P. doi:10.1103/PhysRev.54.899
- Holleman, Arnold F.; Wiberg, Egon; Wiberg, Nils (1985). "Tin". Lehrbuch der Anorganischen Chemie (in German) (91–100 ed.). Walter de Gruyter. pp. 793–800. ISBN 3-11-007511-3
- Pauling, L. (1949). "A Resonating-Valence-Bond Theory of Metals and Intermetallic Compounds". Proceedings of the Royal Society A. 196 (1046): 343. Bibcode:1949RSPSA.196..343P. doi:10.1098/rspa.1949.0032
- Schwartz, Mel (2002). "Tin and Alloys, Properties". Encyclopedia of Materials, Parts and Finishes (2nd ed.). CRC Press. ISBN 1-56676-661-3
- Liu, Gang (2015). "Dynamical equations for the period vectors in a periodic system under constant external stress". Can. J. Phys. 93 (9): 974. Bibcode:2015CaJPh..93..974L. doi:10.1139/cjp-2014-0518

2

Fundamentals of Graphene

Graphene is an allotrope of carbon that forms the fundamental structure of all other allotropes. It is considered to be 200 times stronger than steel and is also a good conductor of heat and electricity. This section will provide an integrated understanding of graphene and its applications.

Graphene

Graphene is an allotrope of carbon in the form of a two-dimensional, atomic-scale, hexagonal lattice in which one atom forms each vertex. It is the basic structural element of other allotropes, including graphite, charcoal, carbon nanotubes and fullerenes. It can be considered as an indefinitely large aromatic molecule, the ultimate case of the family of flat polycyclic aromatic hydrocarbons.

Graphene is an atomic-scale hexagonal lattice made of carbon atoms.

Graphene has many unusual properties. It is about 200 times stronger than the strongest steel. It efficiently conducts heat and electricity and is nearly transparent. Graphene shows a large and nonlinear diamagnetism, greater than graphite and can be levitated by neodymium magnets.

Scientists have theorized about graphene for years. It has unintentionally been produced in small quantities for centuries, through the use of pencils and other similar graphite applications. It was originally observed in electron microscopes in 1962, but it was studied only while supported on metal surfaces. The material was later rediscovered, isolated, and characterized in 2004 by Andre Geim and Konstantin Novoselov at the University of Manchester. Research was informed by existing theoretical descriptions of its composition, structure, and properties. This work resulted in the two winning the Nobel Prize in Physics in 2010 "for groundbreaking experiments regarding the two-dimensional material graphene."

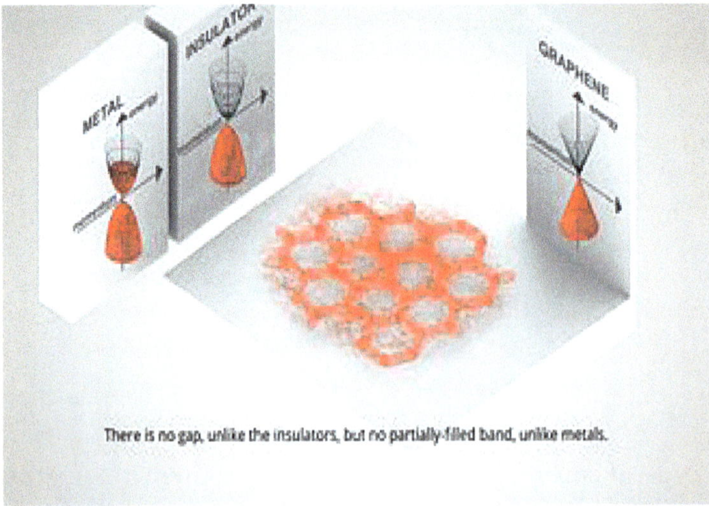

Graphene and its band structure and Dirac cones, effect of a grid on doping

The global market for graphene reached $9 million by 2012 with most sales in the semiconductor, electronics, battery energy, and composites industries.

Definition

"Graphene" is a combination of "graphite" and the suffix -ene, named by Hanns-Peter Boehm, who described single-layer carbon foils in 1962.

The term *graphene* first appeared in 1987 to describe single sheets of graphite as a constituent of graphite intercalation compounds (GICs); conceptually a GIC is a crystalline salt of the intercalant and graphene. The term was also used in early descriptions of carbon nanotubes, as well as for epitaxial graphene and polycyclic aromatic hydrocarbons (PAH). Graphene can be considered an "infinite alternant" (only six-member carbon ring) polycyclic aromatic hydrocarbon.

The IUPAC compendium of technology states: "previously, descriptions such as graphite layers, carbon layers, or carbon sheets have been used for the term graphene... it is incorrect to use for a single layer a term which includes the term graphite, which would imply a three-dimensional structure. The term graphene should be used only when the reactions, structural relations or other properties of individual layers are discussed."

Geim defined "isolated or free-standing graphene" as "graphene is a single atomic plane of graphite, which – and this is essential – is sufficiently isolated from its environment to be considered free-standing." This definition is narrower than the IUPAC definition and refers to cloven, transferred and suspended graphene. Other forms such as graphene grown on various metals, can become free-standing if, for example, suspended or transferred to silicon dioxide (SiO_2) or silicon carbide.

History

A lump of graphite, a graphene transistor, and a tape dispenser. Donated to the Nobel Museum in Stockholm by Andre Geim and Konstantin Novoselov in 2010.

Andre Geim and Konstantin Novoselov, 2010

In 1859 Benjamin Collins Brodie became aware of the highly lamellar structure of thermally reduced graphite oxide.

The structure of graphite was identified in 1916 by the related method of powder diffraction. It was studied in detail by Kohlschütter and Haenni in 1918, who described the properties of graphite oxide paper. Its structure was determined from single-crystal diffraction in 1924.

The theory of graphene was first explored by Wallace in 1947 as a starting point for understanding the electronic properties of 3D graphite. The emergent massless Dirac equation was first pointed out by Semenoff, DiVincenzo and Mele. Semenoff emphasized the occurrence in a magnetic field of an electronic Landau level precisely at the Dirac point. This level is responsible for the anomalous integer quantum Hall effect.

The earliest TEM images of few-layer graphite were published by Ruess and Vogt in 1948. Later, single graphene layers were observed directly by electron microscopy. Before 2004 intercalated graphite compounds were studied under a transmission electron microscope (TEM). Researchers occasionally observed thin graphitic flakes ("few-layer graphene") and possibly even individual layers. An early, detailed study on few-layer graphite dates to 1962 when Boehm reported producing monolayer flakes of reduced graphene oxide.

Starting in the 1970s single layers of graphite were grown epitaxially on top of other materials. This "epitaxial graphene" consists of a single-atom-thick hexagonal lattice of sp^2-bonded carbon atoms, as in free-standing graphene. However, significant charge transfers from the substrate to the epitaxial graphene, and in some cases, the d-orbitals of the substrate atoms hybridize with the π orbitals of graphene, which significantly alters the electronic structure of epitaxial graphene.

Single layers of graphite were observed by TEM within bulk materials, in particular inside soot obtained by chemical exfoliation. Efforts to make thin films of graphite by mechanical exfoliation started in 1990, but nothing thinner than 50 to 100 layers was produced before 2004.

Initial attempts to make atomically thin graphitic films employed exfoliation techniques similar to the drawing method. Multilayer samples down to 10 nm in thickness were obtained. Earlier researchers tried to isolate graphene starting with intercalated compounds, producing very thin graphitic fragments (possibly monolayers). Neither of the earlier observations was sufficient to launch the "graphene gold rush" that awaited macroscopic samples of extracted atomic planes.

One of the first patents pertaining to the production of graphene was filed in October 2002 and granted in 2006. It detailed one of the first large scale graphene production processes. Two years later, in 2004 Geim and Novoselov extracted single-atom-thick crystallites from bulk graphite. They pulled graphene layers from graphite and transferred them onto thin silicon dioxide (SiO_2) on a silicon wafer in a process called either micromechanical cleavage or the Scotch tape technique. The SiO_2 electrically isolated the graphene and weakly interacted with it, providing nearly charge-neutral graphene layers. The silicon beneath the SiO_2 could be used as a "back gate" electrode to vary the charge density in the graphene over a wide range. US patent 6667100, filed in 2002, describes how to process expanded graphite to achieve a graphite thickness of 0.00001" (one hundred-thousandth) of an inch. The key to success was high-throughput visual recognition of graphene on a properly chosen substrate, which provides a small but noticeable optical contrast.

The cleavage technique led directly to the first observation of the anomalous quantum Hall effect in graphene, which provided direct evidence of graphene's theoretically predicted Berry's phase of massless Dirac fermions. The effect was reported by Geim's group and by Kim and Zhang, whose papers appeared in *Nature* in 2005. Before these experiments other researchers had looked for the quantum Hall effect and Dirac fermions in bulk graphite.

Even though graphene on nickel and on silicon carbide have both existed in the laboratory for decades, graphene mechanically exfoliated on SiO_2 provided the first proof of the Dirac fermion nature of electrons.

Geim and Novoselov received awards for their pioneering research on graphene, notably the 2010 Nobel Prize in Physics.

In 2014 National Graphene Institute, a £60m Graphene Engineering Innovation Centre (GEIC), was announced to support applied research and development in partnership with other research organizations and industry.

In 2014 two North East England commercial manufacturers, Applied Graphene Materials and Thomas Swan Limited (with Trinity College, Dublin researchers), began manufacturing. In East Anglia FGV Cambridge Nanosystems operates a graphene powder production facility.

Properties

Graphene has a theoretical specific surface area (SSA) of 2630 m^2/g. This is much larger than that reported to date for carbon black (typically smaller than 900 m^2/g) or for carbon nanotubes (CNTs), from ≈100 to 1000 m^2/g and is similar to activated carbon.

Structure

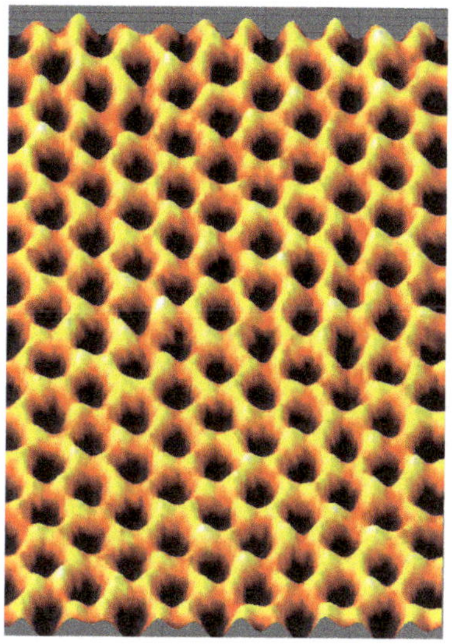

Scanning probe microscopy image of graphene

Graphene is a crystalline allotrope of carbon with 2-dimensional properties. Its carbon atoms are densely packed in a regular atomic-scale chicken wire (hexagonal) pattern.

Each atom has four bonds, one σ bond with each of its three neighbors and one π-bond that is oriented out of plane. The atoms are about 1.42 Å apart.

Graphene's hexagonal lattice can be regarded as two interleaving triangular lattices. This perspective was successfully used to calculate the band structure for a single graphite layer using a tight-binding approximation.

Graphene's stability is due to its tightly packed carbon atoms and a sp^2 orbital hybridization – a combination of orbitals s, p_x and p_y that constitute the σ-bond. The final p_z electron makes up the π-bond. The π-bonds hybridize together to form the π-band and π∗-bands. These bands are responsible for most of graphene's notable electronic properties, via the half-filled band that permits free-moving electrons.

Graphene sheets in solid form usually show evidence in diffraction for graphite's (002) layering. This is true of some single-walled nanostructures. However, unlayered graphene with only (hk0) rings has been found in the core of presolar graphite onions. TEM studies show faceting at defects in flat graphene sheets and suggest a role for two-dimensional crystallization from a melt.

Graphene can self-repair holes in its sheets, when exposed to molecules containing carbon, such as hydrocarbons. Bombarded with pure carbon atoms, the atoms perfectly align into hexagons, completely filling the holes.

The atomic structure of isolated, single-layer graphene was studied by TEM on sheets of graphene suspended between bars of a metallic grid. Electron diffraction patterns showed the expected honeycomb lattice. Suspended graphene showed "rippling" of the flat sheet, with amplitude of about one nanometer. These ripples may be intrinsic to the material as a result of the instability of two-dimensional crystals, or may originate from the ubiquitous dirt seen in all TEM images of graphene. Atomic resolution real-space images of isolated, single-layer graphene on SiO_2 substrates are available via scanning tunneling microscopy (STM). Photoresist residue, which must be removed to obtain atomic-resolution images, may be the "adsorbates" observed in TEM images, and may explain the observed rippling. Rippling on SiO_2 is caused by conformation of graphene to the underlying SiO_2 and is not intrinsic.

Chemical

Graphene is the only form of carbon (or solid material) in which every atom is available for chemical reaction from two sides (due to the 2D structure). Atoms at the edges of a graphene sheet have special chemical reactivity. Graphene has the highest ratio of edge atoms of any allotrope. Defects within a sheet increase its chemical reactivity. The onset temperature of reaction between the basal plane of single-layer graphene and oxygen gas is below 260 °C (530 K). Graphene combusts at 350 °C (620 K). Graphene is commonly modified with oxygen- and nitrogen-containing functional groups and analyzed by infrared spectroscopy and X-ray photoelectron spectroscopy. However, determination of structures of graphene with oxygen- and nitrogen- functional groups requires the structures to be well controlled.

Contrary to the ideal 2D structure of graphene, chemical applications of graphene needs whether structural or chemical irregularities as perfectly flat graphene is chemically inert. In other words, the definition of an ideal graphene is different in chemistry and physics.

Graphene placed on a soda-lime glass (SLG) substrate under ambient conditions exhibited spontaneous n-doping ($1.33 \times 10^{13}\ e/cm^2$) via surface-transfer. On p-type copper indium gallium diselenide (CIGS) semiconductor itself deposited on SLG n-doping reached $2.11 \times 10^{13}\ e/cm^2$.

Electronic

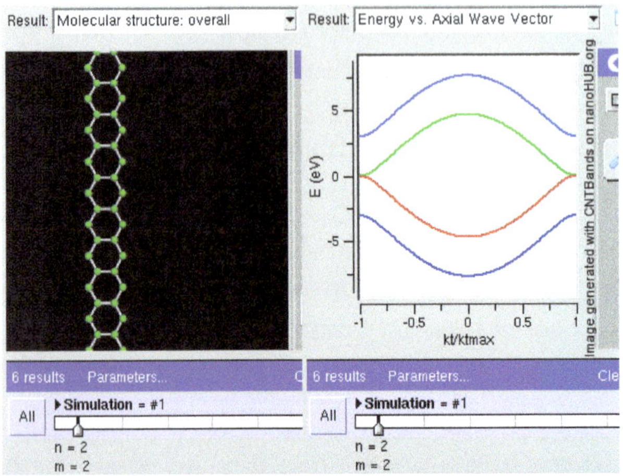

GNR band structure for zig-zag orientation. Tightbinding calculations show that zig-zag orientation is always metallic.

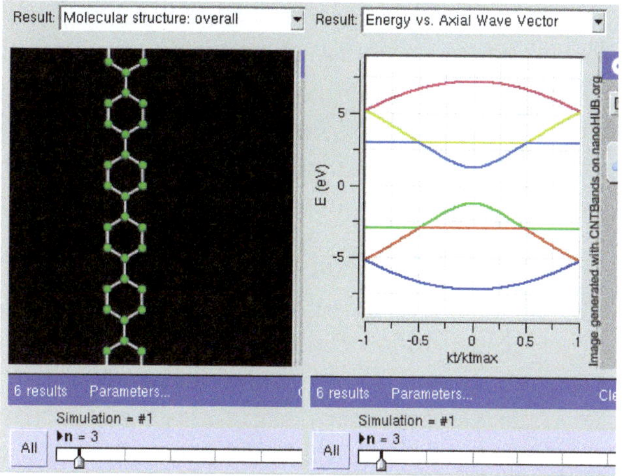

GNR band structure for armchair orientation. Tightbinding calculations show that armchair orientation can be semiconducting or metallic depending on width (chirality).

Graphene is a zero-gap semiconductor, because its conduction and valence bands meet at the Dirac points. The Dirac points are six locations in momentum space, on the edge of the Brillouin zone, divided into two non-equivalent sets of three points. The two sets are labeled K and K'. The sets give graphene a valley degeneracy of $gv = 2$. By contrast, for traditional semiconductors the primary point of interest is generally Γ, where momentum is zero. Four electronic properties separate it from other condensed matter systems.

However, if the in-plane direction is confined, in which case it is referred to as a nanoribbon, its electronic structure is different. If it is "zig-zag", the bandgap is zero. If it is "armchair", the bandgap is non-zero.

Electronic Spectrum

Electrons propagating through graphene's honeycomb lattice effectively lose their mass, producing quasi-particles that are described by a 2D analogue of the Dirac equation rather than the Schrödinger equation for spin-½ particles.

Dispersion Relation

When atoms are placed onto the graphene hexagonal lattice, the overlap between the $p_z(\pi)$ orbitals and the s or the p_x and p_y orbitals is zero by symmetry. The p_z electrons forming the π bands in graphene can be treated independently. Within this π-band approximation, using a conventional tight-binding model, the dispersion relation (restricted to first-nearest-neighbor interactions only) that produces energy of the electrons with wave vector k is

$$E = \pm\sqrt{\gamma_0^2\left(1 + 4\cos^2\frac{3k_y a}{2} + 4\cos\frac{3k_y a}{2}\cdot\cos\frac{k_x\sqrt{3}a}{2}\right)}$$

with the nearest-neighbor (π orbitals) hopping energy $\gamma_0 \approx 2.8$ eV and the lattice constant $a \approx 2.46$ Å. The conduction and valence bands, respectively, correspond to the different signs. With one p_z electron per atom in this model the valence band is fully occupied, while the conduction band is vacant. The two bands touch at the zone corners (the *K* point in the Brillouin zone), where there is a zero density of states but no band gap. The graphene sheet thus displays a semimetallic (or zero-gap semiconductor) character, although not if rolled into a carbon nanotube, due to its curvature. Two of the six Dirac points are independent, while the rest are equivalent by symmetry. In the vicinity of the *K*-points the energy depends *linearly* on the wave vector, similar to a relativistic particle. Since an elementary cell of the lattice has a basis of two atoms, the wave function has an effective 2-spinor structure.

As a consequence, at low energies, even neglecting the true spin, the electrons can be described by an equation that is formally equivalent to the massless Dirac equation. Hence, the electrons and holes are called Dirac fermions. This pseudo-relativistic description is restricted to the chiral limit, i.e., to vanishing rest mass M_0, which leads to additional features:

$$v_F \vec{\sigma} \cdot \nabla \psi(\mathrm{r}) = E\psi(\mathrm{r}).$$

Here $v_F \sim 10^6$ m/s (.003 c) is the Fermi velocity in graphene, which replaces the velocity of light in the Dirac theory; $\vec{\sigma}$ is the vector of the Pauli matrices; $\psi(\mathrm{r})$ is the two-component wave function of the electrons and E is their energy.

The equation describing the electrons' linear dispersion relation is

$$E = \hbar v_F \sqrt{k_x^2 + k_y^2}$$

where the wavevector k is measured from the Dirac points (the zero of energy is chosen here to coincide with the Dirac points). The equation uses a pseudospin matrix formula that describes two sublattices of the honeycomb lattice.

Single-atom Wave Propagation

Electron waves in graphene propagate within a single-atom layer, making them sensitive to the proximity of other materials such as high-κ dielectrics, superconductors and ferromagnetics.

Electron Transport

Graphene displays remarkable electron mobility at room temperature, with reported values in excess of 15000 $cm^2 \cdot V^{-1} \cdot s^{-1}$. Hole and electron mobilities were expected to be nearly identical. The mobility is nearly independent of temperature between 10 K and 100 K, which implies that the dominant scattering mechanism is defect scattering. Scattering by graphene's acoustic phonons intrinsically limits room temperature mobility to 200000 $cm^2 \cdot V^{-1} \cdot s^{-1}$ at a carrier density of 10^{12} cm^{-2}, 10×10^6 times greater than copper.

The corresponding resistivity of graphene sheets would be 10^{-6} Ω·cm. This is less than the resistivity of silver, the lowest otherwise known at room temperature. However, on SiO_2 substrates, scattering of electrons by optical phonons of the substrate is a larger effect than scattering by graphene's own phonons. This limits mobility to 40000 $cm^2 \cdot V^{-1} \cdot s^{-1}$.

Charge transport is affected by adsorption of contaminants such as water and oxygen molecules. This leads to non-repetitive and large hysteresis I-V characteristics. Researchers must carry out electrical measurements in vacuum. Graphene surfaces can be protected by a coating with materials such as SiN, PMMA and h-BN. In January 2015, the first stable graphene device operation in air over several weeks was reported, for graphene whose surface was protected by aluminum oxide. In 2015 lithium-coated graphene was observed to exhibit superconductivity and in 2017 evidence for unconventional superconductivity was demonstrated in single layer graphene placed on the electron-doped (non-chiral) *d*-wave superconductor $Pr_{2-x}Ce_xCuO_4$ (PCCO).

Electrical resistance in 40-nanometer-wide nanoribbons of epitaxial graphene changes in discrete steps. The ribbons' conductance exceeds predictions by a factor of 10. The ribbons can act more like optical waveguides or quantum dots, allowing electrons to flow smoothly along the ribbon edges. In copper, resistance increases in proportion to length as electrons encounter impurities.

Transport is dominated by two modes. One is ballistic and temperature independent, while the other is thermally activated. Ballistic electrons resemble those in cylindrical carbon nanotubes. At room temperature, resistance increases abruptly at a particular length—the ballistic mode at 16 micrometres and the other at 160 nanometres.

Graphene electrons can cover micrometer distances without scattering, even at room temperature.

Despite zero carrier density near the Dirac points, graphene exhibits a minimum conductivity on the order of $4e^2/h$. The origin of this minimum conductivity is unclear. However, rippling of the graphene sheet or ionized impurities in the SiO_2 substrate may lead to local puddles of carriers that allow conduction. Several theories suggest that the minimum conductivity should be $4e^2/(\pi h)$; however, most measurements are of order $4e^2/h$ or greater and depend on impurity concentration.

Near zero carrier density graphene exhibits positive photoconductivity and negative photoconductivity at high carrier density. This is governed by the interplay between photoinduced changes of both the Drude weight and the carrier scattering rate.

Graphene doped with various gaseous species (both acceptors and donors) can be returned to an undoped state by gentle heating in vacuum. Even for dopant concentrations in excess of 10^{12} cm^{-2} carrier mobility exhibits no observable change. Graphene doped with potassium in ultra-high vacuum at low temperature can reduce mobility 20-fold. The mobility reduction is reversible on removing the potassium.

Due to graphene's two dimensions, charge fractionalization (where the apparent charge of individual pseudoparticles in low-dimensional systems is less than a single quantum) is thought to occur. It may therefore be a suitable material for constructing quantum computers using anyonic circuits.

Dirac Fluid

Charged particles in high-purity graphene behave as a strongly interacting, quasi-relativistic plasma. The particles move in a fluid-like manner, traveling along a single path and interacting with high frequency. The behavior was observed in a graphene sheet faced on both sides with a h-BN crystal sheet.

Anomalous Quantum Hall Effect

The quantum Hall effect is a quantum mechanical version of the Hall effect, which is the production of transverse (perpendicular to the main current) conductivity in the presence of a magnetic field. The quantization of the Hall effect σ_{xy} at integer multiples (the "Landau level") of the basic quantity e^2/h (where *e* is the elementary electric charge and *h* is Planck's constant) It can usually be observed only in very clean silicon or gallium arsenide solids at temperatures around 3 K and high magnetic fields.

Graphene shows the quantum Hall effect with respect to conductivity quantization: the effect is *anomalous* in that the sequence of steps is shifted by 1/2 with respect to the standard sequence and with an additional factor of 4. Graphene's Hall conductivity is $\sigma_{xy} = \pm 4 \cdot (N + 1/2) e^2 / h$, where *N* is the Landau level and the double valley and double spin degeneracies give the factor of 4. These anomalies are present at room temperature, i.e. at roughly 20 °C (293 K).

This behavior is a direct result of graphene's massless Dirac electrons. In a magnetic field, their spectrum has a Landau level with energy precisely at the Dirac point. This level is a consequence of the Atiyah–Singer index theorem and is half-filled in neutral graphene, leading to the "+1/2" in the Hall conductivity. Bilayer graphene also shows the quantum Hall effect, but with only one of the two anomalies (i.e. $\sigma_{xy} = \pm 4 \cdot N \cdot e^2 / h$). In the second anomaly, the first plateau at *N*=0 is absent, indicating that bilayer graphene stays metallic at the neutrality point.

Unlike normal metals, graphene's longitudinal resistance shows maxima rather than minima for integral values of the Landau filling factor in measurements of the Shubnikov–de Haas oscillations, whereby the term *integral* quantum Hall effect. These oscillations show a phase shift of π, known as Berry's phase. Berry's phase arises due to the zero effective carrier mass near the Dirac points. The temperature dependence of the oscillations reveals that the carriers have a non-zero cyclotron mass, despite their zero effective mass.

Graphene samples prepared on nickel films, and on both the silicon face and carbon face of silicon carbide, show the anomalous effect directly in electrical measurements. Graphitic layers on the carbon face of silicon carbide show a clear Dirac spectrum in angle-resolved photoemission experiments. The effect is observed in cyclotron resonance and tunneling experiments.

Strong Magnetic Fields

In magnetic fields above 10 tesla or so additional plateaus of the Hall conductivity at $\sigma_{xy} = \nu e^2 / h$ with ν = 0, ±1, ±4 are observed. A plateau at ν = 3 and the fractional quantum Hall effect at $\nu = \frac{1}{3}$ were also reported.

These observations with ν = 0, ±1, ±3, ±4 indicate that the four-fold degeneracy (two valley and two spin degrees of freedom) of the Landau energy levels is partially or completely lifted.

Casimir Effect

The Casimir effect is an interaction between disjoint neutral bodies provoked by the fluctuations of the electrodynamical vacuum. Mathematically it can be explained by considering the normal modes of electromagnetic fields, which explicitly depend on the boundary (or matching) conditions on the interacting bodies' surfaces. Since graphene/electromagnetic field interaction is strong for a one-atom-thick material, the Casimir effect is of interest.

Van der Waals Force

The Van der Waals force (or dispersion force) is also unusual, obeying an inverse cubic, asymptotic power law in contrast to the usual inverse quartic.

'Massive' Electrons

Graphene's unit cell has two identical carbon atoms and two zero-energy states: one in which the electron resides on atom A, the other in which the electron resides on atom B. However, if the two atoms in the unit cell are not identical, the situation changes. Hunt et al. showed that placing hexagonal boron nitride (h-BN) in contact with graphene can alter the potential felt at atom A versus atom B enough that the electrons develop a mass and accompanying band gap of about 30 meV [0.03 Electron Volt(eV)].

The mass can be positive or negative. An arrangement that slightly raises the energy of an electron on atom A relative to atom B gives it a positive mass, while an arrangement that raises the energy of atom B produces a negative electron mass. The two versions behave alike and are indistinguishable via optical spectroscopy. An electron traveling from a positive-mass region to a negative-mass region must cross an intermediate region where its mass once again becomes zero. This region is gapless and therefore metallic. Metallic modes bounding semiconducting regions of opposite-sign mass is a hallmark of a topological phase and display much the same physics as topological insulators.

If the mass in graphene can be controlled, electrons can be confined to massless regions by surrounding them with massive regions, allowing the patterning of quantum dots, wires and other mesoscopic structures. It also produces one-dimensional conductors along the boundary. These wires would be protected against backscattering and could carry currents without dissipation.

Optical

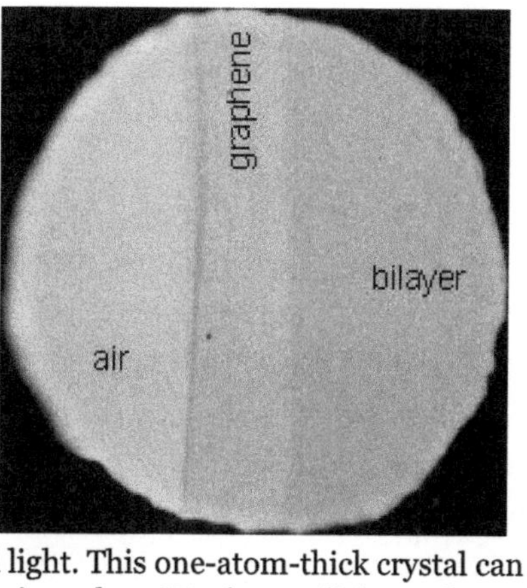

Photograph of graphene in transmitted light. This one-atom-thick crystal can be seen with the naked eye because it absorbs approximately 2.6% of green light, and 2.3% of red light.

Graphene's unique optical properties produce an unexpectedly high opacity for an atomic monolayer in vacuum, absorbing $\pi\alpha \approx 2.3\%$ of red light, where α is the fine-structure constant. This is a consequence of the "unusual low-energy electronic structure of monolayer graphene that features electron and hole conical bands meeting each other at the Dirac point... [which] is qualitatively different from more common quadratic massive bands." Based on the Slonczewski–Weiss–McClure (SWMcC) band model of graphite, the interatomic distance, hopping value and frequency cancel when optical conductance is calculated using Fresnel equations in the thin-film limit.

Although confirmed experimentally, the measurement is not precise enough to improve on other techniques for determining the fine-structure constant.

Multi-Parametric Surface Plasmon Resonance was used to characterize both thickness and refractive index of chemical-vapor-deposition (CVD)-grown graphene films. The measured refractive index and extinction coefficient values at 670 nm wavelength are 3.135 and 0.897, respectively. The thickness was determined as 3.7 Å from a 0.5 mm area, which agrees with 3.35 Å reported for layer-to-layer carbon atom distance of graphite crystals.

The method can be used for real-time label-free interactions of graphene with organic and inorganic substances. The existence of unidirectional surface plasmons in the nonreciprocal graphene-based gyrotropic interfaces has been demonstrated theoretically. By efficiently controlling the chemical potential of graphene, the unidirectional working frequency can be continuously tunable from THz to near-infrared and even visible. Particularly, the unidirectional frequency bandwidth can be 1– 2 orders of magnitude larger than that in metal under the same magnetic field, which arises from the superiority of extremely small effective electron mass in graphene.

Graphene's band gap can be tuned from 0 to 0.25 eV (about 5 micrometre wavelength) by applying voltage to a dual-gate bilayer graphene field-effect transistor (FET) at room temperature. The optical response of graphene nanoribbons is tunable into the terahertz regime by an applied magnetic field. Graphene/graphene oxide systems exhibit electrochromic behavior, allowing tuning of both linear and ultrafast optical properties.

A graphene-based Bragg grating (one-dimensional photonic crystal) demonstrated its capability for excitation of surface electromagnetic waves in the periodic structure using a 633 nm He–Ne laser as the light source.

Saturable Absorption

Such unique absorption could become saturated when the input optical intensity is above a threshold value. This nonlinear optical behavior is termed saturable absorption and the threshold value is called the saturation fluence. Graphene can be saturated readily under strong excitation over the visible to near-infrared region, due to the universal optical absorption and zero band gap. This has relevance for the mode locking of fiber lasers, where fullband mode locking has been achieved by a graphene-based saturable absorber. Due to this special property, graphene has wide application in ultrafast photonics. The optical response of graphene/graphene oxide layers can be tuned electrically. Saturable absorption in graphene could occur at the Microwave and Terahertz bands, owing to its wideband optical absorption property. The microwave saturable absorption in graphene demonstrates the possibility of graphene microwave and terahertz photonics devices, such as a microwave saturable absorber, modulator, polarizer, microwave signal processing and broad-band wireless access networks.

Nonlinear Kerr Effect

Under more intensive laser illumination, graphene could possess a nonlinear phase shift due to the optical nonlinear Kerr effect. Based on a typical open and close aperture z-scan measurement, graphene possesses a nonlinear Kerr coefficient of 10^{-7} $cm^2 \cdot W^{-1}$, almost nine orders of magnitude larger than that of bulk dielectrics. This suggests that graphene may be a powerful nonlinear Kerr medium, with the possibility of observing a variety of nonlinear effects, the most important of which is the soliton. Thermal lens spectroscopy can be used for measuring the thermo-optic coefficient of graphene and inspecting its thermal nonlinearity.

Excitonic

First-principle calculations with quasiparticle corrections and many-body effects explore the electronic and optical properties of graphene-based materials. The approach is described as three stages. With GW calculation, the properties of graphene-based materials are accurately investigated, including bulk graphene, nanoribbons, edge and surface functionalized armchair oribbons, hydrogen saturated armchair ribbons, Josephson effect in graphene SNS junctions with single localized defect and armchair ribbon scaling properties.

Stability

Ab initio calculations show that a graphene sheet is thermodynamically unstable if its size is less than about 20 nm ("graphene is the least stable structure until about 6000 atoms") and becomes the most stable fullerene (as within graphite) only for molecules larger than 24,000 atoms.

Thermal Conductivity

Thermal transport in graphene is an active area of research, which has attracted attention because of the potential for thermal management applications. Early measurements of the thermal conductivity of suspended graphene reported an exceptionally large thermal conductivity of approximately 5300 $W \cdot m^{-1} \cdot K^{-1}$, compared with the thermal conductivity of pyrolytic graphite of approximately 2000 $W \cdot m^{-1} \cdot K^{-1}$ at room temperature. However, later studies have questioned whether this ultrahigh value was overestimated, and instead measured thermal conductivities between 1500 – 2500 $W \cdot m^{-1} \cdot K^{-1}$ for suspended single layer graphene. The large range can be attributed to large measurement uncertainties as well as variations in the graphene quality and processing conditions. In addition, when single-layer graphene is supported on an amorphous material, the thermal conductivity is reduced to about 500 – 600 $W \cdot m^{-1} \cdot K^{-1}$ at room temperature as a result of scattering of graphene lattice waves by the substrate, and can be even lower for few layer graphene encased in amorphous oxide. Likewise, polymeric residue can contribute to a similar decrease for suspended graphene to approximately 500 – 600 $W \cdot m^{-1} \cdot K^{-1}$for bilayer graphene.

It has been suggested that the isotopic composition, the ratio of ^{12}C to ^{13}C, has a significant impact on thermal conductivity. For example, isotopically pure ^{12}C graphene has higher thermal conductivity than either a 50:50 isotope ratio or the naturally occurring 99:1 ratio. It can be shown by using the Wiedemann–Franz law, that the thermal conduction is phonon-dominated. However, for a gated graphene strip, an applied gate bias causing a Fermi energy shift much larger than $k_B T$ can cause the electronic contribution to increase and dominate over the phonon contribution at low temperatures. The ballistic thermal conductance of graphene is isotropic.

Potential for this high conductivity can be seen by considering graphite, a 3D version of graphene that has basal plane thermal conductivity of over a 1000 $W \cdot m^{-1} \cdot K^{-1}$ (comparable to diamond). In graphite, the c-axis (out of plane) thermal conductivity is over a factor of ≈100 smaller due to the weak binding forces between basal planes as well as the larger lattice spacing. In addition, the ballistic thermal conductance of graphene gives the lower limit of the ballistic thermal conductances, per unit circumference and length of carbon nanotubes.

Despite its 2-D nature, graphene has 3 acoustic phonon modes. The two in-plane modes (LA, TA) have a linear dispersion relation, while the out of plane mode (ZA) has a quadratic dispersion relation. Due to this, the T^2 dependent thermal conductivity contribution of the linear modes is dominated at low temperatures by the $T^{1.5}$ contribution of the out-of-plane mode. Some graphene phonon bands display negative Grüneisen parameters (GPs). At low temperatures (where most optical modes with positive GPs are still not excited) the contribution from the negative GPs will be dominant and thermal expansion coefficient (which is directly proportional to GPs) negative. The lowest negative GPs correspond to the lowest transverse acoustic ZA modes. Phonon frequencies for such modes increase with the in-plane lattice parameter since atoms in the layer upon stretching will be less free to move in the z direction. This is similar to the behavior of a stretched string that has vibrations of smaller amplitude and higher frequency. This phenomenon, named "membrane effect," was predicted by Lifshitz in 1952.

Melting Point

An early prediction suggested a melting point of ≈4125 K. Recent, more sophisticated, modelling has increased this temperature to at least 5000 K. At 6000 K (the sun's surface having an effective temperature of 5,777 K) graphene melts into an agglomeration of loosely coupled doubled bonded chains, before becoming a gas.

Mechanical

The carbon–carbon bond length in graphene is about 0.142 nanometers. Graphene sheets stack to form graphite with an interplanar spacing of 0.335 nm.

Graphene is the strongest material ever tested, with an intrinsic tensile strength of 130 GPa and a Young's modulus (stiffness) of 1 TPa (150000000 psi). The Nobel announcement illustrated this by saying that a 1 square meter graphene hammock would support a 4 kg cat but would weigh only as much as one of the cat's whiskers, at 0.77 mg (about 0.001% of the weight of 1 m^2 of paper).

Large-angle-bent graphene monolayer has been achieved with negligible strain, showing mechanical robustness of the two-dimensional carbon nanostructure. Even with extreme deformation, excellent carrier mobility in monolayer graphene can be preserved.

The spring constant of suspended graphene sheets has been measured using an atomic force microscope (AFM). Graphene sheets were suspended over SiO_2 cavities where an atomic force microscope (AFM) tip was used to apply a stress to the sheet to test its mechanical properties. Its spring constant was in the range 1–5 N/m and the stiffness was 0.5 TPa, which differs from that of bulk graphite. These intrinsic properties could lead to applications such as NEMS as pressure sensors and resonators. Due to its large surface energy and out of plane ductility, flat graphene sheets are unstable with respect to scrolling, i.e. bending into a cylindrical shape, which is its lower-energy state.

As is true of all materials, regions of graphene are subject to thermal and quantum fluctuations in relative displacement. Although the amplitude of these fluctuations is bounded in 3D structures (even in the limit of infinite size), the Mermin–Wagner theorem shows that the amplitude of long-wavelength fluctuations grows logarithmically with the scale of a 2D structure and would therefore be unbounded in structures of infinite size. Local deformation and elastic strain are negligibly affected by this long-range divergence in relative displacement. It is believed that a sufficiently large 2D structure, in the absence of applied lateral tension, will bend and crumple to form a fluctuating 3D structure. Researchers have observed ripples in suspended layers of graphene. It has been proposed that the ripples are caused by thermal fluctuations in the material. As a consequence of these dynamical deformations, it is debatable whether graphene is truly a 2D structure. It has recently been shown that these ripples, if amplified through the introduction of vacancy defects, can impart a negative Poisson's ratio into graphene, resulting in the thinnest auxetic material known.

Graphene nanosheets can be incorporated into a nickel matrix through a plating process to form Ni-graphene composites on a target substrate. The enhancement in mechanical properties of the composites is attributed to the high interaction between Ni and graphene and the prevention of the dislocation sliding in the Ni matrix by the graphene.

Fracture Toughness

In 2014, researchers indicated that despite its strength, graphene is also relatively brittle, with a fracture toughness of about 4 MPa√m. This indicates that imperfect graphene is likely to crack in a brittle manner like ceramic materials, as opposed to many metallic materials that have fracture toughnesses in the range of 15–50 MPa√m. Later in 2014, the researchers announced that graphene showed a greater ability to distribute force from an impact than any known material, ten times that of steel per unit weight. The force was transmitted at 22.2 kilometres per second (13.8 mi/s).

Spin Transport

Graphene is claimed to be an ideal material for spintronics due to its small spin-orbit interaction and the near absence of nuclear magnetic moments in carbon (as well as a weak hyperfine interaction). Electrical spin current injection and detection has been demonstrated up to room temperature. Spin coherence length above 1 micrometre at room temperature was observed, and control of the spin current polarity with an electrical gate was observed at low temperature.

Strong Magnetic Fields

In magnetic fields of ≈10 tesla, additional plateaus of Hall conductivity at $\sigma_{xy} = \nu e^2 / h$ with

$\nu = 0, \pm 1, \pm 4$ are observed. The observation of a plateau at $\nu = 3$ and the fractional quantum Hall effect at $\nu = 1/3$ were reported.

These observations with $\nu = 0, \pm 1, \pm 3, \pm 4$ indicate that the four-fold degeneracy (two valley and two spin degrees of freedom) of the Landau energy levels is partially or completely lifted. One hypothesis is that the magnetic catalysis of symmetry breaking is responsible for lifting the degeneracy.

Spintronic and magnetic properties can be present in graphene simultaneously. Low-defect graphene nanomeshes manufactured using a non-lithographic method exhibit large-amplitude ferromagnetism even at room temperature. Additionally a spin pumping effect is found for fields applied in parallel with the planes of few-layer ferromagnetic nanomeshes, while a magnetoresistance hysteresis loop is observed under perpendicular fields.

Magnetic

In 2014 researchers magnetized graphene by placing it on an atomically smooth layer of magnetic yttrium iron garnet. The graphene's electronic properties were unaffected. Prior approaches involved doping. The dopant's presence negatively affected its electronic properties.

Biological

Researchers in 2011 discovered the ability of graphene to accelerate the osteogenic differentiation of human mesenchymal stem cells without the use of biochemical inducers.

In 2015 researchers used graphene to create biosensors with epitaxial graphene on silicon carbide. The sensors bind to 8-hydroxydeoxyguanosine (8-OHdG) and is capable of selective binding with antibodies. The presence of 8-OHdG in blood, urine and saliva is commonly associated with DNA damage. Elevated levels of 8-OHdG have been linked to increased risk of several cancers.

In 2016 researchers revealed that uncoated graphene can be used as neuro-interface electrode without altering or damaging properties such as signal strength or formation of scar tissue. Graphene electrodes in the body stay significantly more stable than electrodes of tungsten or silicon because of properties such as flexibility, bio-compatibility and conductivity.

Substrate

The electronic properties of graphene are significantly influenced by the supporting substrate. The Si(100)/H surface does not perturb graphene's electronic properties, whereas the interaction between it and the clean Si(100) surface changes its electronic states significantly. This effect results from the covalent bonding between C and surface Si atoms, modifying the π-orbital network of the graphene layer. The local density of states shows that the bonded C and Si surface states are highly disturbed near the Fermi energy.

Forms

Monolayer Sheets

In 2013 researchers developed a production unit that produces continuous monolayer sheets of high-strength monolayer graphene (HSMG). The process is based on graphene growth on a liquid metal matrix.

Bilayer

Bilayer graphene displays the anomalous quantum Hall effect, a tunable band gap and potential for excitonic condensation. Bilayer graphene typically can be found either in twisted configurations where the two layers are rotated relative to each other or graphitic Bernal stacked configurations where half the atoms in one layer lie atop half the atoms in the other. Stacking order and orientation govern its optical and electronic properties.

One synthesis method is chemical vapor deposition, which can produce large bilayer regions that almost exclusively conform to a Bernal stack geometry.

Superlattices

Periodically stacked graphene and its insulating isomorph provide a fascinating structural element in implementing highly functional superlattices at the atomic scale, which offers possibilities in designing nanoelectronic and photonic devices. Various types of superlattices can be obtained by stacking graphene and its related forms. The energy band in layer-stacked superlattices is more sensitive to the barrier width than that in conventional III–V semiconductor superlattices. When adding more than one atomic layer to the barrier in each period, the coupling of electronic wavefunctions in neighboring potential wells can be significantly reduced, which leads to the degeneration of continuous subbands into quantized energy levels. When varying the well width, the energy levels in the potential wells along the L-M direction behave distinctly from those along the K-H direction.

Precisely aligned graphene on h-BN always produces giant superlattice known as Moiré pattern. Moiré patterns are observed and the sensitivity of moiré interferometry proves that the graphene grains can align precisely with the underlying h-BN lattice within an error of less than 0.05°. The occurrence of moiré pattern clearly indicates that the graphene locks into h-BN via van der Waals epitaxy with its interfacial stress greatly released.

The existence of the giant Moiré pattern in graphene nanoribbon (GNR) embedded in hBN indicates that the graphene was highly crystalline and precisely aligned with the h-BN underneath. It was noticed that the Moiré pattern appeared to be stretched along the GNR, while it appeared relaxed laterally. This trend differs from regular hexagons with a periodicity of ~14 nm, which have always been observed with well-aligned graphene domains on h-BN. This observation gives a strong indication of the in-plane epitaxy between the graphene and the h-BN at the edges of the trench, where the graphene is stretched by tensile strain along the ribbon, due to a lattice mismatch between the graphene and h-BN.

Nanoribbons

Graphene nanoribbons ("nanostripes" in the "zig-zag" orientation), at low temperatures, show spin-polarized metallic edge currents, which suggest spintronics applications. (In the "armchair" orientation, the edges behave like semiconductors.)

Quantum Dots

Several techniques can prepare nanostructured graphene, e.g., graphene quantum dots (GQDs); these techniques mainly include electron beam lithography, chemical synthesis, electrochemical preparation, graphene oxide (GO) reduction, C60 catalytic transformation, the microwave assisted hydrothermal method (MAH), the Soft-Template method, the hydrothermal method, and the ultrasonic exfoliation method.

Oxide

Using paper-making techniques on dispersed, oxidized and chemically processed graphite in water, monolayer flakes form a single sheet and create strong bonds. These sheets, called graphene oxide paper, have a measured tensile modulus of 32 GPa. The chemical property of graphite oxide is related to the functional groups attached to graphene sheets. These can change the polymerization pathway and similar chemical processes. Graphene oxide flakes in polymers display enhanced photo-conducting properties. Graphene is normally hydrophobic and impermeable to all gases and liquids (vacuum-tight). However, when formed into graphene oxide-based capillary membrane, both liquid water and water vapor flow through as quickly as if the membrane was not present.

Chemical Modification

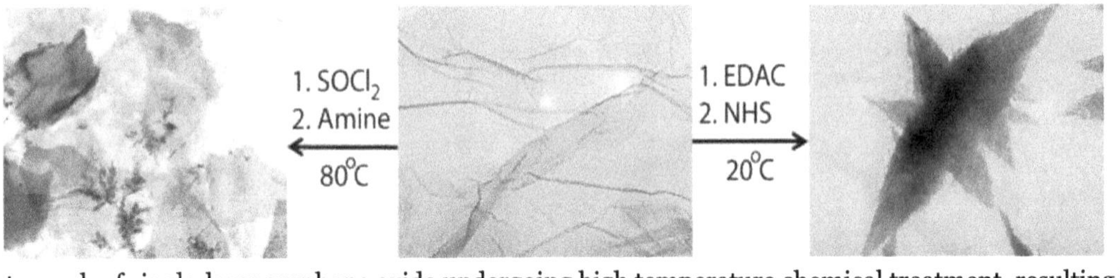

Photograph of single-layer graphene oxide undergoing high temperature chemical treatment, resulting in sheet folding and loss of carboxylic functionality, or through room temperature carbodiimide treatment, collapsing into star-like clusters.

Soluble fragments of graphene can be prepared in the laboratory through chemical modification of graphite. First, microcrystalline graphite is treated with an acidic mixture of sulfuric acid and nitric acid. A series of oxidation and exfoliation steps produce small graphene plates with carboxyl groups at their edges. These are converted to acid chloride groups by treatment with thionyl chloride; next, they are converted to the corresponding graphene amide via treatment with octadecylamine. The resulting material (circular graphene layers of 5.3 angstrom thickness) is soluble in tetrahydrofuran, tetrachloromethane and dichloroethane.

Refluxing single-layer graphene oxide (SLGO) in solvents leads to size reduction and folding of individual sheets as well as loss of carboxylic group functionality, by up to 20%, indicating thermal instabilities of SLGO sheets dependent on their preparation methodology. When using thionyl chloride, acyl chloride groups result, which can then form aliphatic and aromatic amides with a reactivity conversion of around 70–80%.

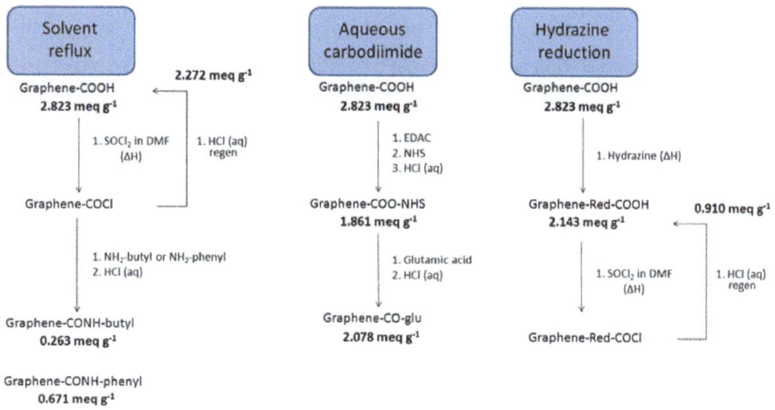

Boehm titration results for various chemical reactions of single-layer graphene oxide, which reveal reactivity of the carboxylic groups and the resultant stability of the SLGO sheets after treatment.

Hydrazine reflux is commonly used for reducing SLGO to SLG(R), but titrations show that only around 20–30% of the carboxylic groups are lost, leaving a significant number available for chemical attachment. Analysis of such SLG(R) reveals that the system is unstable. Using a room temperature stirring with HCl (< 1.0 M) leads to around 60% loss of COOH functionality. Room temperature treatment of SLGO with carbodiimides leads to the collapse of the individual sheets into star-like clusters that exhibited poor subsequent reactivity with amines (c. 3–5% conversion of the intermediate to the final amide). It is apparent that conventional chemical treatment of carboxylic groups on SLGO generates morphological changes of individual sheets that leads to a reduction in chemical reactivity, which may potentially limit their use in composite synthesis. Therefore, chemical reactions types have been explored. SLGO has also been grafted with polyallylamine, cross-linked through epoxy groups. When filtered into graphene oxide paper, these composites exhibit increased stiffness and strength relative to unmodified graphene oxide paper.

Full hydrogenation from both sides of graphene sheet results in graphane, but partial hydrogenation leads to hydrogenated graphene. Similarly, both-side fluorination of graphene (or chemical and mechanical exfoliation of graphite fluoride) leads to fluorographene (graphene fluoride), while partial fluorination (generally halogenation) provides fluorinated (halogenated) graphene.

Ligand/Complex

Graphene can be a ligand to coordinate metals and metal ions by introducing functional groups. Structures of graphene ligands are similar to e.g. metal-porphyrin complex, metal-phthalocyanine complex and metal-phenanthroline complex. Copper and nickel ions can be coordinated with graphene ligands.

Fiber

In 2011, researchers reported making fibers using chemical vapor deposition grown graphene films. The method was scalable and controllable, delivering tunable morphology and pore structure by controlling the evaporation of solvents with suitable surface tension. Flexible all-solid-state supercapacitors based on such fibers were demonstrated in 2013.

In 2015 intercalating small graphene fragments into the gaps formed by larger, coiled graphene sheets after annealing provided pathways for conduction, while the fragments helped reinforce the fibers. The resulting fibers offered better thermal and electrical conductivity and mechanical strength. Thermal conductivity reached 1290 watts per meter per kelvin, while tensile strength reached 1080 megapascals.

In 2016, kilometer-scale continuous graphene fibers with outstanding mechanical properties and excellent electrical conductivity were produced by high-throughput wet-spinning of graphene oxide liquid crystals followed by graphitization through a full-scale synergetic defect-engineering strategy.

3D

In 2013, a three-dimensional honeycomb of hexagonally arranged carbon was termed 3D graphene. Self-supporting 3D graphene was produced that year. Researchers at Stony Brook University have reported a novel radical-initiated crosslinking method to fabricate porous 3D free-standing architectures of graphene and carbon nanotubes using nanomaterials as building blocks without any polymer matrix as support. 3D structures can be fabricated by using either CVD or solution-based methods. A 2016 review summarized the techniques for fabrication of 3D graphene and other related two-dimensional materials. These 3D graphene (all-carbon) scaffolds/foams have potential applications in fields such as energy storage, filtration, thermal management and biomedical devices and implants.

In 2016 box-shaped graphene (BSG) nanostructure resulted from mechanical cleavage of pyrolytic graphite. The discovered nanostructure is a multilayer system of parallel hollow nanochannels located along the surface that displayed quadrangular cross-section. The thickness of the channel walls is approximately equal to 1 nm. Potential applications include: ultra-sensitive detectors, high-performance catalytic cells, nanochannels for DNA sequencing and manipulation, high-performance heat sinking surfaces, rechargeable batteries of enhanced performance, nanomechanical resonators, electron multiplication channels in emission nanoelectronic devices, high-capacity sorbents for safe hydrogen storage.

Three dimensional bilayer graphene was reported in 2012 and 2014.

Gyroid

In 2017 researchers created a graphene gyroid that has five percent of the density of steel, yet is ten times as strong with an enormous surface area to volume ratio. They compressed heated graphene flakes. They then constructed high resolution 3D-printed models of plastic of various configurations – similar to the gyroids that graphene form naturally, though thousands of times larger. These shapes were then tested for tensile strength and compression, and compared to the computer simulations. When then graphene was swapped out for polymers or metals, similar gains in strength were seen.

Molded

A film of graphene soaked in solvent to make it swell and become malleable was overlaid on an underlying substrate "former". The solvent evaporated, leaving behind a layer of graphene that had taken on the shape of the underlying structure. In this way the team was able to produce a range of relatively intricate micro-structured shapes. Features vary from 3.5 to 50 μm. Pure graphene and gold-decorated graphene were each successfully integrated with the substrate.

Aerogel

An aerogel made of graphene layers separated by carbon nanotubes was measured at 0.16 milligrams per cubic centimeter. A solution of graphene and carbon nanotubes in a mold is freeze dried to dehydrate the solution, leaving the aerogel. The material has superior elasticity and absorption. It can recover completely after more than 90% compression, and absorb up to 900 times its weight in oil, at a rate of 68.8 grams per second.

Pillared

Pillared graphene is a hybrid carbon structure consisting of an oriented array of carbon nanotubes connected at each end to a graphene sheet. It was first described theoretically in 2008. Pillared graphene has not been synthesized in the laboratory.

Reinforced

Graphene sheets reinforced with embedded carbon nanotubes ("rebar") are easier to manipulate, while improving the electrical and mechanical qualities of both materials.

Functionalized single- or multiwalled carbon nanotubes are spin-coated on copper foils and then heated and cooled, using the nanotubes as the carbon source. Under heating, the functional carbon groups decompose into graphene, while the nanotubes partially split and form in-plane covalent bonds with the graphene, adding strength. π–π stacking domains add more strength. The nanotubes can overlap, making the material a better conductor than standard CVD-grown graphene. The nanotubes effectively bridge the grain boundaries found in conventional graphene. The technique eliminates the traces of substrate on which later-separated sheets were deposited using epitaxy.

Stacks of a few layers have been proposed as a cost-effective and physically flexible replacement for indium tin oxide (ITO) used in displays and photovoltaic cells.

Nanocoil

In 2015 a coiled form of graphene was discovered in graphitic carbon (coal). The spiraling effect is produced by defects in the material's hexagonal grid that causes it to spiral along its edge, mimicking a Riemann surface, with the graphene surface approximately perpendicular to the axis. When voltage is applied to such a coil, current flows around the spiral, producing a magnetic field. The phenomenon applies to spirals with either zigzag or armchair orientations, although with different current distributions. Computer simulations indicated that a conventional spiral inductor of 205 microns in diameter could be matched by a nanocoil just 70 nanometers wide, with a field strength reaching as much as 1 tesla, about the same as the coils found in typical loudspeakers, about the same field strength as some MRI machines. They found the magnetic field would be strongest in the hollow, nanometer-wide cavity at the spiral's center.

A solenoid made with such a coil behaves as a quantum conductor whose current distribution between the core and exterior varies with applied voltage, resulting in nonlinear inductance.

Production

Multiple production techniques have been developed. Isolated 2D crystals cannot be grown via chemical synthesis beyond small sizes even in principle, because the rapid growth of phonon density with increasing lateral size forces 2D crystallites to bend into the third dimension. In all cases, graphene must bond to a substrate to retain its two-dimensional shape.

Exfoliation

As of 2014, exfoliation produced graphene with the lowest number of defects and highest electron mobility.

Geim and Novoselov initially used adhesive tape to pull graphene sheets away from graphite. Achieving single layers typically requires multiple exfoliation steps. After exfoliation the flakes are deposited on a silicon wafer. Crystallites larger than 1 mm and visible to the naked eye can be obtained.

Alternatively a sharp single-crystal diamond wedge cleave layers from a graphite source.

Rapid heating of graphite oxide and exfoliation yields highly dispersed carbon powder with a few percent of graphene flakes.

Another method is reduction of graphite oxide monolayer films, e.g. by hydrazine with annealing in argon/hydrogen with an almost intact carbon framework that allows efficient removal of functional groups. Measured charge carrier mobility exceeded 1,000 centimetres (393.70 in)/Vs.

Defect-free, unoxidized graphene-containing liquids can be made from graphite using mixers that produce local shear rates greater than 10×10^4.

Burning a graphite oxide coated DVD produced a conductive graphene film (1738 siemens per meter) and specific surface area (1520 square meters per gram) that was highly resistant and malleable.

With definite cleavage parameters, box-shaped graphene (BSG) nanostructure can be prepared on graphite crystal.

Sonication

Dispersing graphite in a liquid medium can produce graphene by sonication followed by centrifugation, producing concentrations of 2.1 mg/mL in N-methylpyrrolidone. Using a suitable ionic liquid as the dispersing liquid medium produced concentrations of 5.33 mg/mL. Restacking is an issue with this technique. Adding a surfactant to a solvent prior to sonication prevents restacking by adsorbing to the graphene's surface. This produces a higher graphene concentration, but removing the surfactant requires chemical treatments.

Sonicating graphite at the interface of two immiscible liquids, most notably heptane and water, produced macro-scale graphene films. The graphene sheets are adsorbed to the high energy interface between the materials and are kept from restacking. The sheets are up to about 95% transparent and conductive.

Molten Salts

Graphite particles can be corroded in molten salts to form a variety of carbon nanostructures including graphene. Hydrogen cations, dissolved in molten lithium chloride, can be discharged on cathodically polarized graphite rods, which then intercalate, peeling graphene sheets. The graphene nanosheets produced displayed a single-crystalline structure with a lateral size of several hundred nanometers and a high degree of crystallinity and thermal stability.

Electrochemical Synthesis

Electrochemical synthesis can exfoliate graphene. Varying a pulsed voltage controls thickness, flake area, number of defects and affects its properties. The process begins by bathing the graphite in a solvent for intercalation. The process can be tracked by monitoring the solution's transparency with an LED and photodiode.

Hydrothermal Self-assembly

Graphene has been prepared by using a sugar (e.g. glucose, sugar, fructose, etc.) This substrate-free "bottom-up" synthesis is safer, simpler and more environmentally friendly than exfoliation. The method can control thickness, ranging from monolayer to multilayers, which is known as "Tang-Lau Method".

Chemical Vapor Deposition

Epitaxy

Epitaxial graphene may be coupled to surfaces weakly enough (by Van der Waals forces) to retain the two dimensional electronic band structure of isolated graphene.

Heating silicon carbide (SiC) to high temperatures (1100 °C) under low pressures (c. 10^{-6} torr) reduces it to graphene.

A normal silicon wafer coated with a layer of germanium (Ge) dipped in dilute hydrofluoric acid strips the naturally forming germanium oxide groups, creating hydrogen-terminated germanium. CVD can coat that with graphene.

A two-step CVD process is shown to grow graphene directly on TiO_2 crystals or exfoliated TiO_2 nanosheets without using a metal catalyst.

Metal Substrates

The atomic structure of metal substrates including ruthenium, iridium, nickel and copper has been used as substrates for graphene production. In 2015 researchers announced the use of commercial copper foils for graphene production, reducing substrate costs by 100-fold.

Dielectric Substrates

The direct growth of high-quality, large single-crystalline domains of graphene on a dielectric substrate is of vital importance for applications in electronics and optoelectronics. Combining the advantages of both catalytic CVD and the ultra-flat dielectric substrate, gaseous catalyst-assisted CVD paves the way for synthesizing high-quality graphene for device applications while avoiding the transfer process.

Sodium Ethoxide Pyrolysis

Gram quantities were produced by the reduction of ethanol by sodium metal, followed by pyrolysis and washing with water.

Roll-to-roll

In 2014 a two-step roll-to-roll manufacturing process was announced. The first roll-to-roll step produces the graphene via chemical vapor deposition. The second step binds the graphene to a substrate.

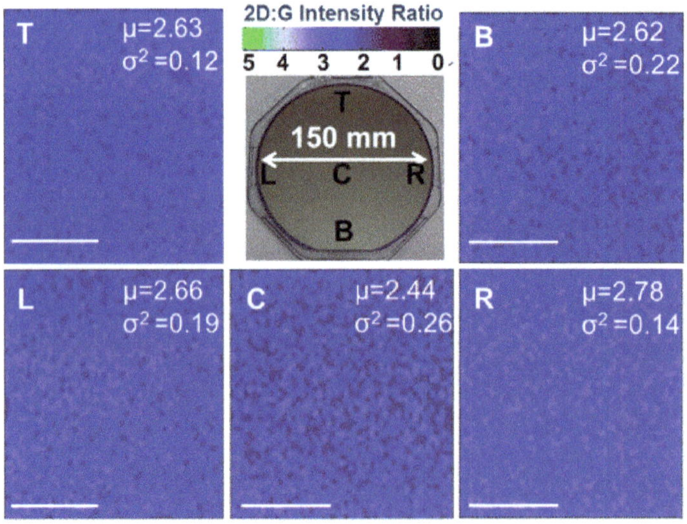

Large-area Raman mapping of CVD graphene on deposited Cu thin film on 150 mm SiO_2/Si wafers reveals >95% monolayer continuity and an average value of ~2.62 for I_{2D}/I_G. The scale bar is 200 μm.

Cold Wall

Growing graphene in an industrial resistive-heating cold wall CVD system was claimed to produce graphene 100 times faster than conventional CVD systems, cut costs by 99% and produce material with enhanced electronic qualities. Cold wall CVD technique can be used to study the underlying surface science involved in graphene nucleation and growth as it allows control of process parameters like gas flow rates, temperature and pressure. A home-built vertical cold wall system used resistive heating by passing direct current through the substrate. It provided insight into a typical surface-mediated nucleation and growth mechanism involved in two-dimensional materials grown using catalytic CVD under conditions sought out in the semiconductor industry.

Wafer Scale CVD

CVD graphene is scalable and has been integrated with ubiquitous CMOS technology via growth on deposited copper thin film catalyst on 100 to 300 mm standard Si/SiO_2 wafers on an Axitron Black Magic system. Monolayer graphene coverage of >95% is achieved on 100 to 300 mm wafer substrates with negligible defects, confirmed by extensive Raman mapping.

Nanotube Slicing

Graphene can be created by opening carbon nanotubes by cutting or etching. In one such method multi-walled carbon nanotubes are cut open in solution by action of potassium permanganate and sulfuric acid.

Carbon Dioxide Reduction

A highly exothermic reaction combusts magnesium in an oxidation–reduction reaction with carbon dioxide, producing carbon nanoparticles including graphene and fullerenes.

Spin Coating

In 2014, carbon nanotube-reinforced graphene was made via spin coating and annealing functionalized carbon nanotubes.

Supersonic Spray

Supersonic acceleration of droplets through a Laval nozzle was used to deposit reduced graphene-oxide on a substrate. The energy of the impact rearranges the carbon atoms into flawless graphene.

Another approach sprays buckyballs at supersonic speeds onto a substrate. The balls cracked open upon impact and the resulting unzipped cages then bond together to form a graphene film.

Laser

A CO_2 infrared laser produced and patterned porous three-dimensional graphene film networks from commercial polymer films. The result exhibits high electrical conductivity.

Microwave-assisted Oxidation

Microwave energy was reported to directly synthesize graphene in one step. This approach avoids use of potassium permanganate in the reaction mixture. Microwave radiation assistance allows graphene oxide with or without holes can be synthesized by controlling microwave time. Microwave heating can dramatically shorten the reaction time from days to seconds.

Ion Implantation

Accelerating carbon ions under an electrical field into a semiconductor made of thin nickel films on a substrate of SiO_2/Si, creates a wafer-scale (4 inches (100 mm)) wrinkle/tear/residue-free graphene layer at a relatively low temperature of 500 °C.

Heated Vegetable Oil

Researchers heated soybean oil in a furnace for ≈30 minutes. The heat decomposed the oil into elemental carbon that deposited on nickel foil as single/few-layer graphene.

Analogs

Analogs (also referred to as "artificial graphene") are two-dimensional systems that exhibit similar properties to graphene. Analogs can be systems in which the physics is easier to observe and to manipulate. In those systems, electrons are not always the chosen particles. They might be optical photons, microwave photons, plasmons, microcavity polaritons or even atoms. Also, the honeycomb structure in which those particles evolve can be of a different nature than carbon atoms in graphene. It can be, respectively, a photonic crystal, an array of metallic rods, metallic nanoparticles, a lattice of coupled microcavities or an optical lattice.

Applications

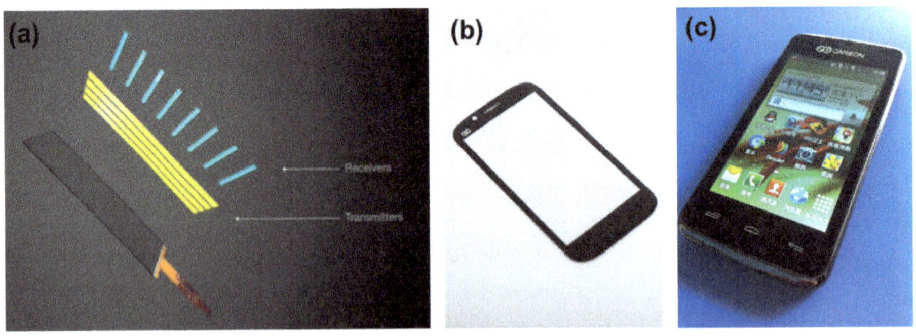

(a) The typical structure of a touch sensor in a touch panel. (Image courtesy of Synaptics, Incorporated.) (b) An actual example of 2D Carbon Graphene Material Co.,Ltd's graphene transparent conductor-based touchscreen that is employed in (c) a commercial smartphone.

Graphene is a transparent and flexible conductor that holds promise for various material/device applications, including solar cells, light-emitting diodes (LED), touch panels and smart windows or phones.

Graphene-based touch panel modules produced by a China-based company (2D Carbon Graphene Material Co., Ltd) have been sold in volume to cell phone, wearable device and home appliance manufacturers. For instance, smart phone products with graphene touch screens are already on the market.

As of 2015 one product was available for commercial use: a graphene-infused printer powder.

In 2016, Adgero announced a regenerative braking system for large trucks that employed a graphene-based supercapacitor.

BAC's 2016 Mono model is said to be made out of graphene as a first of both a street-legal track car and a production car.

Health Risks

The toxicity of graphene has been extensively debated in the literature. The most comprehensive review on graphene toxicity summarized the *in vitro, in vivo*, antimicrobial and environmental effects and highlights the various mechanisms of graphene toxicity. The toxicity of graphene is dependent on factors such as shape, size, purity, post-production processing steps, oxidative state, functional groups, dispersion state, synthesis methods, route, dose of administration and exposure times.

Graphene nanoribbons, graphene nanoplatelets and graphene nano–onions are non-toxic at concentrations up to 50 μg/ml. These nanoparticles do not alter the differentiation of human bone marrow stem cells towards osteoblasts (bone) or adipocytes (fat) suggesting that at low doses graphene nanoparticles are safe for biomedical applications. 10 μm few-layered graphene flakes were able to pierce cell membranes in solution. They were observed to enter initially via sharp and jagged points, allowing graphene to be enter the cell. The physiological effects of this remain uncertain, and this remains a relatively unexplored field.

Properties of Graphene

- Electronic properties
- Thermal properties
- Mechanical properties
- Optical properties
- Relativistic charge carriers
- Anomalous quantum Hall effect

Electronic Properties

- High electron mobility (at room temperature ~ 200.000 $cm^2/(V{\cdot}s)$, ex. Si at RT~ 1400 $cm^2/(V{\cdot}s)$, carbon nanotube: ~ 100.000 $cm^2/(V{\cdot}s)$, organic semiconductors (polymer, oligomer): <10 $cm^2/(V{\cdot}s)$

$$v_d = \mu E$$

Where v_d is the drift velocity in m/s (SI units)

E is the applied electric field in V/m (SI)

μ is the mobility in $m^2/(V.s)$, in SI units

- Resistivity of the graphene sheet ~10^6 Ω·cm, less than the resistivity of silver (Ag), the lowest resistivity substance known at room temperature (electrical resistivity is also as the inverse of the conductivity σ (sigma), of the material, or

$$\rho = \frac{1}{\sigma}$$

Material	Electrical Conductivity (S.m^{-1})	Notes
Graphene	~10^8	
Silver	63.0×10^6	Best electrical conductor of any known metal
Copper	59.6×10^6	Commonly used in electrical wire applications due to very good conductivity and price compared to silver.
Annealed Copper	58.0×10^6	Referred to as 100% IACS or international annealed copper standard. The unit for expressing the conductivity of nonmagnetic meterials by testing using the eddy-current method. Generally used for temper and alloy verification of aluminium.
Gold	45.2×10^6	Gold is commonly used in electrical contacts becouse it does not easily corrode.
Aluminium	37.8×10^6	Commonly used for high voltage electricity distribution cable.
Sea water	4.8	Corresponds to an average salinity of 35 g/kg at 20 °C.
Drinking water	0.0005 to 0.05	This value range is typical of high quality water and not an indicator of water quality.

Mechanical Properties

- High Young's modulus (~1,100 Gpa) High fracture strength (125 Gpa) - Graphene is as the strongest material ever measured, some 200 times stronger than structural steel	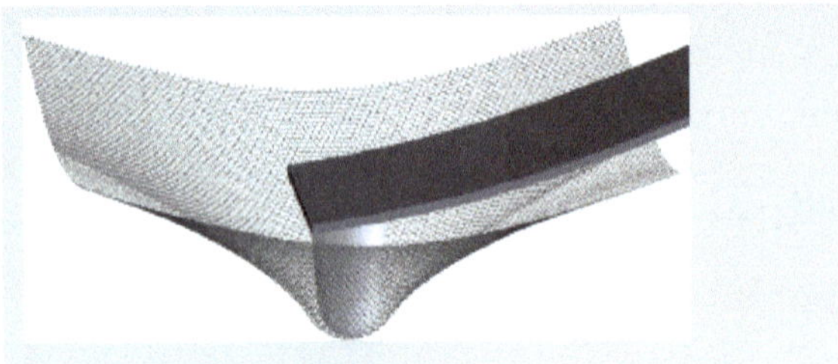 A representation of a diamond tip with a two nanometer radius indenting into a single atomic sheet of graphene (*Science*, **321** (5887): 385)

Optical Properties

Monolayer graphene absorbs $\pi\alpha \approx$ 2.3% of white light (97.7 % transmittance), where α is the fine-structure constant.

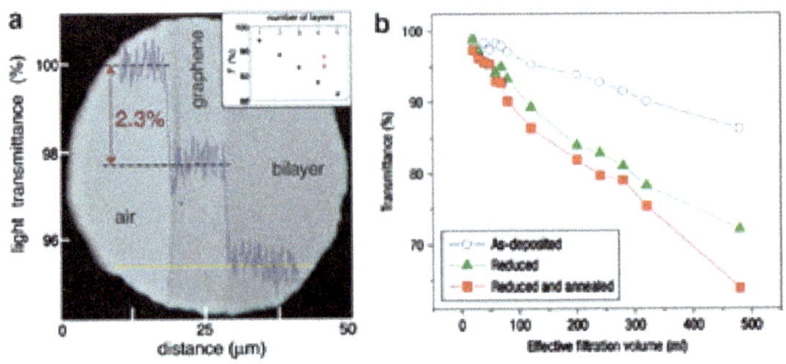

Brief History of Graphene

The term graphene first appeared in 1987 to describe single sheets of graphite as one of the constituents of graphite intercalation compounds (GICs). Larger graphene molecules or sheets (so that they can be considered as true isolated 2D crystals) cannot be grown even in principle. In the 1930s, Landau and Peierls (and Mermin, later) showed thermodynamics prevented 2-d crystals in free state, a n article in Physics Today reads:

"Fundamental forces place seemingly insurmountable barriers in the way of creating [2D crystals] ... Nascent 2D crystallites try to minimize their surface energy and inevitably morph into one of the rich variety of stable 3D structures that occur in soot. But there is a way around the problem. Interactions with 3D structures stabilize 2D crystals during growth. So one can make 2D crystals sandwiched between or placed on top of the atomic planes of a bulk crystal. In that respect, graphene already exists within graphite ... One can then hope to fool Nature and extract single-atom-thick crystallites at a low enough temperature that they remain in the quenched state prescribed by the original higher-temperature 3D growth."

In 2004: Andre Geim and Kostya Novoselov at Manchester University managed to extract single-atom-thick crystallites (graphene) from bulk graphite: Pulled out graphene layers from graphite and transferred them onto thin silicon dioxide on a silicon wafer in a process sometimes called micromechanical cleavage or, simply, the Scotch tape technique. Since 2004, an explosion in the investigation of graphene in term of synthesis, characterization, properties as well as specifical potential application were reported.

Preparation and Characterization Graphene

Preparation methods

Top-down approach
(From graphite)

- Micromechanical exfoliation of graphite (Scotch tape or peel-off method)
- Creation of colloidal suspensions from graphite oxide or graphite intercalation compounds (GICs)

Bottom up approach
(from carbon precursors)

- By chemical vapour deposition (CVD) of hydrocarbon
- By epitaxial growth on electrically insulating surfaces such as SiC
- Total Organic Synthesis

Advantages and disadvantages for techniques currently used to produce graphene.

	Advantages	Disadvantages
Mechanical exfoliation	Low-cost and easy No special equipment needed, SiO_2 thickness is tuned for better contrast	Serendipitous Uneven films Labor intensive (not suitable for large-scale production)
Epitaxial growth	Most even films (of any method) Large scale area	Difficult control of morphology and adsorption energy High-temperature process
Graphene oxide	Straightforward up-scaling Versatile handling of the suspension Rapid process	Fragile stability of the colloidal dispersion Reduction to graphene is only partial

Characterization Methods

Scanning Probe Microscopy (SPM):	Raman Spectroscopy	Transmission electron Microscopy (TEM)	X-ray diffraction (XRD)

- Atomic force microscopes (AFMs)

- Scanning tunneling microscopy (STM)

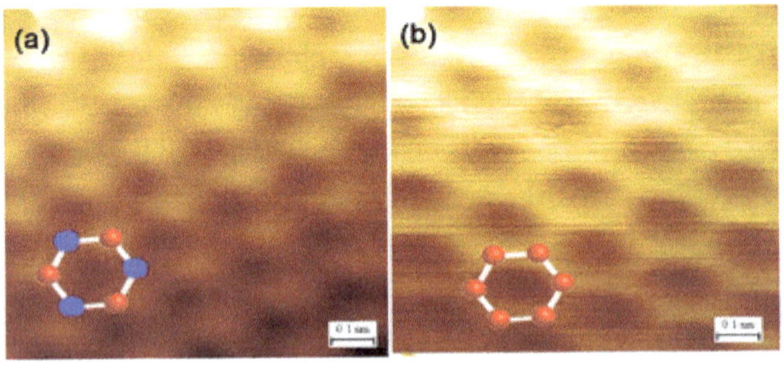

(a) STM image of graphite showing only the three carbons that eclipse a neighbour in the sheet directly below.
(b) In contrest, all six carbons are equivalent and thus visible in mechanically exforliated single-layer graphen.

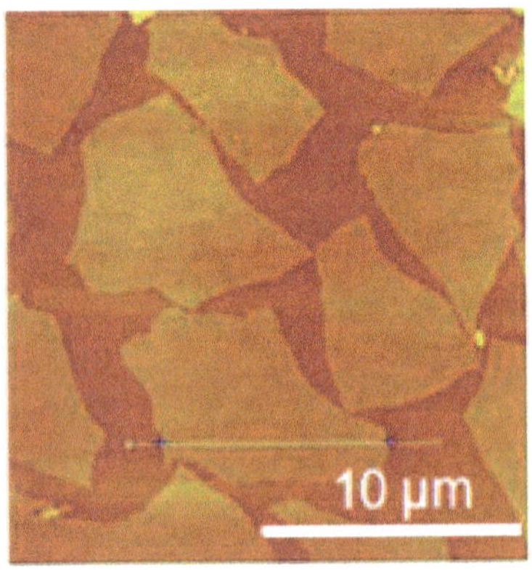

Atomic force microscopy images of a graphite oxide film deposited by Langmuir-Blodgett assembly

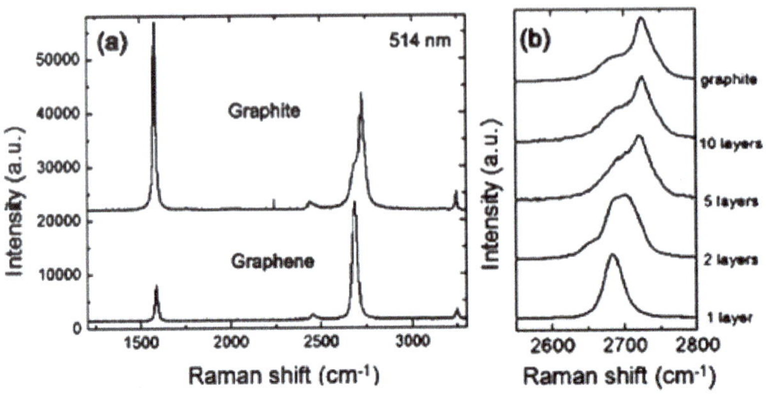

Raman Spectroscopy

Raman Spectroscopy is a powerful diagnostic tool for the study of graphene. Both the G (1584 cm^{-1}) and G' (2700 cm^{-1}) bands undergo Significant changes due to the thickness of AB stacked flakes, as produced by mechanical exfoliation

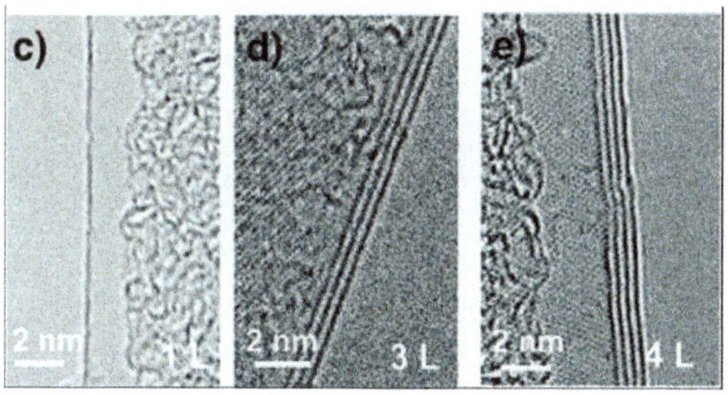

Transmission electron Microscopy (TEM)

TEM images show the nucleation of (c) one, (d) three, or (e) four layers during the growth process

Preparation Methods and Discussions

Top-down approach (From graphite)

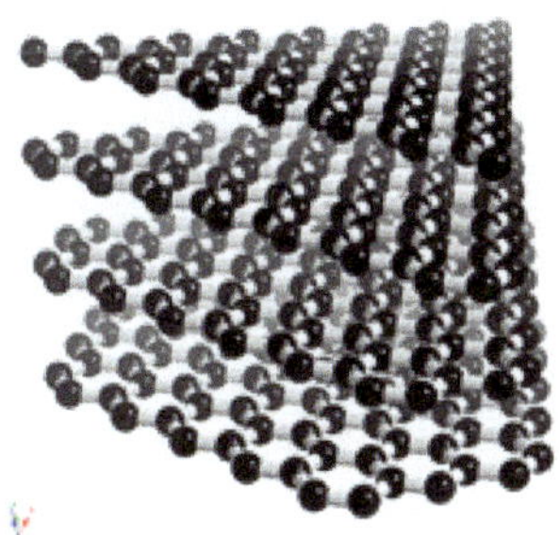

Direct exfoliation of graphite

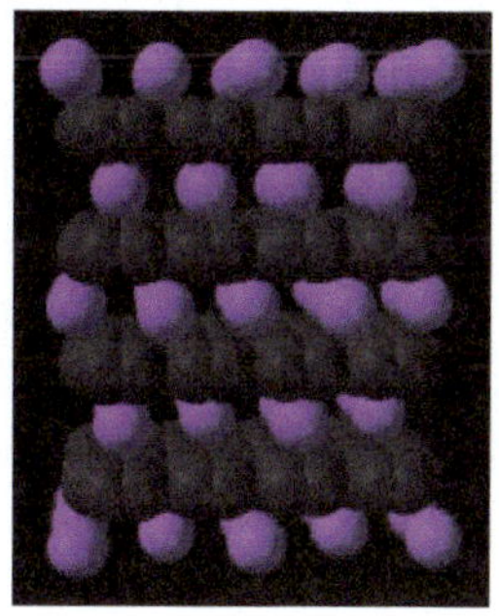

Graphite intercalation compound

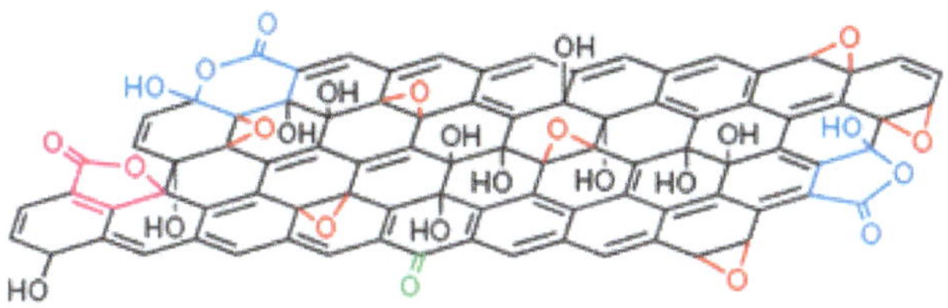

Graphite oxide method

Graphite Oxide Method (Most Common and High Yield Method)

Oxidation (Hummers' Method)

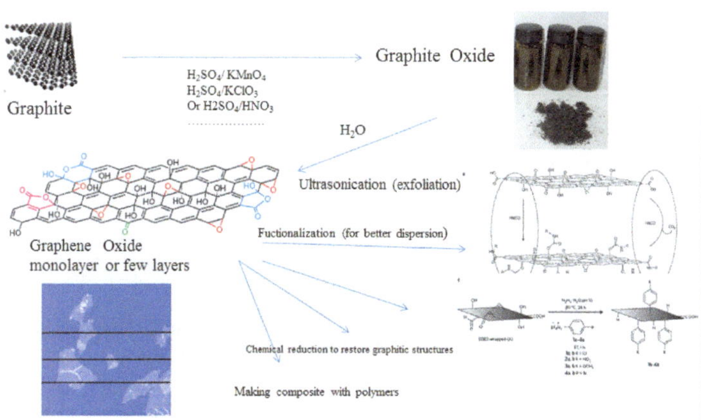

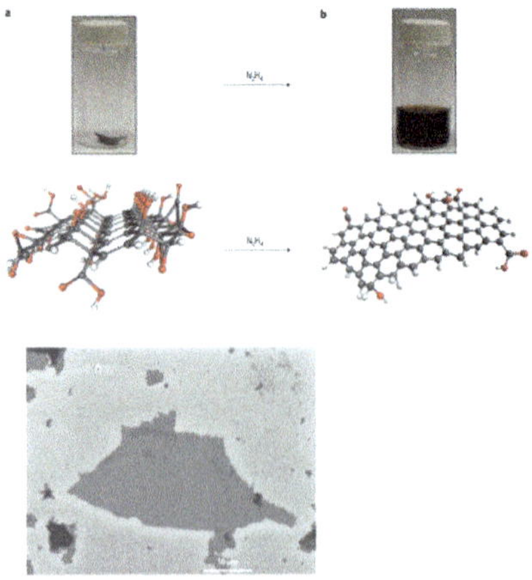

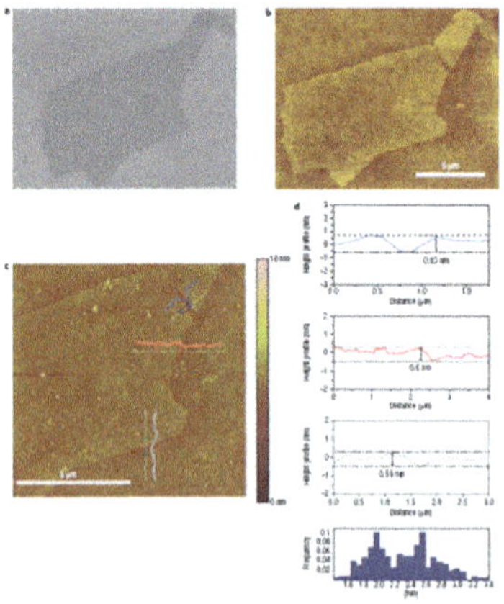

Tung, V. C., Allen, M. J., Yang, Y. & Kaner , R. B. High-throughput solution processing of large-scale graphene

Potential application of graphene

- Single molecule gas detection
- Graphene transistors
- Integrated circuits
- Transparent conducting electrodes for the replacement of ITO
- Ultracapacitors
- Graphene biodevices
- Reinforcement for polymer nanocomposites: Electrical, thermally conductive nanocomposites, antistatic coating, transparent conductive composites..etc

Electrical, Thermally Conductive Nanocomposites

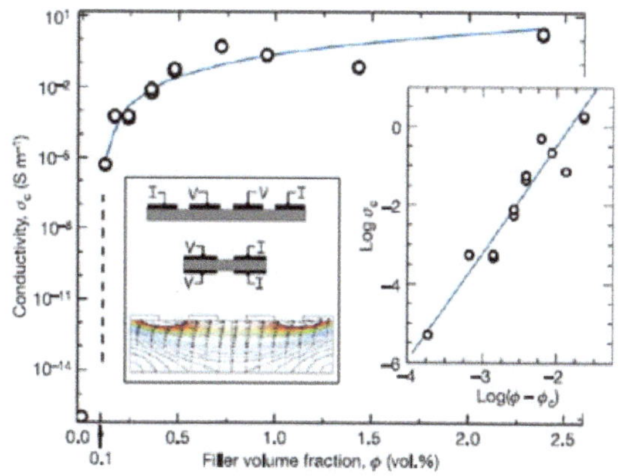

Electrical conductivity of the polystyrene-graphene composites as a function of filler volum fraction. Main figure, composite conductivity, σ_c plotted against filler volume fraction, ϕ Right inset, log σ_c plotted against log $(\phi - \phi_c)$, where ϕ_c is the percolation threshold. Solid lines in both graphs are calculated conductivities based on the fitting (inset, log-log plot) of the experimental data to the effective conductivity equation described in the text. Fitted parameters are: $t = 2.74 \pm 0.20$, $\sigma_f = 10^{4.92 \pm 0.52}\,\mathrm{S m^{-1}}$ and $\phi_c = 0.1\,\mathrm{vol.\%}$ Left inset: top and midlle diagrams show the four-probe setup for in-plane and transverse measurements, respectively; bottom diagram ,one of the computed distributions of the current density (contour lines) with local directions and magnitude (show by arrows) in a specimen for the following conditions- the sample thickness is twice the electrode width and the gap between them, and the in-plane resistivity is 10 times lower than the transverse resistivity.

Graphene has an interesting history, but many now wonder about its future. The subject of considerable scholarly debate, it does seem reasonable to assert a few things looking ahead:

First, the quality and availability of “synthetic” graphene will continue to improve. Whether high quality material comes in the form of an alternative chemical route to the complete exfoliation of graphite or from optimization of the thermal processes required for substrate-based methods, there is no sign that synthetic techniques are nearing their upper limit. This means that device engineers will have ample access to improved materials for developing novel structures and finding ways to integrate graphene into present-day electronic devices.

Second, chemical modification of graphene’s basal plane or its edges will substantially influence graphene -based devices. For electronic applications, one can imagine the attachment of functional groups aimed at self-assembly of simple circuits or the incorporation of chemical dopants to limit leakage current under zero gate bias. For sensors, lock and-key type binding sites could provide selective sensitivity to a wide variety of analytes. These might include chemical warfare agents or even biological species.

Third, industrial use of graphene as a transparent conductor could have huge implications for the solar industry. As synthetic routes improve, the prospect of replacing ITO with a low-cost carbon-based coating seems feasible. This would not only remove significant uncertainty about the availability and cost of indium but also enable non evaporative roll-to roll processing of transparent conductors.

Graphene Nanoribbons

Graphene nanoribbons (GNRs, also called nano-graphene ribbons or nano-graphite ribbons), are strips of graphene with <50 nm width. Graphene ribbons were introduced as a theoretical model by Mitsutaka Fujita and coauthors to examine the edge and nanoscale size effect in graphene.

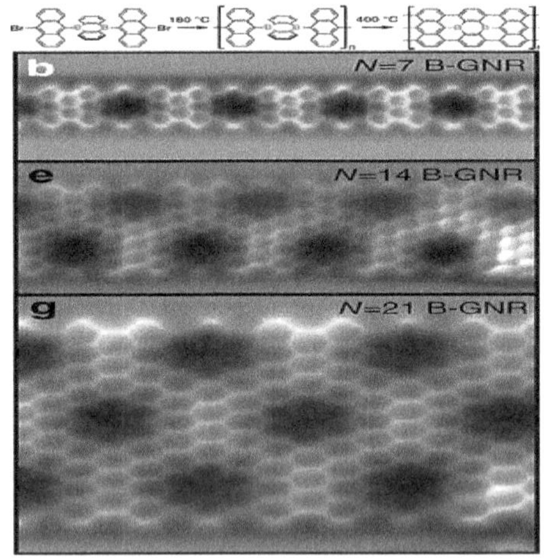

Atomic Force Microscopy (AFM) images of graphene nanoribbons having periodic width and boron doping pattern. The polymerization reaction used for their synthesis is shown on top.

Production

Nanotomy

Large quantities of width-controlled GNRs can be produced via graphite nanotomy, where applying a sharp diamond knife on graphite produces graphite nanoblocks, which can then be exfoliated to produce GNRs. GNRs can also be produced by "unzipping" or axially cutting nanotubes. In one such method multi-walled carbon nanotubes were unzipped in solution by action of potassium permanganate and sulfuric acid. In another method GNRs were produced by plasma etching of nanotubes partly embedded in a polymer film. More recently, graphene nanoribbons were grown onto silicon carbide (SiC) substrates using ion implantation followed by vacuum or laser annealing. The latter technique allows any pattern to be written on SiC substrates with 5 nm precision.

Epitaxy

GNRs were grown on the edges of three-dimensional structures etched into silicon carbide wafers. When the wafers are heated to approximately 1,000 °C (1,270 K; 1,830 °F), silicon is preferentially driven off along the edges, forming nanoribbons whose structure is determined by the pattern of the three-dimensional surface. The ribbons had perfectly smooth edges, annealed by the fabrication process. Electron mobility measurements surpassing one million correspond to a sheet resistance of one ohm per square— two orders of magnitude lower than in two-dimensional graphene.

Chemical Vapor Deposition

Nanoribbons narrower than 10 nm grown on a germanium wafer act like semiconductors, exhibiting a band gap. Inside a reaction chamber, using chemical vapor deposition, methane is used to deposit hydrocarbons on the wafer surface, where they react with each other to produce long, smooth-edged ribbons. The ribbons were used to create prototype transistors. At a very slow growth rate, the graphene crystals naturally grow into long nanoribbons on a specific germanium crystal facet. By controlling the growth rate and growth time, the researchers achieved control over the nanoribbon width.

Recently, researchers from SIMIT(Shanghai Institute of Microsystem and Information Technology,Chinese Academy of Sciences) reported on a strategy to grow graphene nanoribbons with controlled widths and smooth edges directly onto dielectric hexagonal boron nitride (h-BN) substrates. The team use nickel nanoparticles to etch monolayer-deep, nanometre-wide trenches into h-BN, and subsequently fill them with graphene using chemical vapour deposition. Modifying the etching parameters allows the width of the trench to be tuned to less than 10 nm, and the resulting sub-10-nm ribbons display bandgaps of almost 0.5 eV. Integrating these nanoribbons into field effect transistor devices reveals on–off ratios of greater than 10^4 at room temperature, as well as high carrier mobilities of ~750 $cm^2 V^{-1} s^{-1}$.

Multistep Nanoribbon Synthesis

A bottom-up approach was investigated. In 2017 dry contact transfer was used to press a fiberglass applicator coated with a powder of atomically precise graphene nanoribbons on a hydrogen-passivated Si(100) surface under vacuum. 80 of 115 GNRs visibly obscured the substrate lattice with an average apparent height of 0.30 nm. The GNRs do not align to the Si lattice, indicating a weak coupling. The average bandgap over 21 GNRs was 2.85 eV with a standard deviation of 0.13 eV.

The method unintentionally overlapped some nanoribbons, allowing the study of multilayer GNRs. Such overlaps could be formed deliberately by manipulation with a scanning tunneling microscope. Hydrogen depassivation left no band-gap. Covalent bonds between the Si surface and the GNR leads to metallic behavior. The Si surface atoms move outward, and the GNR changes from flat to distorted, with some C atoms moving in toward the Si surface.

Electronic Structure

The electronic states of GNRs largely depend on the edge structures (armchair or zigzag). In zigzag edges each successive edge segment is at the opposite angle to the previous. In armchair edges, each pair of segments is a 120/-120 degree rotation of the prior pair. Zigzag edges provide the edge localized state with non-bonding molecular orbitals near the Fermi energy. They are expected to have large changes in optical and electronic properties from quantization.

Calculations based on tight binding theory predict that zigzag GNRs are always metallic while armchairs can be either metallic or semiconducting, depending on their width. However, density functional theory (DFT) calculations show that armchair nanoribbons are semiconducting with an energy gap scaling with the inverse of the GNR width. Experiments verified that energy gaps increase with decreasing GNR width. Claims that graphene nanoribbons with controlled edge orientation have been fabricated by scanning tunneling microscope (STM) lithography are highly controversial and some of the data might have been fabricated, as one of the authors has been recognized guilty of scientific misconduct by an official commission of the Max Planck Gesellschaft. Energy gaps up to 0.5 eV in a 2.5 nm wide armchair ribbon were reported.

Zigzag nanoribbons are semiconducting and present spin polarized edges. Their gap opens thanks to an unusual antiferromagnetic coupling between the magnetic moments at opposite edge carbon atoms. This gap size is inversely proportional to the ribbon width and its behavior can be traced back to the spatial distribution properties of edge-state wave functions, and the mostly local character of the exchange interaction that originates the spin polarization. Therefore, the quantum confinement, inter-edge superexchange, and intra-edge direct exchange interactions in zigzag GNR are important for its magnetism and band gap. The edge magnetic moment and band gap of zigzag GNR are reversely proportional to the electron/hole concentration and they can be controlled by alkaline adatoms.

Their 2D structure, high electrical and thermal conductivity and low noise also make GNRs a possible alternative to copper for integrated circuit interconnects. Research is exploring the creation of quantum dots by changing the width of GNRs at select points along the ribbon, creating quantum confinement.

Graphene nanoribbons possess semiconductive properties and may be a technological alternative to silicon semiconductors capable of sustaining microprocessor clock speeds in the vicinity of 1 THz field-effect transistors less than 10 nm wide have been created with GNR – "GNRFETs" – with an I_{on}/I_{off} ratio >10^6 at room temperature.

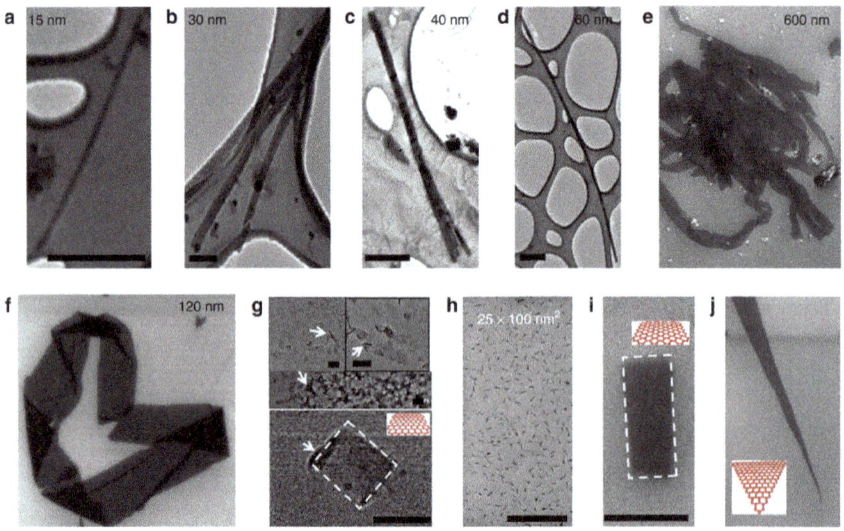

TEM micrographs of GNRs of (a) w=15, (b) w=30, (c) w=40 (exfoliating), and (d) w=60 nm deposited on 400 mesh lacey carbon grids and (e) FESEM micrograph of 600 nm ribbon. (f) Electron microscope images of a 120-nm graphene ribbons (FESEM), (g) 50 nm square GQDs (FESEM), (h,i) 25×100 nm² rectangular GQDs (FESEM), and (j) 8°-angled tapered GNR (or triangular GQD) (FESEM)). The large densities of square and rectangular GQDs (g) showed extensive folding (white arrows). Bar sizes=(a) 250 nm, (b,g,i) 50 nm, (c,d) 500 nm, and (h) 1 μm.

Exciton Properties

First-principle calculations with quasiparticle corrections and many-body effects explored the electronic and optical properties of graphene-based materials. With GW calculation, the properties of graphene-based materials are accurately investigated, including graphene nanoribbons, edge and surface functionalized armchair graphene nanoribbons and scaling properties in armchair graphene nanoribbons.

Applications

Polymeric Nanocomposites

Graphene nanoribbons and their oxidized counterparts called graphene oxide nanoribbons have been investigated as nano-fillers to improve the mechanical properties of polymeric nanocomposites. Increases in the mechanical properties of epoxy composites on loading of graphene nanoribbons were observed. An increase in the mechanical properties of biodegradable polymeric nanocomposites of poly(propylene fumarate) at low weight% was achieved by loading of oxidized graphene nanoribbons, fabricated for bone tissue engineering applications.

Contrast Agent for Bioimaging

Hybrid imaging modalities, such as photoacoustic (PA) tomography (PAT) and thermoacoustic (TA) tomography (TAT) have been developed for bioimaging applications. PAT/TAT combines advantages of pure ultrasound and pure optical imaging/radio frequency (RF), providing good spatial resolution, great penetration depth and high soft-tissue contrast. GNR synthesized by unzipping single- and multi-walled carbon nanotubes have been reported as contrast agents for photoacoustic and thermoacoustic imaging and tomography.

Graphene Production Techniques

A rapidly increasing list of graphene production techniques have been developed to enable graphene's use in commercial applications.

Isolated 2D crystals cannot be grown via chemical synthesis beyond small sizes even in principle, because the rapid growth of phonon density with increasing lateral size forces 2D crystallites to bend into the third dimension. However, other routes to 2d materials exist:

Fundamental forces place seemingly insurmountable barriers in the way of creating [2D crystals]... The nascent 2D crystallites try to minimize their surface energy and inevitably morph into one of the rich variety of stable 3D structures that occur in soot.

But there is a way around the problem. Interactions with 3D structures stabilize 2D crystals during growth. So one can make 2D crystals sandwiched between or placed on top of the atomic planes of a bulk crystal. In that respect, graphene already exists within graphite. One can then hope to fool Nature and extract single-atom-thick crystallites at a low enough temperature that they remain in the quenched state prescribed by the original higher-temperature 3D growth.

The early approaches of cleaving multi-layer graphite into single layers or growing it epitaxially by depositing a layer of carbon onto another material have been supplemented by numerous alternatives. In all cases, the graphite must bond to some substrate to retain its 2d shape.

Exfoliation

As of 2014 exfoliation produced graphene with the lowest number of defects and highest electron mobility.

Adhesive Tape

Andre Geim and Konstantin Novoselov initially used adhesive tape to split graphite into graphene. Achieving single layers typically requires multiple exfoliation steps, each producing a slice with fewer layers, until only one remains. After exfoliation the flakes are deposited on a silicon wafer. Crystallites larger than 1 mm and visible to the naked eye can be obtained.

Wedge-based

In this method, a sharp single-crystal diamond wedge penetrates onto the graphite source to exfoliate layers. This method uses highly ordered pyrolytic graphite (HOPG) as the starting material. The experiments were supported by molecular dynamic simulations.

Graphite Oxide Reduction

P. Boehm reported producing monolayer flakes of reduced graphene oxide in 1962. Rapid heating of graphite oxide and exfoliation yields highly dispersed carbon powder with a few percent of graphene flakes. Reduction of graphite oxide monolayer films, e.g. by hydrazine with annealing in argon/hydrogen also yielded graphene films. Later the oxidation protocol was enhanced to yield graphene oxide with an almost intact carbon framework that allows efficient removal of functional groups, neither of which was originally possible. The measured charge carrier mobility exceeded 1,000 centimetres (393.70 in)/Vs. Spectroscopic analysis of reduced graphene oxide has been conducted.

Shearing

In 2014 defect-free, unoxidized graphene-containing liquids were made from graphite using mixers that produce local shear rates greater than 10×10^4. The method was claimed to be applicable to other 2D materials, including boron nitride, Molybdenum disulfide and other layered crystals.

Sonication

Solvent-aided

Dispersing graphite in a proper liquid medium can produce graphene by sonication. Graphene is separated from graphite by centrifugation, producing graphene concentrations initially up to 0.01 mg/mL in N-methylpyrrolidone (NMP) and later to 2.1 mg/mL in NMP. Using a suitable ionic liquid as the dispersing liquid medium produced concentrations of 5.33 mg/mL. Graphene concentration produced by this method is very low, because nothing prevents the sheets from restacking due to van der Waals forces. The maximum concentrations achieved are the points at which the van der Waals forces overcome the interactive forces between the graphene sheets and the solvent molecules.

Adding a surfactant to a solvent prior to sonication prevents restacking by adsorbing to the graphene's surface. This produces a higher graphene concentration, but removing the surfactant requires chemical treatments.

Immiscible Liquids

Sonicating graphite at the interface of two immiscible liquids, most notably heptane and water, produced macro-scale graphene films. The graphene sheets are adsorbed to the high energy interface between the heptane and the water, where they are kept from restacking. The graphene remained at the interface even when exposed to force in excess of 300,000 g. The solvents may then be evaporated. The sheets are up to ~95% transparent and conductive.

Molten Salts

Graphite particles can be corroded in molten salts to form a variety of carbon nanostructures including graphene. Hydrogen cations, dissolved in molten Lithium chloride, can be discharged on cathodically polarized graphite rods, which then intercalate into the graphite structure, peeling graphite to produce graphene. The graphene nanosheets produced displayed a single-crystalline structure with a lateral size of several hundred nanometers and a high degree of crystallinity and thermal stability.

Electrochemical Synthesis

Electrochemical synthesis can exfoliate graphene. Varying a pulsed voltage controls thickness, flake area, number of defects and affects its properties. The process begins by bathing the graphite in a solvent for intercalation. The process can be tracked by monitoring the solution's transparency with an LED and photodiode.

Hydrothermal Self-assembly

Graphene has been prepared by using a sugar (e.g. glucose, fructose, etc.) This substrate-free "bottom-up" synthesis is safer, simpler and more environmentally friendly than exfoliation. The method can control thickness, ranging from monolayer to multilayers.

Chemical Vapor Deposition

Epitaxy

Epitaxy refers to the deposition of a crystalline overlayer on a crystalline substrate, where there is registry between the two. In some cases epitaxial graphene layers are coupled to surfaces weakly enough (by Van der Waals forces) to retain the two dimensional electronic band structure of isolated graphene. An example of this weak coupling is epitaxial graphene on SiC.

Silicon Carbide

Heating silicon carbide (SiC) to high temperatures (>1100 °C) under low pressures (~10^{-6} torr) reduces it to graphene. This process produces epitaxial graphene with dimensions dependent upon the size of the wafer. The face of the SiC used for graphene formation, silicon- or carbon-terminated, highly influences the thickness, mobility and carrier density.

Graphene's electronic band-structure (so-called Dirac cone structure) was first visualized in this material. Weak anti-localization is observed in this material, but not in exfoliated graphene produced by the drawing method. Large, temperature-independent mobilities approach those in exfoliated graphene placed on silicon oxide, but lower than mobilities in suspended graphene produced by the drawing method. Even without transfer, graphene on SiC exhibits massless Dirac fermions. The graphene–substrate interaction can be further passivated.

The weak van der Waals force that coheres multilayer stacks does not always affect the individual layers' electronic properties. That is, while the electronic properties of certain multilayered epitaxial graphenes are identical to that of a single layer, other properties are affected, as they are in bulk graphite. This effect is well understood theoretically and is related to the symmetry of the interlayer interactions.

Epitaxial graphene on SiC can be patterned using standard microelectronics methods. A band gap can be created and tuned by laser irradiation.

Silicon/germanium/hydrogen

A normal silicon wafer coated with a layer of germanium (Ge) dipped in dilute hydrofluoric acid strips the naturally forming germanium oxide groups, creating hydrogen-terminated germanium. Chemical vapor deposition deposits a layer of graphene on top. The graphene can be peeled from the wafer using a dry process and is then ready for use. The wafer can be reused. The graphene is wrinkle-free, high quality and low in defects.

Metal Substrates

The atomic structure of metal substrates can seed the growth of graphene.

Ruthenium

Graphene grown on ruthenium does not typically produce uniform layer thickness. Bonding between the bottom graphene layer and the substrate may affect layer properties.

Iridium

Graphene grown on iridium is weakly bonded, uniform in thickness, albeit rippled, and can be highly ordered. Due to the long-range order of these ripples, minigaps in the electronic band-structure (Dirac cone) become visible.

Nickel

High-quality sheets of few-layer graphene exceeding 1 cm^2 (0.2 sq in) in area have been synthesized via CVD on thin nickel films using multiple techniques. First the film is exposed to Argon gas at 900–1000 degrees Celsius. Methane is then mixed into the gas, and the methane's disassociated carbon is absorbed into the film. The solution is then cooled and the carbon diffuses out of the nickel to form graphene films.

Another method used temperatures compatible with conventional CMOS processing, using a nickel-based alloy with a gold catalyst. This process dissolves carbon atoms inside a transition metal melt at a certain temperature and then precipitates the dissolved carbon at lower temperatures as single layer graphene (SLG).

The metal is first melted in contact with a carbon source, possibly a graphite crucible inside which the melt is carried out or graphite powder/chunks that are placed in the melt. Keeping the melt in contact with the carbon at a specific temperature dissolves the carbon atoms, saturating the melt based on the metal–carbon binary phase diagram. Lowering the temperature decreases carbon's solubility and the excess carbon precipitates onto the melt. The floating layer can be either skimmed or frozen for later removal.

Using different morphology, including thick graphite, few layer graphene (FLG) and SLG were observed on metal substrate. Raman spectroscopy proved that SLG had grown on nickel substrate. The SLG Raman spectrum featured no D and D′ band, indicating its pristine nature. Since nickel is not Raman active, direct Raman spectroscopy of graphene layers on top of the nickel is achievable.

Another approach covered a sheet of silicon dioxide glass (the substrate) on one side with a nickel film. Graphene deposited via chemical vapor deposition formed into layers on both sides of the film, one on the exposed top side, and one on the underside, sandwiched between nickel and glass. Peeling the nickel and the top layer of graphene left an intervening layer of graphene on the glass. While the top graphene layer could be harvested from the foil as in earlier methods, the bottom layer was already in place on the glass. The quality and purity of the attached layer was not assessed.

Copper

Copper foil, at room temperature and very low pressure and in the presence of small amounts of methane produces high quality graphene. The growth automatically stops after a single layer forms. Arbitrarily large films can be created. The single layer growth is due to the low concentration of carbon in methane. The process is surface-based rather than relying on absorption into the metal and then diffusion of carbon into graphene layers on the surface. The room temperature process eliminates the need for postproduction steps and reduces production from a ten-hour/nine- to ten-step procedure to a single step that takes five minutes. A chemical reaction between the hydrogen plasma formed from the methane and ordinary air molecules in the chamber generates cyano radicals—carbon–nitrogen molecules without electrons. These charged molecules scour away surface imperfections, providing a pristine substrate. The graphene deposits form lines that merge into each other, forming a seamless sheet that contributes to mechanical and electrical integrity.

Larger hydrocarbons such as ethane and propane produce bilayer coatings. Atmospheric pressure CVD growth produces multilayer graphene on copper (similar to nickel).

The material has fewer defects, which in higher temperature processes result from thermal expansion/contraction. Ballistic transport was observed in the resulting material.

Sodium Ethoxide Pyrolysis

Gram-quantities were produced by the reduction of ethanol by sodium metal, followed by pyrolysis of the ethoxide product and washing with water to remove sodium salts.

Roll-to-roll

Large scale roll-to-roll production of graphene based on chemical vapor deposition, was first demonstrated in 2010. In 2014 a two-step roll-to-roll manufacturing process was announced. The first roll-to-roll step produces the graphene via chemical vapor deposition, and the second step binds the graphene to a substrate.

Cold Wall

Growing graphene in an industrial resistive-heating cold wall CVD system was claimed to produce graphene 100 times faster than conventional CVD systems, cuts costs by 99 percent and produce material with enhanced electronic qualities.

Cold wall CVD technique can be used to study the underlying surface science involved in graphene nucleation and growth as it allows unprecedented control of process parameters like gas flow rates, temperature and pressure as demonstrated in a recent study. The study was carried out in a home-built vertical cold wall system utilizing resistive heating by passing direct current through the substrate. It provided conclusive insight into a typical surface-mediated nucleation and growth mechanism involved in two-dimensional materials grown using catalytic CVD under conditions sought out in the semiconductor industry.

Nanotube Slicing

Graphene can be created by cutting open carbon nanotubes. In one such method multi-walled carbon nanotubes are cut open in solution by action of potassium permanganate and sulfuric acid. In another method graphene nanoribbons were produced by plasma etching of nanotubes partly embedded in a polymer film.

Carbon Dioxide Reduction

A highly exothermic reaction combusts magnesium in an oxidation–reduction reaction with carbon dioxide, producing a variety of carbon nanoparticles including graphene and fullerenes. The carbon dioxide reactant may be either solid (dry-ice) or gaseous. The products of this reaction are carbon and magnesium oxide. US patent 8377408 was issued for this process.

Spin Coating

In 2014, carbon nanotube-reinforced graphene was made via spin coating and annealing functionalized carbon nanotubes. The resulting material was stronger, flexible and more conductive than conventional graphene.

Supersonic Spray

Supersonic acceleration of droplets through a Laval nozzle was used to deposit small droplets of reduced graphene-oxide in suspension on a substrate. The droplets disperse evenly, evaporate rapidly and display reduced flake aggregations. In addition, the topological defects (Stone-Wales defect and C_2 vacancies) originally in the flakes disappeared. The result was a higher quality graphene layer. The energy of the impact stretches the graphene and rearranges its carbon atoms into flawless hexagonal graphene with no need for post-treatment. The high amount of energy also allows the graphene droplets to heal any defects in the graphene layer that occur during this process.

Another approach sprays buckyballs at supersonic speeds onto a substrate. The balls cracked open upon impact, and the resulting unzipped cages then bond together to form a graphene film. The buckyballs are released into a helium or hydrogen gas, which expands at supersonic speeds, carrying the carbon balls with it. The buckyballs achieve energies of around 40 keV without changing their internal dynamics. This material contains hexagons and pentagons that come from the original structures. The pentagons could introduce a band gap.

Intercalation

Producing graphene via intercalation splits graphite into single layer graphene by inserting guest molecules/ions between the graphite layers. Graphite was first intercalated in 1841 using a strong oxidizing or reducing agent that damaged the material's desirable properties. Kovtyukhova developed a widely used oxidative intercalation method in 1999. In 2014, she was able to achieve intercalation using non-oxidizing Brønsted acids (phosphoric, sulfuric, dichloroacetic and alkylsulfonic acids), but without oxidizing agents. The new method has yet to achieve output sufficient for commercialization.

Laser

In 2014 a laser-based single-step, scalable approach to graphene production was announced. The technique produced and patterned porous three-dimensional graphene film networks from commercial polymer films. The system used a CO_2 infrared laser. The sp^3-carbon atoms were photothermally converted to sp^2-carbon atoms by pulsed laser irradiation. The result exhibits high electrical conductivity. The material can produce interdigitated electrodes for in-plane microsupercapacitors with specific capacitances of >4 mF cm^{-2} and power densities of ~9 mW cm^{-2}. Laser-induced production appeared to allow roll-to-roll manufacturing processes and provides a route to electronic and energy storage devices.

Applying a layer of graphite oxide film to a DVD and burning it in a DVD writer produced a thin graphene film with high electrical conductivity (1738 siemens per meter) and specific surface area (1520 square meters per gram) that was highly resistant and malleable.

Microwave-assisted Oxidation

In 2012, a microwave-assisted, scalable approach was reported to directly synthesize graphene with different size from graphite in one step. The resulting graphene does not need any post reduction treatment as it contains very little oxygen. This approach avoids use of potassium permanganate in the reaction mixture. It was also reported that by microwave radiation assistance, graphene oxide with or without holes can be synthesized by controlling microwave time. This method uses a recipe similar to Hummer's method, but uses microwave heating instead of traditional heating. Microwave heating can dramatically shorten the reaction time from days to seconds.

Ion Implantation

Accelerating carbon ions under an electrical field into a semiconductor made of thin Ni films on a substrate of SiO2/Si, creates a wafer-scale (4 inches (100 mm)) wrinkle/tear/residue-free graphene layer that changes the semiconductor's physical, chemical and electrical properties. The process uses 20 keV and a dose of 1×10^{15} cm^{-2} at a relatively low temperature of 500 °C. This was followed by high-temperature activation annealing (600–900 °C) to form an sp^2-bonded structure.

References

- Gibson, Robert (10 June 2014). "Consett firm Thomas Swan sees export success with grapheme". The Journal. Retrieved 23 July 2014
- Y S. Touloukian (1970). Thermophysical Properties of Matter: Thermal conductivity : nonmetallic solids. IFI/Plenum. ISBN 978-0-306-67020-6
- Geim, A. K. (2012). "Graphene Prehistory". Physica Scripta. T146: 014003. Bibcode:2012PhST..146a4003G. doi:10.1088/0031-8949/2012/T146/014003
- "Global breakthrough: Irish scientists discover how to mass produce 'wonder material' graphene". The Journal.ie. 20 April 2014. Retrieved 20 December 2014
- Burn-Callander, Rebecca (1 July 2014). "Graphene maker aims to build British, billion-pound venture". Daily Telegraph. Retrieved 24 July 2014
- Kohlschütter, V.; Haenni, P. (1919). "Zur Kenntnis des Graphitischen Kohlenstoffs und der Graphitsäure". Zeitschrift für anorganische und allgemeine Chemie (in German). 105 (1): 121–144. doi:10.1002/zaac.19191050109
- Bernatowicz; T. J.; et al. (1996). "Constraints on stellar grain formation from presolar graphite in the Murchison meteorite". Astrophysical Journal. 472 (2): 760–782. Bibcode:1996ApJ...472..760B. doi:10.1086/178105
- Borghino, Dario (February 15, 2016). "Liquid-like graphene could be the key to understanding black holes". New Atlas. Retrieved February 18, 2017
- Ishigami, Masa; et al. (2007). "Atomic Structure of Graphene on SiO_2". Nano Letters. 7 (6): 1643–1648. Bibcode:2007NanoL...7.1643I. doi:10.1021/nl070613a. PMID 17497819
- Fraundorf, P.; Wackenhut, M. (2002). "The core structure of presolar graphite onions". Astrophysical Journal Letters. 578 (2): L153–156. arXiv:astro-ph/0110585 . Bibcode:2002ApJ...578L.153F. doi:10.1086/344633
- Dorrieron, Jason (4 December 2014). "Graphene Armor Would Be Light, Flexible and Far Stronger Than Steel". Singularity Hub. Retrieved 6 October 2016
- Avouris, P.; Chen, Z.; Perebeinos, V. (2007). "Carbon-based electronics". Nature Nanotechnology. 2 (10): 605–15. Bibcode:2007NatNa...2..605A. doi:10.1038/nnano.2007.300. PMID 18654384
- Denis, P. A.; Iribarne, F. (2013). "Comparative Study of Defect Reactivity in Graphene". Journal of Physical Chemistry C. 117 (37): 19048–19055. doi:10.1021/jp4061945
- Coxworth, Ben (1 December 2014). "Graphene could find use in lightweight ballistic body armor". Gizmag. Retrieved 6 October 2016
- Fuhrer, Michael S. (2009). "A physicist peels back the layers of excitement about graphene". Nature. 459 (7250): 1037. Bibcode:2009Natur.459.1037F. doi:10.1038/4591037e. PMID 19553953
- Yamada, Y.; Murota, K; Fujita, R; Kim, J; et al. (2014). "Subnanometer vacancy defects introduced on graphene by oxygen gas". Journal of the American Chemical Society. 136 (6): 2232–2235. doi:10.1021/ja4117268. PMID 24460150
- Saito, K.; Nakamura, J.; Natori, A. (2007). "Ballistic thermal conductance of a graphene sheet". Physical Review B. 76 (11): 115409. Bibcode:2007PhRvB..76k5409S. doi:10.1103/PhysRevB.76.115409
- Eftekhari, A.; Jafarkhani, P. (2013). "Curly Graphene with Specious Interlayers Displaying Superior Capacity for Hydrogen Storage". Journal of Physical Chemistry C. 117 (48): 25845–25851. doi:10.1021/jp410044v
- "Samsung's graphene breakthrough could finally put the wonder material into real-world devices". ExtremeTech. 7 April 2014. Retrieved 13 April 2014
- Ho, Sungjae; Chen, Yung-Fu; Fuhrer, Michael S. (2007). "Gate-tunable Graphene Spin Valve". Applied Physics Letters. 91 (12): 123105. arXiv:0706.1597 . Bibcode:2007ApPhL..91l3105C. doi:10.1063/1.2784934

Nanoscience and Nanotechnology

Nanotechnology is the science that deals with modification of elements and particles at their molecular or atomic state. It also includes manufacturing products that are very small in size. One nanometre can be defined as one billionth of a metre. This section elucidates the crucial theories and principles of nanotechnology and nanoscience.

Nanoscale Science

Nanoscale science can be defined as the chemistry and physics of structures that are on the length scale of 1-100 nm (1nm = 10^{-9} m or 10 Å), or require tolerances below 100 nm.

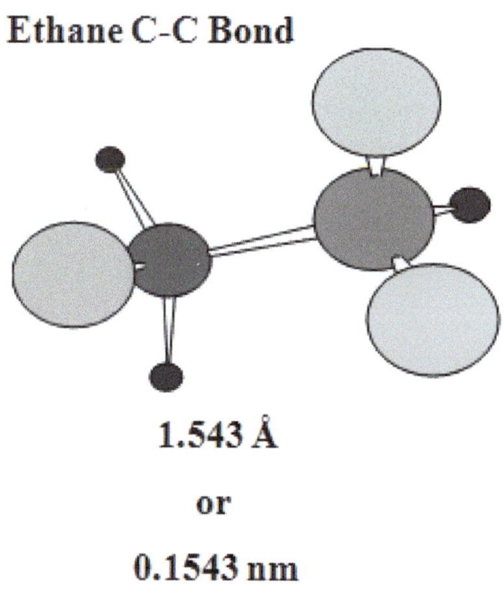

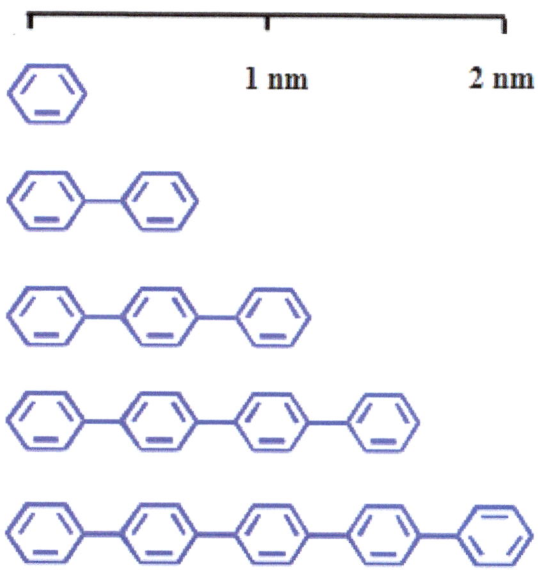

Early Nano History

When viewed in reflected light the minute metallic particles are just coarse enough to reflect enough of the light without eliminating the transmission. In transmitted light the fine particles scatter the blue end of the spectrum more effectively than the red end, resulting in red transmission, and this is the color observed.

A **nanostructured material** (or **nanomaterial**) is defined as a solid material characterized by at least one dimension in the nanometer range

MOLECULES

	Size (approx.)	*Materials*
Nanocrystals and clusters (quantum dots)	1 – 10 nm (diam.)	Metals, semiconductors, magnetic materials
Other nanoparticles	1 – 100 nm (diam.)	Ceramic oxides
Nanowires	1 – 100 nm (diam.)	Metals, semiconductors, oxides, sulfides, nitrides
Nanotubes	1 – 100 nm (diam.)	Carbon, layered metal chalcogenides
Nanoporous solids	0.5 – 10 nm (pore diam.)	Zeolites, phosphates, etc.
2-D arrays (of nanoparticles)	Several nm^2 – μm^2	Metals, semiconductors, magnetic mater.
Surfaces and thin films	1 – 1000 nm (thickness)	A variety of materials
3-D structures (superlattices)	Several nm in the three dimensions	Metals, semiconductors, magnetic materials

BULK MATERIALS

Properties of a Material Vary with the Size of the Material

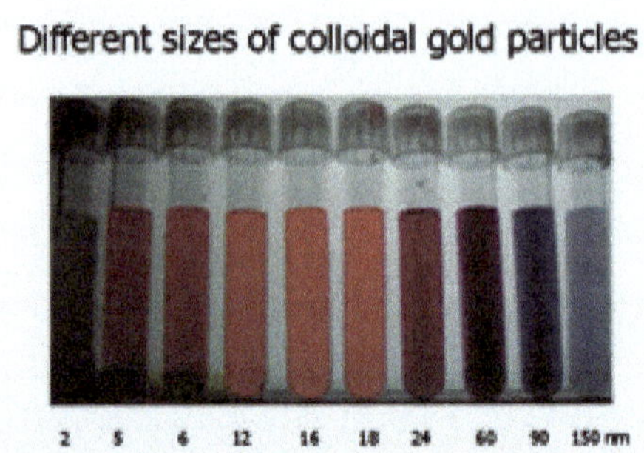

(Bulk) Gold is a shiny yellow metal Nanoscopic gold, i.e. clusters of gold atoms measuring 1 nm across, appears red Bulk gold does not exhibit catalytic properties Au nanocrystal is an excellent low temperature catalyst.Therefore, if we can control the processes that make a nanoscopic material, then we can control the material's properties.

Nanotechnology

Nanotechnology ("nanotech") is manipulation of matter on an atomic, molecular, and supramolecular scale. The earliest, widespread description of nanotechnology referred to the particular technological goal of precisely manipulating atoms and molecules for fabrication of macroscale products, also now referred to as molecular nanotechnology. A more generalized description of nanotechnology was subsequently established by the National Nanotechnology Initiative, which defines nanotechnology as the manipulation of matter with at least one dimension sized from 1 to 100 nanometers. This definition reflects the fact that quantum mechanical effects are important at this quantum-realm scale, and so the definition shifted from a particular technological goal to a research category inclusive of all types of research and technologies that deal with the special properties of matter which occur below the given size threshold. It is therefore common to see the plural form "nanotechnologies" as well as "nanoscale technologies" to refer to the broad range of research and applications whose common trait is size. Because of the variety of potential applications (including industrial and military), governments have invested billions of dollars in nanotechnology research. Until 2012, through its National Nanotechnology Initiative, the USA has invested 3.7 billion dollars, the European Union has invested 1.2 billion and Japan 750 million dollars.

Nanotechnology as defined by size is naturally very broad, including fields of science as diverse as surface science, organic chemistry, molecular biology, semiconductor physics, microfabrication, molecular engineering, etc. The associated research and applications are equally diverse, ranging from extensions of conventional device physics to completely new approaches based upon molecular self-assembly, from developing new materials with dimensions on the nanoscale to direct control of matter on the atomic scale.

Scientists currently debate the future implications of nanotechnology. Nanotechnology may be able to create many new materials and devices with a vast range of applications, such as in nanomedicine, nanoelectronics, biomaterials energy production, and consumer products. On the other hand, nanotechnology raises many of the same issues as any new technology, including concerns about the toxicity and environmental impact of nanomaterials, and their potential effects on global economics, as well as speculation about various doomsday scenarios. These concerns have led to a debate among advocacy groups and governments on whether special regulation of nanotechnology is warranted.

Origins

The concepts that seeded nanotechnology were first discussed in 1959 by renowned physicist Richard Feynman in his talk *There's Plenty of Room at the Bottom*, in which he described the possibility of synthesis via direct manipulation of atoms. The term "nano-technology" was first used by Norio Taniguchi in 1974, though it was not widely known.

Inspired by Feynman's concepts, K. Eric Drexler used the term "nanotechnology" in his 1986 book *Engines of Creation: The Coming Era of Nanotechnology*, which proposed the idea of a nanoscale "assembler" which would be able to build a copy of itself and of other items of arbitrary complexity with atomic control. Also in 1986, Drexler co-founded The Foresight Institute (with which he is no longer affiliated) to help increase public awareness and understanding of nanotechnology concepts and implications.

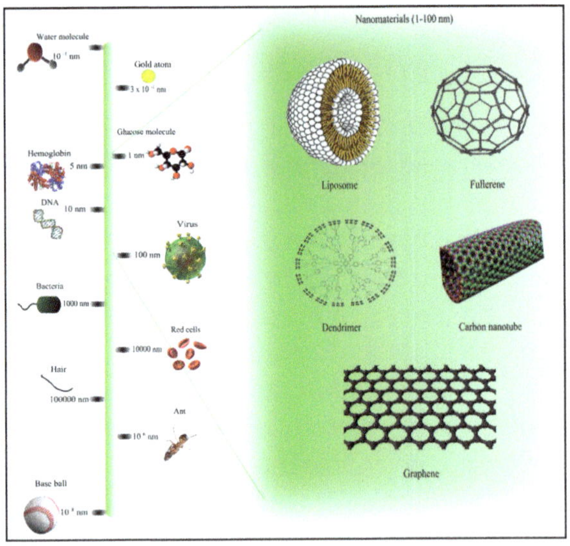

Comparison of Nanomaterials Sizes

Thus, emergence of nanotechnology as a field in the 1980s occurred through convergence of Drexler's theoretical and public work, which developed and popularized a conceptual framework for nanotechnology, and high-visibility experimental advances that drew additional wide-scale attention to the prospects of atomic control of matter. In the 1980s, two major breakthroughs sparked the growth of nanotechnology in modern era.

First, the invention of the scanning tunneling microscope in 1981 which provided unprecedented visualization of individual atoms and bonds, and was successfully used to manipulate individual atoms in 1989. The microscope's developers Gerd Binnig and Heinrich Rohrer at IBM Zurich Research Laboratory received a Nobel Prize in Physics in 1986. Binnig, Quate and Gerber also invented the analogous atomic force microscope that year.

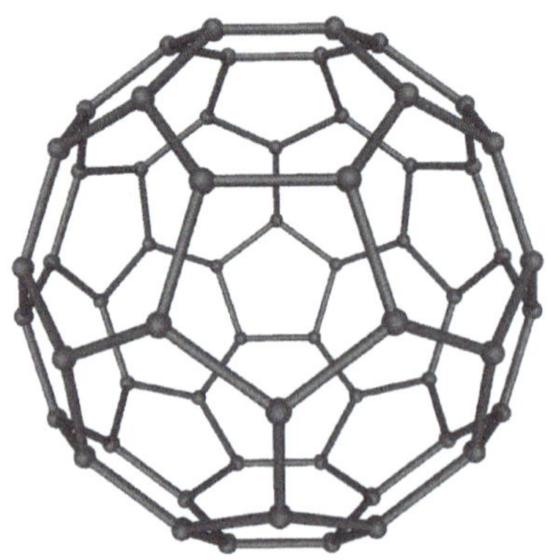

Buckminsterfullerene C_{60}, also known as the buckyball, is a representative member of the carbon structures known as fullerenes. Members of the fullerene family are a major subject of research falling under the nanotechnology umbrella.

Second, Fullerenes were discovered in 1985 by Harry Kroto, Richard Smalley, and Robert Curl, who together won the 1996 Nobel Prize in Chemistry. C_{60} was not initially described as nanotechnology; the term was used regarding subsequent work with related graphene tubes (called carbon nanotubes and sometimes called Bucky tubes) which suggested potential applications for nanoscale electronics and devices.

In the early 2000s, the field garnered increased scientific, political, and commercial attention that led to both controversy and progress. Controversies emerged regarding the definitions and potential implications of nanotechnologies, exemplified by the Royal Society's report on nanotechnology. Challenges were raised regarding the feasibility of applications envisioned by advocates of molecular nanotechnology, which culminated in a public debate between Drexler and Smalley in 2001 and 2003.

Meanwhile, commercialization of products based on advancements in nanoscale technologies began emerging. These products are limited to bulk applications of nanomaterials and do not involve atomic control of matter. Some examples include the Silver Nano platform for using silver nanoparticles as an antibacterial agent, nanoparticle-based transparent sunscreens, carbon fiber strengthening using silica nanoparticles, and carbon nanotubes for stain-resistant textiles.

Governments moved to promote and fund research into nanotechnology, such as in the U.S. with the National Nanotechnology Initiative, which formalized a size-based definition of nanotechnology and established funding for research on the nanoscale, and in Europe via the European Framework Programmes for Research and Technological Development.

By the mid-2000s new and serious scientific attention began to flourish. Projects emerged to produce nanotechnology roadmaps which center on atomically precise manipulation of matter and discuss existing and projected capabilities, goals, and applications.

Fundamental Concepts

Nanotechnology is the engineering of functional systems at the molecular scale. This covers both current work and concepts that are more advanced. In its original sense, nanotechnology refers to the projected ability to construct items from the bottom up, using techniques and tools being developed today to make complete, high performance products.

One nanometer (nm) is one billionth, or 10^{-9}, of a meter. By comparison, typical carbon-carbon bond lengths, or the spacing between these atoms in a molecule, are in the range 0.12–0.15 nm, and a DNA double-helix has a diameter around 2 nm. On the other hand, the smallest cellular life-forms, the bacteria of the genus *Mycoplasma*, are around 200 nm in length. By convention, nanotechnology is taken as the scale range 1 to 100 nm following the definition used by the National Nanotechnology Initiative in the US. The lower limit is set by the size of atoms (hydrogen has the smallest atoms, which are approximately a quarter of a nm diameter) since nanotechnology must build its devices from atoms and molecules. The upper limit is more or less arbitrary but is around the size below which phenomena not observed in larger structures start to become apparent and can be made use of in the nano device. These new phenomena make nanotechnology distinct from devices which are merely miniaturised versions of an equivalent macroscopic device; such devices are on a larger scale and come under the description of microtechnology.

To put that scale in another context, the comparative size of a nanometer to a meter is the same as that of a marble to the size of the earth. Or another way of putting it: a nanometer is the amount an average man's beard grows in the time it takes him to raise the razor to his face.

Two main approaches are used in nanotechnology. In the "bottom-up" approach, materials and devices are built from molecular components which assemble themselves chemically by principles of molecular recognition. In the "top-down" approach, nano-objects are constructed from larger entities without atomic-level control.

Areas of physics such as nanoelectronics, nanomechanics, nanophotonics and nanoionics have evolved during the last few decades to provide a basic scientific foundation of nanotechnology.

Larger to Smaller: A Materials Perspective

Image of reconstruction on a clean Gold(100) surface, as visualized using scanning tunneling microscopy. The positions of the individual atoms composing the surface are visible.

Several phenomena become pronounced as the size of the system decreases. These include statistical mechanical effects, as well as quantum mechanical effects, for example the "quantum size effect" where the electronic properties of solids are altered with great reductions in particle size. This effect does not come into play by going from macro to micro dimensions. However, quantum effects can become significant when the nanometer size range is reached, typically at distances of 100 nanometers or less, the so-called quantum realm. Additionally, a number of physical (mechanical, electrical, optical, etc.) properties change when compared to macroscopic systems. One example is the increase in surface area to volume ratio altering mechanical, thermal and catalytic properties of materials. Diffusion and reactions at nanoscale, nanostructures materials and nanodevices with fast ion transport are generally referred to nanoionics. *Mechanical* properties of nanosystems are of interest in the nanomechanics research. The catalytic activity of nanomaterials also opens potential risks in their interaction with biomaterials.

Materials reduced to the nanoscale can show different properties compared to what they exhibit on a macroscale, enabling unique applications. For instance, opaque substances can become transparent (copper); stable materials can turn combustible (aluminium); insoluble materials may become soluble (gold). A material such as gold, which is chemically inert at normal scales, can serve as a potent chemical catalyst at nanoscales. Much of the fascination with nanotechnology stems from these quantum and surface phenomena that matter exhibits at the nanoscale.

Simple to Complex: A Molecular Perspective

Modern synthetic chemistry has reached the point where it is possible to prepare small molecules to almost any structure. These methods are used today to manufacture a wide variety of useful chemicals such as pharmaceuticals or commercial polymers. This ability raises the question of extending this kind of control to the next-larger level, seeking methods to assemble these single molecules into supramolecular assemblies consisting of many molecules arranged in a well defined manner.

These approaches utilize the concepts of molecular self-assembly and/or supramolecular chemistry to automatically arrange themselves into some useful conformation through a bottom-up approach. The concept of molecular recognition is especially important: molecules can be designed so that a specific configuration or arrangement is favored due to non-covalent intermolecular forces. The Watson–Crick basepairing rules are a direct result of this, as is the specificity of an enzyme being targeted to a single substrate, or the specific folding of the protein itself. Thus, two or more components can be designed to be complementary and mutually attractive so that they make a more complex and useful whole.

Such bottom-up approaches should be capable of producing devices in parallel and be much cheaper than top-down methods, but could potentially be overwhelmed as the size and complexity of the desired assembly increases. Most useful structures require complex and thermodynamically unlikely arrangements of atoms. Nevertheless, there are many examples of self-assembly based on molecular recognition in biology, most notably Watson–Crick basepairing and enzyme-substrate interactions. The challenge for nanotechnology is whether these principles can be used to engineer new constructs in addition to natural ones.

Molecular Nanotechnology: A Long-term View

Molecular nanotechnology, sometimes called molecular manufacturing, describes engineered nanosystems (nanoscale machines) operating on the molecular scale. Molecular nanotechnology is especially associated with the molecular assembler, a machine that can produce a desired structure or device atom-by-atom using the principles of mechanosynthesis. Manufacturing in the context of productive nanosystems is not related to, and should be clearly distinguished from, the conventional technologies used to manufacture nanomaterials such as carbon nanotubes and nanoparticles.

When the term "nanotechnology" was independently coined and popularized by Eric Drexler (who at the time was unaware of an earlier usage by Norio Taniguchi) it referred to a future manufacturing technology based on molecular machine systems. The premise was that molecular scale biological analogies of traditional machine components demonstrated molecular machines were possible: by the countless examples found in biology, it is known that sophisticated, stochastically optimised biological machines can be produced.

It is hoped that developments in nanotechnology will make possible their construction by some other means, perhaps using biomimetic principles. However, Drexler and other researchers have proposed that advanced nanotechnology, although perhaps initially implemented by biomimetic means, ultimately could be based on mechanical engineering principles, namely, a manufacturing technology based on the mechanical functionality of these components (such as gears, bearings, motors, and structural members) that would enable programmable, positional assembly to atomic specification. The physics and engineering performance of exemplar designs were analyzed in Drexler's book *Nanosystems*.

In general it is very difficult to assemble devices on the atomic scale, as one has to position atoms on other atoms of comparable size and stickiness. Another view, put forth by Carlo Montemagno, is that future nanosystems will be hybrids of silicon technology and biological molecular machines. Richard Smalley argued that mechanosynthesis are impossible due to the difficulties in mechanically manipulating individual molecules.

This led to an exchange of letters in the ACS publication Chemical & Engineering News in 2003. Though biology clearly demonstrates that molecular machine systems are possible, non-biological molecular machines are today only in their infancy. Leaders in research on non-biological molecular machines are Dr. Alex Zettl and his colleagues at Lawrence Berkeley Laboratories and UC Berkeley. They have constructed at least three distinct molecular devices whose motion is controlled from the desktop with changing voltage: a nanotube nanomotor, a molecular actuator, and a nanoelectromechanical relaxation oscillator.

An experiment indicating that positional molecular assembly is possible was performed by Ho and Lee at Cornell University in 1999. They used a scanning tunneling microscope to move an individual carbon monoxide molecule (CO) to an individual iron atom (Fe) sitting on a flat silver crystal, and chemically bound the CO to the Fe by applying a voltage.

Current Research

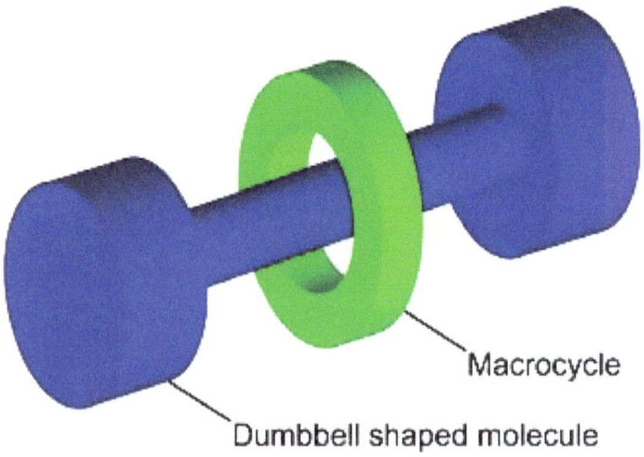

Graphical representation of a rotaxane, useful as a molecular switch.

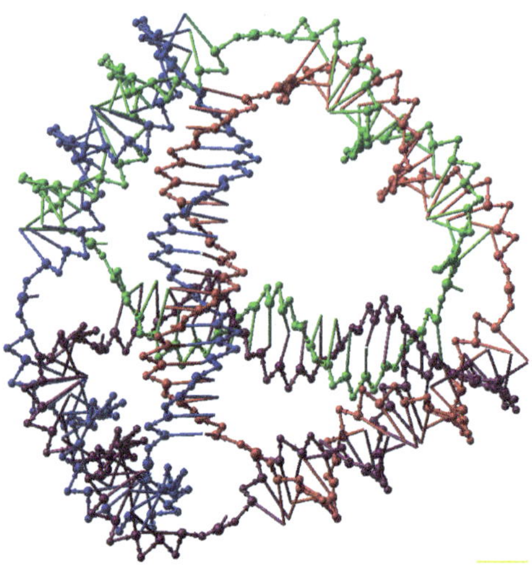

This DNA tetrahedron is an artificially designed nanostructure of the type made in the field of DNA nanotechnology. Each edge of the tetrahedron is a 20 base pair DNA double helix, and each vertex is a three-arm junction.

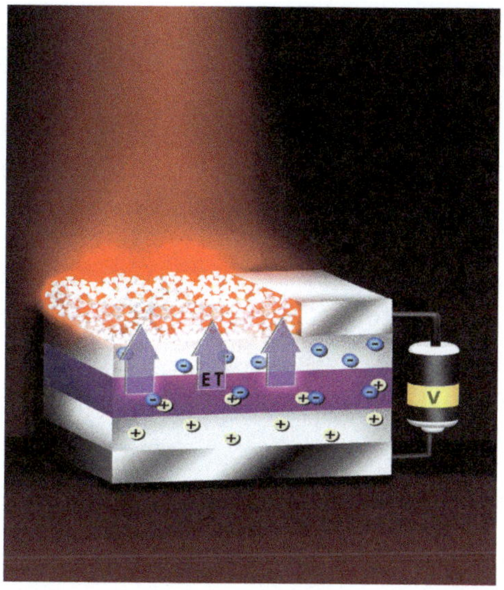

This device transfers energy from nano-thin layers of quantum wells to nanocrystals above them, causing the nanocrystals to emit visible light.

Nanomaterials

The nanomaterials field includes subfields which develop or study materials having unique properties arising from their nanoscale dimensions.

- Interface and colloid science has given rise to many materials which may be useful in nanotechnology, such as carbon nanotubes and other fullerenes, and various nanoparticles and nanorods. Nanomaterials with fast ion transport are related also to nanoionics and nanoelectronics.
- Nanoscale materials can also be used for bulk applications; most present commercial applications of nanotechnology are of this flavor.
- Progress has been made in using these materials for medical applications.
- Nanoscale materials such as nanopillars are sometimes used in solar cells which combats the cost of traditional silicon solar cells.
- Development of applications incorporating semiconductor nanoparticles to be used in the next generation of products, such as display technology, lighting, solar cells and biological imaging.
- Recent application of nanomaterials include a range of biomedical applications, such as tissue engineering, drug delivery, and biosensors.

Bottom-up Approaches

These seek to arrange smaller components into more complex assemblies.

- DNA nanotechnology utilizes the specificity of Watson–Crick basepairing to construct well-defined structures out of DNA and other nucleic acids.

- Approaches from the field of "classical" chemical synthesis (Inorganic and organic synthesis) also aim at designing molecules with well-defined shape (e.g. bis-peptides).
- More generally, molecular self-assembly seeks to use concepts of supramolecular chemistry, and molecular recognition in particular, to cause single-molecule components to automatically arrange themselves into some useful conformation.
- Atomic force microscope tips can be used as a nanoscale "write head" to deposit a chemical upon a surface in a desired pattern in a process called dip pen nanolithography. This technique fits into the larger subfield of nanolithography.

Top-down Approaches

These seek to create smaller devices by using larger ones to direct their assembly.

- Many technologies that descended from conventional solid-state silicon methods for fabricating microprocessors are now capable of creating features smaller than 100 nm, falling under the definition of nanotechnology. Giant magnetoresistance-based hard drives already on the market fit this description, as do atomic layer deposition (ALD) techniques. Peter Grünberg and Albert Fert received the Nobel Prize in Physics in 2007 for their discovery of Giant magnetoresistance and contributions to the field of spintronics.
- Solid-state techniques can also be used to create devices known as nanoelectromechanical systems or NEMS, which are related to microelectromechanical systems or MEMS.
- Focused ion beams can directly remove material, or even deposit material when suitable precursor gasses are applied at the same time. For example, this technique is used routinely to create sub-100 nm sections of material for analysis in Transmission electron microscopy.
- Atomic force microscope tips can be used as a nanoscale "write head" to deposit a resist, which is then followed by an etching process to remove material in a top-down method.

Functional Approaches

These seek to develop components of a desired functionality without regard to how they might be assembled.

- Magnetic assembly for the synthesis of anisotropic superparamagnetic materials such as recently presented magnetic nanochains.
- Molecular scale electronics seeks to develop molecules with useful electronic properties. These could then be used as single-molecule components in a nanoelectronic device. For an example see rotaxane.
- Synthetic chemical methods can also be used to create synthetic molecular motors, such as in a so-called nanocar.

Biomimetic Approaches

- Bionics or biomimicry seeks to apply biological methods and systems found in nature, to the study and design of engineering systems and modern technology. Biomineralization is one example of the systems studied.
- Bionanotechnology is the use of biomolecules for applications in nanotechnology, including use of viruses and lipid assemblies. Nanocellulose is a potential bulk-scale application.

Speculative

These subfields seek to anticipate what inventions nanotechnology might yield, or attempt to propose an agenda along which inquiry might progress. These often take a big-picture view of nanotechnology, with more emphasis on its societal implications than the details of how such inventions could actually be created.

- Molecular nanotechnology is a proposed approach which involves manipulating single molecules in finely controlled, deterministic ways. This is more theoretical than the other subfields, and many of its proposed techniques are beyond current capabilities.
- Nanorobotics centers on self-sufficient machines of some functionality operating at the nanoscale. There are hopes for applying nanorobots in medicine, but it may not be easy to do such a thing because of several drawbacks of such devices. Nevertheless, progress on innovative materials and methodologies has been demonstrated with some patents granted about new nanomanufacturing devices for future commercial applications, which also progressively helps in the development towards nanorobots with the use of embedded nanobioelectronics concepts.
- Productive nanosystems are “systems of nanosystems” which will be complex nanosystems that produce atomically precise parts for other nanosystems, not necessarily using novel nanoscale-emergent properties, but well-understood fundamentals of manufacturing. Because of the discrete (i.e. atomic) nature of matter and the possibility of exponential growth, this stage is seen as the basis of another industrial revolution. Mihail Roco, one of the architects of the USA’s National Nanotechnology Initiative, has proposed four states of nanotechnology that seem to parallel the technical progress of the Industrial Revolution, progressing from passive nanostructures to active nanodevices to complex nanomachines and ultimately to productive nanosystems.
- Programmable matter seeks to design materials whose properties can be easily, reversibly and externally controlled though a fusion of information science and materials science.
- Due to the popularity and media exposure of the term nanotechnology, the words picotechnology and femtotechnology have been coined in analogy to it, although these are only used rarely and informally.

Dimensionality in Nanomaterials

Nanomaterials can be classified in 0D, 1D, 2D and 3D nanomaterials. The dimensionality play a major role in determining the characteristic of nanomaterials including physical, chemical and biological characteristics. With the decrease in dimensionality, an increase in surface-to-volume ratio is observed. This indicate that smaller dimensional nanomaterials have higher surface area compared to 3D nanomaterials. Recently, two dimensional (2D) nanomaterials are extensively investigated for electronic, biomedical, drug delivery and biosensor applications.

Tools and Techniques

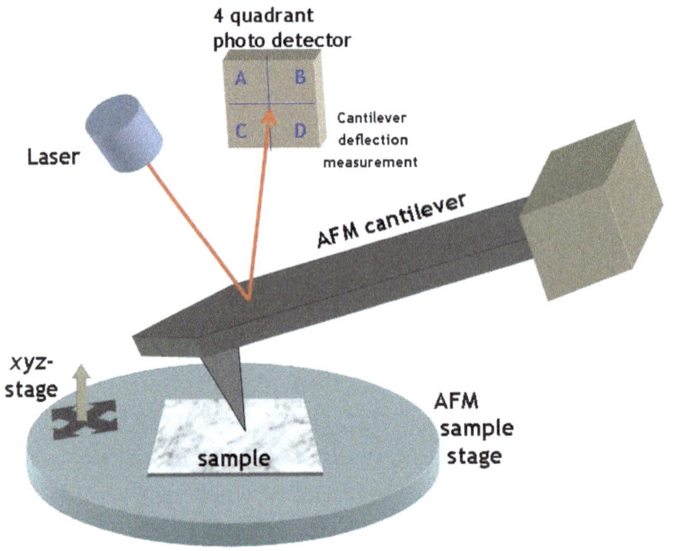

Typical AFM setup. A microfabricated cantilever with a sharp tip is deflected by features on a sample surface, much like in a phonograph but on a much smaller scale. A laser beam reflects off the backside of the cantilever into a set of photodetectors, allowing the deflection to be measured and assembled into an image of the surface.

There are several important modern developments. The atomic force microscope (AFM) and the Scanning Tunneling Microscope (STM) are two early versions of scanning probes that launched nanotechnology. There are other types of scanning probe microscopy. Although conceptually similar to the scanning confocal microscope developed by Marvin Minsky in 1961 and the scanning acoustic microscope (SAM) developed by Calvin Quate and coworkers in the 1970s, newer scanning probe microscopes have much higher resolution, since they are not limited by the wavelength of sound or light.

The tip of a scanning probe can also be used to manipulate nanostructures (a process called positional assembly). Feature-oriented scanning methodology may be a promising way to implement these nanomanipulations in automatic mode. However, this is still a slow process because of low scanning velocity of the microscope.

Various techniques of nanolithography such as optical lithography, X-ray lithography dip pen nanolithography, electron beam lithography or nanoimprint lithography were also developed. Lithography is a top-down fabrication technique where a bulk material is reduced in size to nanoscale pattern.

Another group of nanotechnological techniques include those used for fabrication of nanotubes and nanowires, those used in semiconductor fabrication such as deep ultraviolet lithography, electron beam lithography, focused ion beam machining, nanoimprint lithography, atomic layer deposition, and molecular vapor deposition, and further including molecular self-assembly techniques such as those employing di-block copolymers. The precursors of these techniques preceded the nanotech era, and are extensions in the development of scientific advancements rather than techniques which were devised with the sole purpose of creating nanotechnology and which were results of nanotechnology research.

The top-down approach anticipates nanodevices that must be built piece by piece in stages, much as manufactured items are made. Scanning probe microscopy is an important technique both for characterization and synthesis of nanomaterials. Atomic force microscopes and scanning tunneling microscopes can be used to look at surfaces and to move atoms around. By designing different tips for these microscopes, they can be used for carving out structures on surfaces and to help guide self-assembling structures. By using, for example, feature-oriented scanning approach, atoms or molecules can be moved around on a surface with scanning probe microscopy techniques. At present, it is expensive and time-consuming for mass production but very suitable for laboratory experimentation.

In contrast, bottom-up techniques build or grow larger structures atom by atom or molecule by molecule. These techniques include chemical synthesis, self-assembly and positional assembly. Dual polarisation interferometry is one tool suitable for characterisation of self assembled thin films. Another variation of the bottom-up approach is molecular beam epitaxy or MBE. Researchers at Bell Telephone Laboratories like John R. Arthur. Alfred Y. Cho, and Art C. Gossard developed and implemented MBE as a research tool in the late 1960s and 1970s. Samples made by MBE were key to the discovery of the fractional quantum Hall effect for which the 1998 Nobel Prize in Physics was awarded. MBE allows scientists to lay down atomically precise layers of atoms and, in the process, build up complex structures. Important for research on semiconductors, MBE is also widely used to make samples and devices for the newly emerging field of spintronics.

However, new therapeutic products, based on responsive nanomaterials, such as the ultradeformable, stress-sensitive Transfersome vesicles, are under development and already approved for human use in some countries.

Applications

As of August 21, 2008, the Project on Emerging Nanotechnologies estimates that over 800 manufacturer-identified nanotech products are publicly available, with new ones hitting the market at a pace of 3–4 per week. The project lists all of the products in a publicly accessible online database. Most applications are limited to the use of "first generation" passive nanomaterials which includes titanium dioxide in sunscreen, cosmetics, surface coatings, and some food products; Carbon allotropes used to produce gecko tape; silver in food packaging, clothing, disinfectants and household appliances; zinc oxide in sunscreens and cosmetics, surface coatings, paints and outdoor furniture varnishes; and cerium oxide as a fuel catalyst.

Further applications allow tennis balls to last longer, golf balls to fly straighter, and even bowling balls to become more durable and have a harder surface. Trousers and socks have been infused with nanotechnology so that they will last longer and keep people cool in the summer. Bandages are being infused with silver nanoparticles to heal cuts faster. Video game consoles and personal computers may become cheaper, faster, and contain more memory thanks to nanotechnology. Nanotechnology may have the ability to make existing medical applications cheaper and easier to use in places like the general practitioner's office and at home. Cars are being manufactured with nanomaterials so they may need fewer metals and less fuel to operate in the future.

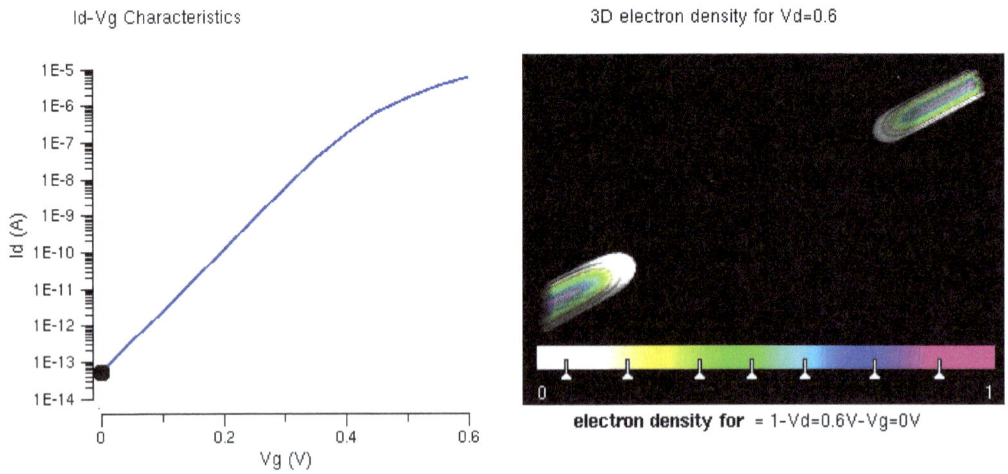

One of the major applications of nanotechnology is in the area of nanoelectronics with MOSFET's being made of small nanowires ~10 nm in length. Here is a simulation of such a nanowire.

Scientists are now turning to nanotechnology in an attempt to develop diesel engines with cleaner exhaust fumes. Platinum is currently used as the diesel engine catalyst in these engines. The catalyst is what cleans the exhaust fume particles. First a reduction catalyst is employed to take nitrogen atoms from NOx molecules in order to free oxygen. Next the oxidation catalyst oxidizes the hydrocarbons and carbon monoxide to form carbon dioxide and water. Platinum is used in both the reduction and the oxidation catalysts. Using platinum though, is inefficient in that it is expensive and unsustainable. Danish company InnovationsFonden invested DKK 15 million in a search for new catalyst substitutes using nanotechnology. The goal of the project, launched in the autumn of 2014, is to maximize surface area and minimize the amount of material required. Objects tend to minimize their surface energy; two drops of water, for example, will join to form one drop and decrease surface area. If the catalyst's surface area that is exposed to the exhaust fumes is maximized, efficiency of the catalyst is maximized. The team working on this project aims to create nanoparticles that will not merge. Every time the surface is optimized, material is saved. Thus, creating these nanoparticles will increase the effectiveness of the resulting diesel engine catalyst—in turn leading to cleaner exhaust fumes—and will decrease cost. If successful, the team hopes to reduce platinum use by 25%.

Nanotechnology also has a prominent role in the fast developing field of Tissue Engineering. When designing scaffolds, researchers attempt to the mimic the nanoscale features of a Cell's microenvironment to direct its differentiation down a suitable lineage. For example, when creating scaffolds to support the growth of bone, researchers may mimic osteoclast resorption pits.

Researchers have successfully used DNA origami-based nanobots capable of carrying out logic functions to achieve targeted drug delivery in cockroaches. It is said that the computational power of these nanobots can be scaled up to that of a Commodore 64.

Implications

An area of concern is the effect that industrial-scale manufacturing and use of nanomaterials would have on human health and the environment, as suggested by nanotoxicology research. For these reasons, some groups advocate that nanotechnology be regulated by governments. Others counter that overregulation would stifle scientific research and the development of beneficial innovations. Public health research agencies, such as the National Institute for Occupational Safety and Health are actively conducting research on potential health effects stemming from exposures to nanoparticles.

Some nanoparticle products may have unintended consequences. Researchers have discovered that bacteriostatic silver nanoparticles used in socks to reduce foot odor are being released in the wash. These particles are then flushed into the waste water stream and may destroy bacteria which are critical components of natural ecosystems, farms, and waste treatment processes.

Public deliberations on risk perception in the US and UK carried out by the Center for Nanotechnology in Society found that participants were more positive about nanotechnologies for energy applications than for health applications, with health applications raising moral and ethical dilemmas such as cost and availability.

Experts, including director of the Woodrow Wilson Center's Project on Emerging Nanotechnologies David Rejeski, have testified that successful commercialization depends on adequate oversight, risk research strategy, and public engagement. Berkeley, California is currently the only city in the United States to regulate nanotechnology; Cambridge, Massachusetts in 2008 considered enacting a similar law, but ultimately rejected it. Relevant for both research on and application of nanotechnologies, the insurability of nanotechnology is contested. Without state regulation of nanotechnology, the availability of private insurance for potential damages is seen as necessary to ensure that burdens are not socialised implicitly.

Health and Environmental Concerns

Nanofibers are used in several areas and in different products, in everything from aircraft wings to tennis rackets. Inhaling airborne nanoparticles and nanofibers may lead to a number of pulmonary diseases, e.g. fibrosis. Researchers have found that when rats breathed in nanoparticles, the particles settled in the brain and lungs, which led to significant increases in biomarkers for inflammation and stress response and that nanoparticles induce skin aging through oxidative stress in hairless mice.

A two-year study at UCLA's School of Public Health found lab mice consuming nano-titanium dioxide showed DNA and chromosome damage to a degree "linked to all the big killers of man, namely cancer, heart disease, neurological disease and aging".

A major study published more recently in Nature Nanotechnology suggests some forms of carbon nanotubes – a poster child for the "nanotechnology revolution" – could be as harmful as asbestos if inhaled in sufficient quantities. Anthony Seaton of the Institute of Occupational Medicine in Edinburgh, Scotland, who contributed to the article on carbon nanotubes said "We know that some of them probably have the potential to cause mesothelioma. So those sorts of materials need to be handled very carefully." In the absence of specific regulation forthcoming from governments, Paull and Lyons (2008) have called for an exclusion of engineered nanoparticles in food. A newspaper article reports that workers in a paint factory developed serious lung disease and nanoparticles were found in their lungs.

Regulation

Calls for tighter regulation of nanotechnology have occurred alongside a growing debate related to the human health and safety risks of nanotechnology. There is significant debate about who is responsible for the regulation of nanotechnology. Some regulatory agencies currently cover some nanotechnology products and processes (to varying degrees) – by "bolting on" nanotechnology to existing regulations – there are clear gaps in these regimes. Davies (2008) has proposed a regulatory road map describing steps to deal with these shortcomings.

Stakeholders concerned by the lack of a regulatory framework to assess and control risks associated with the release of nanoparticles and nanotubes have drawn parallels with bovine spongiform encephalopathy ("mad cow" disease), thalidomide, genetically modified food, nuclear energy, reproductive technologies, biotechnology, and asbestosis. Dr. Andrew Maynard, chief science advisor to the Woodrow Wilson Center's Project on Emerging Nanotechnologies, concludes that there is insufficient funding for human health and safety research, and as a result there is currently limited understanding of the human health and safety risks associated with nanotechnology. As a result, some academics have called for stricter application of the precautionary principle, with delayed marketing approval, enhanced labelling and additional safety data development requirements in relation to certain forms of nanotechnology.

The Royal Society report identified a risk of nanoparticles or nanotubes being released during disposal, destruction and recycling, and recommended that "manufacturers of products that fall under extended producer responsibility regimes such as end-of-life regulations publish procedures outlining how these materials will be managed to minimize possible human and environmental exposure" (p. xiii).

The Center for Nanotechnology in Society has found that people respond to nanotechnologies differently, depending on application – with participants in public deliberations more positive about nanotechnologies for energy than health applications – suggesting that any public calls for nano regulations may differ by technology sector.

Nanomaterials

Nanomaterials describe, in principle, materials of which a single unit is sized (in at least one dimension) between 1 and 1000 nanometres (10^{-9} meter) but is usually 1—100 nm (the usual definition of nanoscale).

Nanomaterials research takes a materials science-based approach to nanotechnology, leveraging advances in materials metrology and synthesis which have been developed in support of microfabrication research. Materials with structure at the nanoscale often have unique optical, electronic, or mechanical properties.

Nanomaterials are slowly becoming commercialized and beginning to emerge as commodities.

Types

Natural Nanomaterials

Biological systems often feature natural, functional nanomaterials. The structure of foraminifera (mainly chalk) and viruses (protein, capsid), the wax crystals covering a lotus or nasturtium leaf, spider and spider-mite silk, the blue hue of tarantulas, the "spatulae" on the bottom of gecko feet, some butterfly wing scales, natural colloids (milk, blood), horny materials (skin, claws, beaks, feathers, horns, hair), paper, cotton, nacre, corals, and even our own bone matrix are all natural *organic* nanomaterials.

Natural *inorganic* nanomaterials occur through crystal growth in the diverse chemical conditions of the Earth's crust. For example, clays display complex nanostructures due to anisotropy of their underlying crystal structure, and volcanic activity can give rise to opals, which are an instance of a naturally occurring photonic crystals due to their nanoscale structure. Fires represent particularly complex reactions and can produce pigments, cement, fumed silica etc.

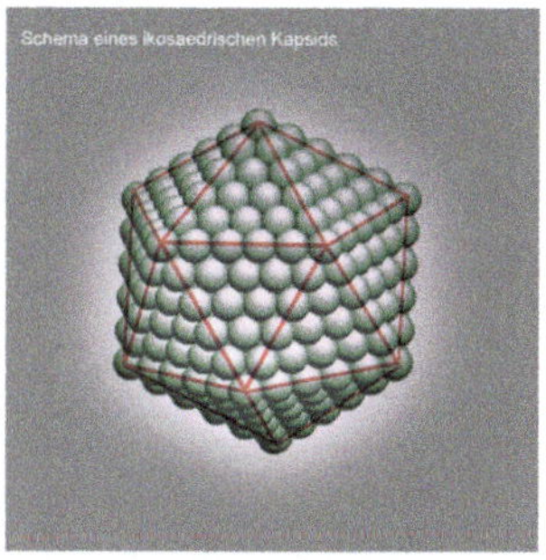

Viral capsid

"Lotus effect", hydrophobic effect with self-cleaning ability

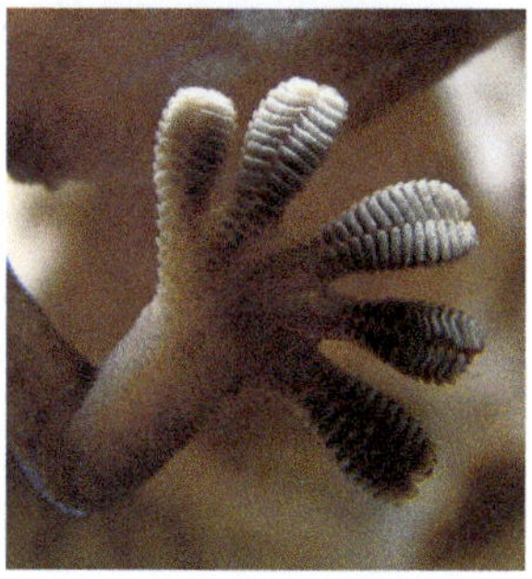
Close-up of the underside of a gecko's foot as it walks on a glass wall. (spatula: 200 × 10-15 nm).

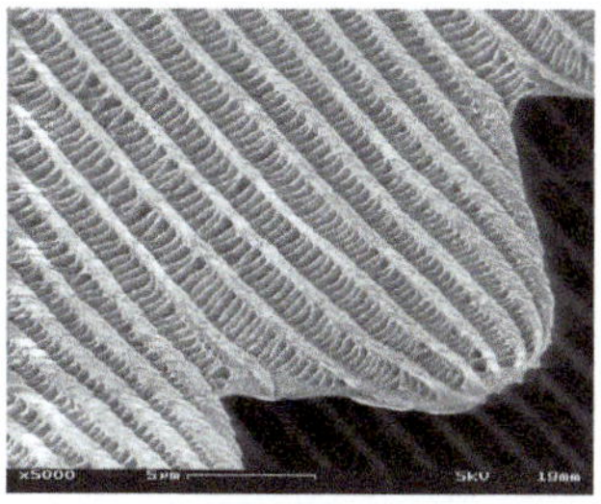
SEM micrograph of a butterfly wing scale (× 5000)

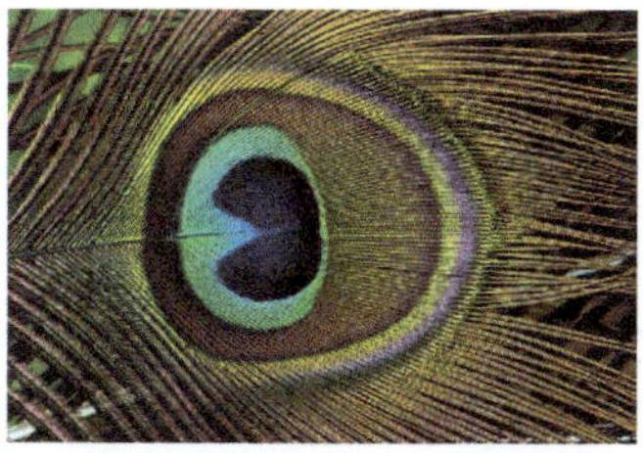
Peacock feather (detail)

Blue hue of a species of tarantula (450 nm ± 20 nm)

Lycurgus Cup, glass, 4th century, Roman. Nanoparticles (70 nm) of gold and silver, dispersed in colloidal form, are responsible for the dichroic effect (red/green).

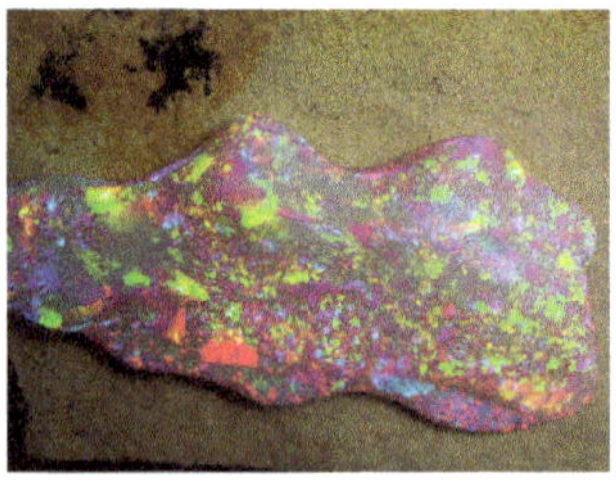

Brazilian Crystal Opal. The play of color is caused by the interference and diffraction of light between silica spheres (150 - 300 nm in diameter).

Fullerenes

The fullerenes are a class of allotropes of carbon which conceptually are graphene sheets rolled into tubes or spheres. These include the carbon nanotubes (or silicon nanotubes) which are of interest both because of their mechanical strength and also because of their electrical properties.

The first fullerene molecule to be discovered, and the family's namesake, buckminsterfullerene (C_{60}), was prepared in 1985 by Richard Smalley, Robert Curl, James Heath, Sean O'Brien, and Harold Kroto at Rice University. The name was a homage to Buckminster Fuller, whose geodesic domes it resembles. Fullerenes have since been found to occur in nature. More recently, fullerenes have been detected in outer space.

For the past decade, the chemical and physical properties of fullerenes have been a hot topic in the field of research and development, and are likely to continue to be for a long time. In April 2003, fullerenes were under study for potential medicinal use: binding specific antibiotics to the structure of resistant bacteria and even target certain types of cancer cells such as melanoma. The October 2005 issue of Chemistry and Biology contains an article describing the use of fullerenes as light-activated antimicrobial agents. In the field of nanotechnology, heat resistance and superconductivity are among the properties attracting intense research.

A common method used to produce fullerenes is to send a large current between two nearby graphite electrodes in an inert atmosphere. The resulting carbon plasma arc between the electrodes cools into sooty residue from which many fullerenes can be isolated.

There are many calculations that have been done using ab-initio Quantum Methods applied to fullerenes. By DFT and TDDFT methods one can obtain IR, Raman and UV spectra. Results of such calculations can be compared with experimental results.

Graphene Nanostructures

2D materials are crystalline materials consisting of a two-dimensional single layer of atoms. The most important representative graphene was discovered in 2004.

Box-shaped graphene (BSG) nanostructure is an example of 3D nanomaterial. BSG nanostructure has appeared after mechanical cleavage of pyrolytic graphite. This nanostructure is a multilayer system of parallel hollow nanochannels located along the surface and having quadrangular cross-section. The thickness of the channel walls is approximately equal to 1 nm. The typical width of channel facets makes about 25 nm.

Nanoparticles

Inorganic nanomaterials, (e.g. quantum dots, nanowires and nanorods) because of their interesting optical and electrical properties, could be used in optoelectronics. Furthermore, the optical and electronic properties of nanomaterials which depend on their size and shape can be tuned via synthetic techniques. There are the possibilities to use those materials in organic material based optoelectronic devices such as Organic solar cells, OLEDs etc. The operating principles of such devices are governed by photoinduced processes like electron transfer and energy transfer. The performance of the devices depends on the efficiency of the photoinduced process responsible for their functioning. Therefore, better understanding of those photoinduced processes in organic/inorganic nanomaterial composite systems is necessary in order to use them in organic optoelectronic devices.

Nanoparticles or nanocrystals made of metals, semiconductors, or oxides are of particular interest for their mechanical, electrical, magnetic, optical, chemical and other properties. Nanoparticles have been used as quantum dots and as chemical catalysts such as nanomaterial-based catalysts. Recently, a range of nanoparticles are extensively investigated for biomedical applications including tissue engineering, drug delivery, biosensor.

Nanoparticles are of great scientific interest as they are effectively a bridge between bulk materials and atomic or molecular structures. A bulk material should have constant physical properties regardless of its size, but at the nano-scale this is often not the case. Size-dependent properties are observed such as quantum confinement in semiconductor particles, surface plasmon resonance in some metal particles and superparamagnetism in magnetic materials.

Nanoparticles exhibit a number of special properties relative to bulk material. For example, the bending of bulk copper (wire, ribbon, etc.) occurs with movement of copper atoms/clusters at about the 50 nm scale. Copper nanoparticles smaller than 50 nm are considered super hard materials that do not exhibit the same malleability and ductility as bulk copper. The change in properties is not always desirable. Ferroelectric materials smaller than 10 nm can switch their magnetisation direction using room temperature thermal energy, thus making them useless for memory storage. Suspensions of nanoparticles are possible because the interaction of the particle surface with the solvent is strong enough to overcome differences in density, which usually result in a material either sinking or floating in a liquid. Nanoparticles often have unexpected visual properties because they are small enough to confine their electrons and produce quantum effects. For example, gold nanoparticles appear deep red to black in solution.

The often very high surface area to volume ratio of nanoparticles provides a tremendous driving force for diffusion, especially at elevated temperatures. Sintering is possible at lower temperatures and over shorter durations than for larger particles. This theoretically does not affect the density of the final product, though flow difficulties and the tendency of nanoparticles to agglomerate do complicate matters. The surface effects of nanoparticles also reduces the incipient melting temperature.

Nanozymes

Nanozymes are nanomaterials with enzyme-like characteristics. They are an emerging type of artificial enzyme, which have been used for wide applications in such as biosensing, bioimaging, tumor diagnosis and therapy, antibiofouling, etc.

Synthesis

The goal of any synthetic method for nanomaterials is to yield a material that exhibits properties that are a result of their characteristic length scale being in the nanometer range (~1 – 100 nm). Accordingly, the synthetic method should exhibit control of size in this range so that one property or another can be attained. Often the methods are divided into two main types "Bottom Up" and "Top Down."

Bottom up Methods

Bottom up methods involve the assembly of atoms or molecules into nanostructured arrays. In these methods the raw material sources can be in the form of gases, liquids or solids. The latter requiring some sort of disassembly prior to their incorporation onto a nanostructure. Bottom methods generally fall into two categories: chaotic and controlled.

Chaotic Processes

Chaotic processes involve elevating the constituent atoms or molecules to a chaotic state and then suddenly changing the conditions so as to make that state unstable. Through the clever manipulation of any number of parameters, products form largely as a result of the insuring kinetics. The collapse from the chaotic state can be difficult or impossible to control and so ensemble statistics often govern the resulting size distribution and average size. Accordingly, nanoparticle formation is controlled through manipulation of the end state of the products.

Examples of Chaotic Processes are: Laser ablation, Exploding wire, Arc, Flame pyrolysis, Combustion, Precipitation synthesis techniques.

Controlled Processes

Controlled Processes involve the controlled delivery of the constituent atoms or molecules to the site(s) of nanoparticle formation such that the nanoparticle can grow to a prescribed sizes in a controlled manner. Generally the state of the constituent atoms or molecules are never far from that needed for nanoparticle formation. Accordingly, nanoparticle formation is controlled through the control of the state of the reactants.

Examples of controlled processes are self-limiting growth solution, self-limited chemical vapor deposition, shaped pulse femtosecond laser techniques, and molecular beam epitaxy.

Characterization

Novel effects can occur in materials when structures are formed with sizes comparable to any one of many possible length scales, such as the de Broglie wavelength of electrons, or the optical wavelengths of high energy photons. In these cases quantum mechanical effects can dominate material properties. One example is quantum confinement where the electronic properties of solids are altered with great reductions in particle size. The optical properties of nanoparticles, e.g. fluorescence, also become a function of the particle diameter. This effect does not come into play by going from macrosocopic to micrometer dimensions, but becomes pronounced when the nanometer scale is reached.

In addition to optical and electronic properties, the novel mechanical properties of many nanomaterials is the subject of nanomechanics research. When added to a bulk material, nanoparticles can strongly influence the mechanical properties of the material, such as the stiffness or elasticity. For example, traditional polymers can be reinforced by nanoparticles (such as carbon nanotubes) resulting in novel materials which can be used as lightweight replacements for metals. Such composite materials may enable a weight reduction accompanied by an increase in stability and improved functionality.

Finally, nanostructured materials with small particle size such as zeolites, and asbestos, are used as catalysts in a wide range of critical industrial chemical reactions. The further development of such catalysts can form the basis of more efficient, environmentally friendly chemical processes.

The first observations and size measurements of nano-particles were made during the first decade of the 20th century. Zsigmondy made detailed studies of gold sols and other nanomaterials with sizes down to 10 nm and less. He published a book in 1914. He used an ultramicroscope that employs a *dark field* method for seeing particles with sizes much less than light wavelength.

There are traditional techniques developed during 20th century in Interface and Colloid Science for characterizing nanomaterials. These are widely used for *first generation* passive nanomaterials specified.

These methods include several different techniques for characterizing particle size distribution. This characterization is imperative because many materials that are expected to be nano-sized are actually aggregated in solutions. Some of methods are based on light scattering. Others apply ultrasound, such as ultrasound attenuation spectroscopy for testing concentrated nano-dispersions and microemulsions.

There is also a group of traditional techniques for characterizing surface charge or zeta potential of nano-particles in solutions. This information is required for proper system stabilzation, preventing its aggregation or flocculation. These methods include microelectrophoresis, electrophoretic light scattering and electroacoustics. The last one, for instance colloid vibration current method is suitable for characterizing concentrated systems.

Uniformity

The chemical processing and synthesis of high performance technological components for the private, industrial and military sectors requires the use of high purity ceramics, polymers, glass-ceramics and material composites. In condensed bodies formed from fine powders, the irregular sizes and shapes of nanoparticles in a typical powder often lead to non-uniform packing morphologies that result in packing density variations in the powder compact.

Uncontrolled agglomeration of powders due to attractive van der Waals forces can also give rise to in microstructural inhomogeneities. Differential stresses that develop as a result of non-uniform drying shrinkage are directly related to the rate at which the solvent can be removed, and thus highly dependent upon the distribution of porosity. Such stresses have been associated with a plastic-to-brittle transition in consolidated bodies, and can yield to crack propagation in the unfired body if not relieved.

In addition, any fluctuations in packing density in the compact as it is prepared for the kiln are often amplified during the sintering process, yielding inhomogeneous densification. Some pores and other structural defects associated with density variations have been shown to play a detrimental role in the sintering process by growing and thus limiting end-point densities. Differential stresses arising from inhomogeneous densification have also been shown to result in the propagation of internal cracks, thus becoming the strength-controlling flaws.

It would therefore appear desirable to process a material in such a way that it is physically uniform with regard to the distribution of components and porosity, rather than using particle size distributions which will maximize the green density. The containment of a uniformly dispersed assembly of strongly interacting particles in suspension requires total control over particle-particle interactions. It should be noted here that a number of dispersants such as ammonium citrate (aqueous) and imidazoline or oleyl alcohol (nonaqueous) are promising solutions as possible additives for enhanced dispersion and deagglomeration. Monodisperse nanoparticles and colloids provide this potential.

Monodisperse powders of colloidal silica, for example, may therefore be stabilized sufficiently to ensure a high degree of order in the colloidal crystal or polycrystalline colloidal solid which results from aggregation. The degree of order appears to be limited by the time and space allowed for longer-range correlations to be established. Such defective polycrystalline colloidal structures would appear to be the basic elements of sub-micrometer colloidal materials science, and, therefore, provide the first step in developing a more rigorous understanding of the mechanisms involved in microstructural evolution in high performance materials and components.

Legal Definition

On 18 October 2011, the European Commission adopted the following definition of a nanomaterial: "A natural, incidental or manufactured material containing particles, in an unbound state or as an aggregate or as an agglomerate and where, for 50% or more of the particles in the number size distribution, one or more external dimensions is in the size range 1 nm – 100 nm. In specific cases and where warranted by concerns for the environment, health, safety or competitiveness the number size distribution threshold of 50% may be replaced by a threshold between 1 and 50%."

However, this differs from the definition adopted by the International Organization for Standardization (ISO), which is: "Material with any external dimension in the nanoscale or having internal structure in the nanoscale."

"Nanoscale" is, in turn, defined as: "Size range from approximately 1 nm to 100 nm."

It is not currently known which of these, if any, will prevail in courts of law.

Safety

Health Impact

While nanomaterials and nanotechnologies are expected to yield numerous health and health care advances, such as more targeted methods of delivering drugs, new cancer therapies, and methods of early detection of diseases, they also may have unwanted effects. Increased rate of absorption is the main concern associated with manufactured nanoparticles.

When materials are made into nanoparticles, their surface area to volume ratio increases. The greater specific surface area (surface area per unit weight) may lead to increased rate of absorption through the skin, lungs, or digestive tract and may cause unwanted effects to the lungs as well as other organs. However, the particles must be absorbed in sufficient quantities in order to pose health risks.

Nanoparticles created adventitiously (e.g., through the rubbing of prostheses) have long been known to be a health hazard, but as the use of nanomaterials increases worldwide, concerns for worker and user safety are mounting. To address such concerns, the Swedish Karolinska Institute conducted a study in which various nanoparticles were introduced to human lung epithelial cells. The results, released in 2008, showed that iron oxide nanoparticles caused little DNA damage and were non-toxic. Zinc oxide nanoparticles were slightly worse. Titanium dioxide caused only DNA damage. Carbon nanotubes caused DNA damage at low levels. Copper oxide was found to be the worst offender, and was the only nanomaterial identified by the researchers as a clear health risk. Though nanomaterials are not confirmed as a health risk to workers who produce them, NIOSH recommends that exposure precautions and personal protective equipment be used to protect workers until the risks of nanomaterial manufacture are better understood.

Occupational Safety

Beyond normal chemical safety practices such as using personal protective equipment, The United States National Institute for Occupational Safety and Health recommends the use of HEPA filtration on local exhaust ventilation such as fume hoods. General laboratory ventilation is often not sufficient to effectively clear nanomaterials released into the general room air over a 30-minute period; therefore, researchers should leave the hood fan on even after synthesis is complete. Nanomaterials in dry powder form have the most risk of airborne dust generation and dermal contact, followed by liquid suspensions and then by nanomaterials embedded in a solid matrix.

Nanoparticles behave differently than other similarly sized particles. It is therefore necessary to develop specialized approaches to testing and monitoring their effects on human health and on the environment. The OECD Chemicals Committee has established the Working Party on Manufactured Nanomaterials to address this issue and to study the practices of OECD member countries in regards to nanomaterial safety.

Approaches to make Structures on the Nanoscale

The bottom-up approach:

whereby structures are made atom-by-atom and molecule-by molecule, harnessing covalent, ionic, metallic or non-covalent bonds. This approach represents how nature self-assembles functioning nanostructures, such as enzymes and viruses.

The top-down approach:

whereby structures are etched into bulk materials such as silicon. This approach represents how silicon chips are fabricated.

Methods of Synthesis of Nanostructured Materials

- Colloidal Synthesis
- Reduction
- Sono-chemical synthesis
- Microwave synthesis
- Micro-emulsion synthesis
- Precipitation From Homogenous Solutions (PFHS)
- Hydrothermal Synthesis

Nanofabrication Techniques: Grouped According to the form of Products

Nanoparticles:

colloidal processing

flame combustion

phase segregation

Nanorods or Nanowires:

template-based electroplating

solution-liquid-solid growth (SLS)

Thin Films:

Chemical vapour deposition (CBD and MOCVD)

Molecular beam epitaxy

Atomic layer deposition

Nanostructured Bulk Materials:

Photonic bandgap crystals by self-assembly of nanosized particles

Molecular Nanotechnology

Molecular nanotechnology (MNT) is a technology based on the ability to build structures to complex, atomic specifications by means of mechanosynthesis. This is distinct from nanoscale materials. Based on Richard Feynman's vision of miniature factories using nanomachines to build complex products (including additional nanomachines), this advanced form of nanotechnology (or *molecular manufacturing*) would make use of positionally-controlled mechanosynthesis guided by molecular machine systems. MNT would involve combining physical principles demonstrated by biophysics, chemistry, other nanotechnologies, and the molecular machinery of life with the systems engineering principles found in modern macroscale factories.

Introduction

While conventional chemistry uses inexact processes obtaining inexact results, and biology exploits inexact processes to obtain definitive results, molecular nanotechnology would employ original definitive processes to obtain definitive results. The desire in molecular nanotechnology would be to balance molecular reactions in positionally-controlled locations and orientations to obtain desired chemical reactions, and then to build systems by further assembling the products of these reactions.

A roadmap for the development of MNT is an objective of a broadly based technology project led by Battelle (the manager of several U.S. National Laboratories) and the Foresight Institute. The roadmap was originally scheduled for completion by late 2006, but was released in January 2008. The Nanofactory Collaboration is a more focused ongoing effort involving 23 researchers from 10 organizations and 4 countries that is developing a practical research agenda specifically aimed at positionally-controlled diamond mechanosynthesis and diamondoid nanofactory development. In August 2005, a task force consisting of 50+ international experts from various fields was organized by the Center for Responsible Nanotechnology to study the societal implications of molecular nanotechnology.

Projected Applications and Capabilities

Smart Materials and Nanosensors

One proposed application of MNT is so-called smart materials. This term refers to any sort of material designed and engineered at the nanometer scale for a specific task. It encompasses a wide variety of possible commercial applications. One example would be materials designed to respond differently to various molecules; such a capability could lead, for example, to artificial drugs which would recognize and render inert specific viruses. Another is the idea of self-healing structures, which would repair small tears in a surface naturally in the same way as self-sealing tires or human skin.

A MNT nanosensor would resemble a smart material, involving a small component within a larger machine that would react to its environment and change in some fundamental, intentional way. A very simple example: a photosensor might passively measure the incident light and discharge its absorbed energy as electricity when the light passes above or below a specified threshold, sending a signal to a larger machine. Such a sensor would supposedly cost less and use less power than a conventional sensor, and yet function usefully in all the same applications — for example, turning on parking lot lights when it gets dark.

While smart materials and nanosensors both exemplify useful applications of MNT, they pale in comparison with the complexity of the technology most popularly associated with the term: the replicating nanorobot.

Replicating Nanorobots

MNT nanofacturing is popularly linked with the idea of swarms of coordinated nanoscale robots working together, a popularization of an early proposal by K. Eric Drexler in his 1986 discussions of MNT, but superseded in 1992. In this early proposal, sufficiently capable nanorobots would construct more nanorobots in an artificial environment containing special molecular building blocks.

Critics have doubted both the feasibility of self-replicating nanorobots and the feasibility of control if self-replicating nanorobots could be achieved: they cite the possibility of mutations removing any control and favoring reproduction of mutant pathogenic variations. Advocates address the first doubt by pointing out that the first macroscale autonomous machine replicator, made of Lego blocks, was built and operated experimentally in 2002. While there are sensory advantages present at the macroscale compared to the limited sensorium available at the nanoscale, proposals for positionally controlled nanoscale mechanosynthetic fabrication systems employ dead reckoning of tooltips combined with reliable reaction sequence design to ensure reliable results, hence a limited sensorium is no handicap; similar considerations apply to the positional assembly of small nanoparts. Advocates address the second doubt by arguing that bacteria are (of necessity) evolved to evolve, while nanorobot mutation could be actively prevented by common error-correcting techniques. Similar ideas are advocated in the Foresight Guidelines on Molecular Nanotechnology, and a map of the 137-dimensional replicator design space recently published by Freitas and Merkle provides numerous proposed methods by which replicators could, in principle, be safely controlled by good design.

However, the concept of suppressing mutation raises the question: How can design evolution occur at the nanoscale without a process of random mutation and deterministic selection? Critics argue that MNT advocates have not provided a substitute for such a process of evolution in this nanoscale arena where conventional sensory-based selection processes are lacking. The limits of the sensorium available at the nanoscale could make it difficult or impossible to winnow successes from failures. Advocates argue that design evolution should occur deterministically and strictly under human control, using the conventional engineering paradigm of modeling, design, prototyping, testing, analysis, and redesign.

In any event, since 1992 technical proposals for MNT do not include self-replicating nanorobots, and recent ethical guidelines put forth by MNT advocates prohibit unconstrained self-replication.

Medical Nanorobots

One of the most important applications of MNT would be medical nanorobotics or nanomedicine, an area pioneered by Robert Freitas in numerous books and papers. The ability to design, build, and deploy large numbers of medical nanorobots would, at a minimum, make possible the rapid elimination of disease and the reliable and relatively painless recovery from physical trauma. Medical nanorobots might also make possible the convenient correction of genetic defects, and help to ensure a greatly expanded lifespan. More controversially, medical nanorobots might be used to augment natural human capabilities.

Utility Fog

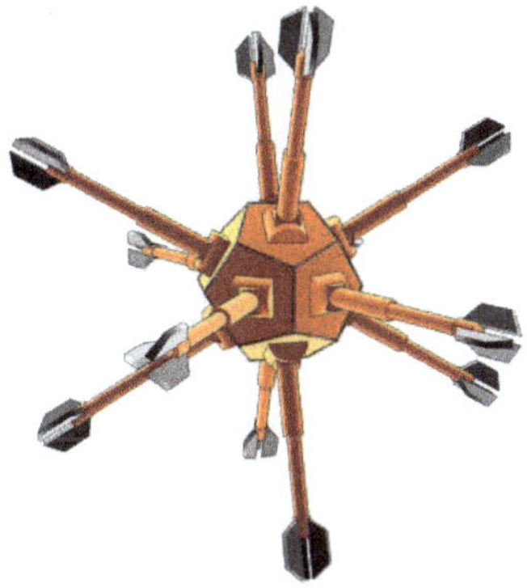

Diagram of a 100 micrometer foglet

Another proposed application of molecular nanotechnology is "utility fog" — in which a cloud of networked microscopic robots (simpler than assemblers) would change its shape and properties to form macroscopic objects and tools in accordance with software commands. Rather than modify the current practices of consuming material goods in different forms, utility fog would simply replace many physical objects.

Phased-array Optics

Yet another proposed application of MNT would be phased-array optics (PAO). However, this appears to be a problem addressable by ordinary nanoscale technology. PAO would use the principle of phased-array millimeter technology but at optical wavelengths. This would permit the duplication of any sort of optical effect but virtually. Users could request holograms, sunrises and sunsets, or floating lasers as the mood strikes. PAO systems were described in BC Crandall's *Nanotechnology: Molecular Speculations on Global Abundance* in the Brian Wowk article "Phased-Array Optics."

Potential Social Impacts

Molecular manufacturing is a potential future subfield of nanotechnology that would make it possible to build complex structures at atomic precision. Molecular manufacturing requires significant advances in nanotechnology, but once achieved could produce highly advanced products at low costs and in large quantities in nanofactories weighing a kilogram or more. When nanofactories gain the ability to produce other nanofactories production may only be limited by relatively abundant factors such as input materials, energy and software.

The products of molecular manufacturing could range from cheaper, mass-produced versions of known high-tech products to novel products with added capabilities in many areas of application. Some applications that have been suggested are advanced smart materials, nanosensors, medical nanorobots and space travel. Additionally, molecular manufacturing could be used to cheaply produce highly advanced, durable weapons, which is an area of special concern regarding the impact of nanotechnology. Being equipped with compact computers and motors these could be increasingly autonomous and have a large range of capabilities.

According to Chris Phoenix and Mike Treder from the Center for Responsible Nanotechnology as well as Anders Sandberg from the Future of Humanity Institute molecular manufacturing is the application of nanotechnology that poses the most significant global catastrophic risk. Several nanotechnology researchers state that the bulk of risk from nanotechnology comes from the potential to lead to war, arms races and destructive global government. Several reasons have been suggested why the availability of nanotech weaponry may with significant likelihood lead to unstable arms races (compared to e.g. nuclear arms races): (1) A large number of players may be tempted to enter the race since the threshold for doing so is low; (2) the ability to make weapons with molecular manufacturing will be cheap and easy to hide; (3) therefore lack of insight into the other parties' capabilities can tempt players to arm out of caution or to launch preemptive strikes; (4) molecular manufacturing may reduce dependency on international trade, a potential peace-promoting factor; (5) wars of aggression may pose a smaller economic threat to the aggressor since manufacturing is cheap and humans may not be needed on the battlefield.

Since self-regulation by all state and non-state actors seems hard to achieve, measures to mitigate war-related risks have mainly been proposed in the area of international cooperation. International infrastructure may be expanded giving more sovereignty to the international level. This could help coordinate efforts for arms control. International institutions dedicated specifically to nanotechnology (perhaps analogously to the International Atomic Energy Agency IAEA) or general arms control may also be designed. One may also jointly make differential technological progress on defensive technologies, a policy that players should usually favour. The Center for Responsible Nanotechnology also suggest some technical restrictions. Improved transparency regarding technological capabilities may be another important facilitator for arms-control.

A grey goo is another catastrophic scenario, which was proposed by Eric Drexler in his 1986 book *Engines of Creation*, has been analyzed by Freitas in "Some Limits to Global Ecophagy by Biovorous Nanoreplicators, with Public Policy Recommendations" and has been a theme in mainstream media and fiction. This scenario involves tiny self-replicating robots that consume the entire biosphere using it as a source of energy and building blocks. Nanotech experts including Drexler now discredit the scenario. According to Chris Phoenix a "So-called grey goo could only be the product of a deliberate and difficult engineering process, not an accident". With the advent of nano-biotech, a different scenario called green goo has been forwarded. Here, the malignant substance is not nanobots but rather self-replicating biological organisms engineered through nanotechnology.

Benefits

Nanotechnology (or molecular nanotechnology to refer more specifically to the goals discussed here) will let us continue the historical trends in manufacturing right up to the fundamental limits imposed by physical law. It will let us make remarkably powerful molecular computers. It will let us make materials over fifty times lighter than steel or aluminium alloy but with the same strength. We'll be able to make jets, rockets, cars or even chairs that, by today's standards, would be remarkably light, strong, and inexpensive. Molecular surgical tools, guided by molecular computers and injected into the blood stream could find and destroy cancer cells or invading bacteria, unclog arteries, or provide oxygen when the circulation is impaired.

Nanotechnology will replace our entire manufacturing base with a new, radically more precise, radically less expensive, and radically more flexible way of making products. The aim is not simply to replace today's computer chip making plants, but also to replace the assembly lines for cars, televisions, telephones, books, surgical tools, missiles, bookcases, airplanes, tractors, and all the rest. The objective is a pervasive change in manufacturing, a change that will leave virtually no product untouched. Economic progress and military readiness in the 21st Century will depend fundamentally on maintaining a competitive position in nanotechnology.

Despite the current early developmental status of nanotechnology and molecular nanotechnology, much concern surrounds MNT's anticipated impact on economics and on law. Whatever the exact effects, MNT, if achieved, would tend to reduce the scarcity of manufactured goods and make many more goods (such as food and health aids) manufacturable.

It is generally considered that future citizens of a molecular-nanotechnological society would still need money, in the form of unforgeable digital cash or physical specie (in special circumstances). They might use such money to buy goods and services that are unique, or limited within the solar system. These might include: matter, energy, information, real estate, design services, entertainment services, legal services, fame, political power, or the attention of other people to one's political/religious/philosophical message. Furthermore, futurists must consider war, even between prosperous states, and non-economic goals.

If MNT were realized, some resources would remain limited, because unique physical objects are limited (a plot of land in the real Jerusalem, mining rights to the larger near-earth asteroids) or because they depend on the goodwill of a particular person (the love of a famous person, a live audience in a musical concert). Demand will always exceed supply for some things, and a political economy may continue to exist in any case. Whether the interest in these limited resources would diminish with the advent of virtual reality, where they could be easily substituted, is yet unclear. One reason why it might not is a hypothetical preference for "the real thing", although such an opinion could easily be mollified if virtual reality were to develop to a certain level of quality.

MNT should make possible nanomedical capabilities able to cure any medical condition not already cured by advances in other areas. Good health would be common, and poor health of any form would be as rare as smallpox and scurvy are today. Even cryonics would be feasible, as cryopreserved tissue could be fully repaired.

Risks

Molecular nanotechnology is one of the technologies that some analysts believe could lead to a Technological Singularity. Some feel that molecular nanotechnology would have daunting risks. It conceivably could enable cheaper and more destructive conventional weapons. Also, molecular nanotechnology might permit weapons of mass destruction that could self-replicate, as viruses and cancer cells do when attacking the human body. Commentators generally agree that, in the event molecular nanotechnology were developed, its self-replication should be permitted only under very controlled or "inherently safe" conditions.

A fear exists that nanomechanical robots, if achieved, and if designed to self-replicate using naturally occurring materials (a difficult task), could consume the entire planet in their hunger for raw materials, or simply crowd out natural life, out-competing it for energy (as happened historically when blue-green algae appeared and outcompeted earlier life forms). Some commentators have referred to this situation as the "grey goo" or "ecophagy" scenario. K. Eric Drexler considers an accidental "grey goo" scenario extremely unlikely and says so in later editions of *Engines of Creation*.

In light of this perception of potential danger, the Foresight Institute (founded by K. Eric Drexler to prepare for the arrival of future technologies) has drafted a set of guidelines for the ethical development of nanotechnology. These include the banning of free-foraging self-replicating pseudo-organisms on the Earth's surface, at least, and possibly in other places.

Technical Issues and Criticism

The feasibility of the basic technologies analyzed in *Nanosystems* has been the subject of a formal scientific review by U.S. National Academy of Sciences, and has also been the focus of extensive debate on the internet and in the popular press.

Study and Recommendations by the U.S. National Academy of Sciences

In 2006, U.S. National Academy of Sciences released the report of a study of molecular manufacturing as part of a longer report, *A Matter of Size: Triennial Review of the National Nanotechnology Initiative* The study committee reviewed the technical content of *Nanosystems*, and in its conclusion states that no current theoretical analysis can be considered definitive regarding several questions of potential system performance, and that optimal paths for implementing high-performance systems cannot be predicted with confidence. It recommends experimental research to advance knowledge in this area:

> "Although theoretical calculations can be made today, the eventually attainable range of chemical reaction cycles, error rates, speed of operation, and thermodynamic efficiencies of such bottom-up manufacturing systems cannot be reliably predicted at this time. Thus, the eventually attainable perfection and complexity of manufactured products, while they can be calculated in theory, cannot be predicted with confidence. Finally, the optimum research paths that might lead to systems which greatly exceed the thermodynamic efficiencies and other capabilities of biological systems cannot be reliably predicted at this time. Research funding that is based on the ability of investigators to produce experimental demonstrations that link to abstract models and guide long-term vision is most appropriate to achieve this goal."

Assemblers Versus Nanofactories

A section heading in Drexler's *Engines of Creation* reads "Universal Assemblers", and the following text speaks of multiple types of assemblers which, collectively, could hypothetically "build almost anything that the laws of nature allow to exist." Drexler's colleague Ralph Merkle has noted that, contrary to widespread legend, Drexler never claimed that assembler systems could build absolutely any molecular structure. The endnotes in Drexler's book explain the qualification "almost":

"For example, a delicate structure might be designed that, like a stone arch, would self-destruct unless all its pieces were already in place. If there were no room in the design for the placement and removal of a scaffolding, then the structure might be impossible to build. Few structures of practical interest seem likely to exhibit such a problem, however."

In 1992, Drexler published *Nanosystems: Molecular Machinery, Manufacturing, and Computation*, a detailed proposal for synthesizing stiff covalent structures using a table-top factory. Diamondoid structures and other stiff covalent structures, if achieved, would have a wide range of possible applications, going far beyond current MEMS technology. An outline of a path was put forward in 1992 for building a table-top factory in the absence of an assembler. Other researchers have begun advancing tentative, alternative proposed paths for this in the years since Nanosystems was published.

Hard Versus Soft Nanotechnology

In 2004 Richard Jones wrote Soft Machines (nanotechnology and life), a book for lay audiences published by Oxford University. In this book he describes radical nanotechnology (as advocated by Drexler) as a deterministic/mechanistic idea of nano engineered machines that does not take into account the nanoscale challenges such as wetness, stickness, Brownian motion, and high viscosity. He also explains what is soft nanotechnology or more appropriatelly biomimetic nanotechnology which is the way forward, if not the best way, to design functional nanodevices that can cope with all the problems at a nanoscale. One can think of soft nanotechnology as the development of nanomachines that uses the lessons learned from biology on how things work, chemistry to precisely engineer such devices and stochastic physics to model the system and its natural processes in detail.

The Smalley-Drexler Debate

Several researchers, including Nobel Prize winner Dr. Richard Smalley (1943–2005), attacked the notion of universal assemblers, leading to a rebuttal from Drexler and colleagues, and eventually to an exchange of letters. Smalley argued that chemistry is extremely complicated, reactions are hard to control, and that a universal assembler is science fiction. Drexler and colleagues, however, noted that Drexler never proposed universal assemblers able to make absolutely anything, but instead proposed more limited assemblers able to make a very wide variety of things. They challenged the relevance of Smalley's arguments to the more specific proposals advanced in *Nanosystems*. Also, Smalley argued that nearly all of modern chemistry involves reactions that take place in a solvent (usually water), because the small molecules of a solvent contribute many things, such as lowering binding energies for transition states. Since nearly all known chemistry requires a solvent, Smalley felt that Drexler's proposal to use a high vacuum environment was not feasible. However, Drexler addresses this in Nanosystems by showing mathematically that well designed catalysts can provide the effects of a solvent and can fundamentally be made even more efficient than a solvent/enzyme reaction could ever be. It is noteworthy that, contrary to Smalley's opinion that enzymes require water, "Not only do enzymes work vigorously in anhydrous organic media, but in this unnatural milieu they acquire remarkable properties such as greatly enhanced stability, radically altered substrate and enantiomeric specificities, molecular memory, and the ability to catalyse unusual reactions.

Design Issues

For the future, some means have to be found for MNT design evolution at the nanoscale which mimics the process of biological evolution at the molecular scale. Biological evolution proceeds by random variation in ensemble averages of organisms combined with culling of the less-successful variants and reproduction of the more-successful variants, and macroscale engineering design also proceeds by a process of design evolution from simplicity to complexity as set forth somewhat satirically by John Gall: "A complex system that works is invariably found to have evolved from a simple system that worked. . . . A complex system designed from scratch never works and can not be patched up to make it work. You have to start over, beginning with a system that works." A breakthrough in MNT is needed which proceeds from the simple atomic ensembles which can be built with, e.g., an STM to complex MNT systems via a process of design evolution. A handicap in this process is the difficulty of seeing and manipulation at the nanoscale compared to the macroscale which makes deterministic selection of successful trials difficult; in contrast biological evolution proceeds via action of what Richard Dawkins has called the "blind watchmaker" comprising random molecular variation and deterministic reproduction/extinction.

At present in 2007 the practice of nanotechnology embraces both stochastic approaches (in which, for example, supramolecular chemistry creates waterproof pants) and deterministic approaches wherein single molecules (created by stochastic chemistry) are manipulated on substrate surfaces (created by stochastic deposition methods) by deterministic methods comprising nudging them with STM or AFM probes and causing simple binding or cleavage reactions to occur. The dream of a complex, deterministic molecular nanotechnology remains elusive. Since the mid-1990s, thousands of surface scientists and thin film technocrats have latched on to the nanotechnology bandwagon and redefined their disciplines as nanotechnology. This has caused much confusion in the field and has spawned thousands of "nano"-papers on the peer reviewed literature. Most of these reports are extensions of the more ordinary research done in the parent fields.

The Feasibility of the Proposals in *Nanosystems*

The feasibility of Drexler's proposals largely depends, therefore, on whether designs like those in *Nanosystems* could be built in the absence of a universal assembler to build them and would work as described. Supporters of molecular nanotechnology frequently claim that no significant errors have been discovered in *Nanosystems* since 1992. Even some critics concede that "Drexler has carefully considered a number of physical principles underlying the 'high level' aspects of the nanosystems he proposes and, indeed, has thought in some detail" about some issues.

Other critics claim, however, that *Nanosystems* omits important chemical details about the low-level 'machine language' of molecular nanotechnology. They also claim that much of the other low-level chemistry in *Nanosystems* requires extensive further work, and that Drexler's higher-level designs therefore rest on speculative foundations. Recent such further work by Freitas and Merkle is aimed at strengthening these foundations by filling the existing gaps in the low-level chemistry.

Drexler argues that we may need to wait until our conventional nanotechnology improves before solving these issues: "Molecular manufacturing will result from a series of advances in molecular machine systems, much as the first Moon landing resulted from a series of advances in liquid-fuel rocket systems. We are now in a position like that of the British Interplanetary Society of the 1930s which described how multistage liquid-fueled rockets could reach the Moon and pointed to early rockets as illustrations of the basic principle." However, Freitas and Merkle argue that a focused effort to achieve diamond mechanosynthesis (DMS) can begin now, using existing technology, and might achieve success in less than a decade if their "direct-to-DMS approach is pursued rather than a more circuitous development approach that seeks to implement less efficacious nondiamondoid molecular manufacturing technologies before progressing to diamondoid".

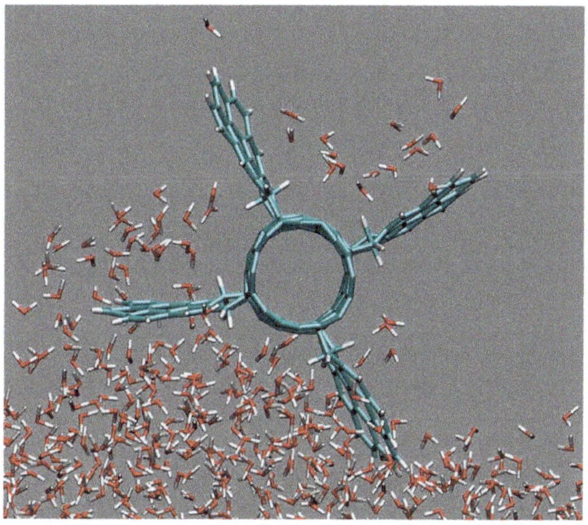

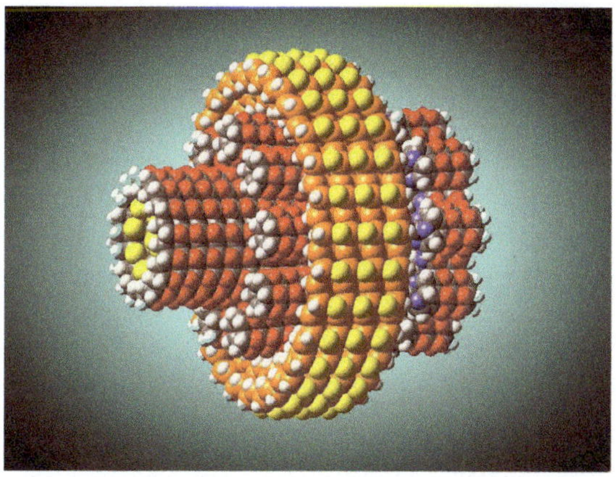

Top, a molecular propellor. Bottom, a molecular planetary gear system.
The feasibility of devices like these has been questioned.

To summarize the arguments against feasibility: First, critics argue that a primary barrier to achieving molecular nanotechnology is the lack of an efficient way to create machines on a molecular/atomic scale, especially in the absence of a well-defined path toward a self-replicating assembler or diamondoid nanofactory. Advocates respond that a preliminary research path leading to a diamondoid nanofactory is being developed.

A second difficulty in reaching molecular nanotechnology is design. Hand design of a gear or bearing at the level of atoms might take a few to several weeks. While Drexler, Merkle and others have created designs of simple parts, no comprehensive design effort for anything approaching the complexity of a Model T Ford has been attempted. Advocates respond that it is difficult to undertake a comprehensive design effort in the absence of significant funding for such efforts, and that despite this handicap much useful design-ahead has nevertheless been accomplished with new software tools that have been developed, e.g., at Nanorex.

In the latest report *A Matter of Size: Triennial Review of the National Nanotechnology Initiative* put out by the National Academies Press in December 2006 (roughly twenty years after Engines of Creation was published), no clear way forward toward molecular nanotechnology could yet be seen, as per the conclusion on page 108 of that report: "Although theoretical calculations can be made today, the eventually attainable range of chemical reaction cycles, error rates, speed of operation, and thermodynamic efficiencies of such bottom-up manufacturing systems cannot be reliably predicted at this time. Thus, the eventually attainable perfection and complexity of manufactured products, while they can be calculated in theory, cannot be predicted with confidence. Finally, the optimum research paths that might lead to systems which greatly exceed the thermodynamic efficiencies and other capabilities of biological systems cannot be reliably predicted at this time. Research funding that is based on the ability of investigators to produce experimental demonstrations that link to abstract models and guide long-term vision is most appropriate to achieve this goal." This call for research leading to demonstrations is welcomed by groups such as the Nanofactory Collaboration who are specifically seeking experimental successes in diamond mechanosynthesis. The "Technology Roadmap for Productive Nanosystems" aims to offer additional constructive insights.

It is perhaps interesting to ask whether or not most structures consistent with physical law can in fact be manufactured. Advocates assert that to achieve most of the vision of molecular manufacturing it is not necessary to be able to build "any structure that is compatible with natural law." Rather, it is necessary to be able to build only a sufficient (possibly modest) subset of such structures—as is true, in fact, of any practical manufacturing process used in the world today, and is true even in biology. In any event, as Richard Feynman once said, "It is scientific only to say what's more likely or less likely, and not to be proving all the time what's possible or impossible."

Existing Work on Diamond Mechanosynthesis

There is a growing body of peer-reviewed theoretical work on synthesizing diamond by mechanically removing/adding hydrogen atoms and depositing carbon atoms (a process known as mechanosynthesis). This work is slowly permeating the broader nanoscience community and is being critiqued. For instance, Peng et al. (2006) (in the continuing research effort by Freitas, Merkle and their collaborators) reports that the most-studied mechanosynthesis tooltip motif (DCB6Ge) successfully places a C_2 carbon dimer on a C(110) diamond surface at both 300 K (room temperature) and 80 K (liquid nitrogen temperature), and that the silicon variant (DCB6Si) also works at 80 K but not at 300 K. Over 100,000 CPU hours were invested in this latest study. The DCB6 tooltip motif, initially described by Merkle and Freitas at a Foresight Conference in 2002, was the first complete tooltip ever proposed for diamond mechanosynthesis and remains the only tooltip motif that has been successfully simulated for its intended function on a full 200-atom diamond surface.

The tooltips modeled in this work are intended to be used only in carefully controlled environments (e. g., vacuum). Maximum acceptable limits for tooltip translational and rotational misplacement errors are reported in Peng et al. (2006) -- tooltips must be positioned with great accuracy to avoid bonding the dimer incorrectly. Peng et al. (2006) reports that increasing the handle thickness from 4 support planes of C atoms above the tooltip to 5 planes decreases the resonance frequency of the entire structure from 2.0 THz to 1.8 THz. More importantly, the vibrational footprints of a DCB6Ge tooltip mounted on a 384-atom handle and of the same tooltip mounted on a similarly constrained but much larger 636-atom "crossbar" handle are virtually identical in the non-crossbar directions. Additional computational studies modeling still bigger handle structures are welcome, but the ability to precisely position SPM tips to the requisite atomic accuracy has been repeatedly demonstrated experimentally at low temperature, or even at room temperature constituting a basic existence proof for this capability.

Further research to consider additional tooltips will require time-consuming computational chemistry and difficult laboratory work.

A working nanofactory would require a variety of well-designed tips for different reactions, and detailed analyses of placing atoms on more complicated surfaces. Although this appears a challenging problem given current resources, many tools will be available to help future researchers: Moore's Law predicts further increases in computer power, semiconductor fabrication techniques continue to approach the nanoscale, and researchers grow ever more skilled at using proteins, ribosomes and DNA to perform novel chemistry.

Works of Fiction

- In The Diamond Age by Neal Stephenson diamond can be constructed by simply building it out of carbon atoms. Also all sorts of devices from dust size detection devices to giant diamond zeppelins are constructed atom by atom using only carbon, oxygen, nitrogen and chlorine atoms.
- In the novel *Tomorrow* by Andrew Saltzman (ISBN 1-4243-1027-X), a scientist uses nanorobotics to create a liquid that when inserted into the bloodstream, renders one nearly invincible given that the microscopic machines repair tissue almost instantaneously after it is damaged.
- In the roleplaying game Splicers by Palladium Books, humanity has succumbed to a "nanobot plague" that causes any object made of a non-precious metal to twist and change shape (sometimes into a type of robot) moments after being touched by a human. The object will then proceed to attack the human. This has forced humanity to develop "biotechnological" devices to replace those previously made of metal.
- On the television show Mystery Science Theater 3000, the Nanites (voiced variously by Kevin Murphy, Paul Chaplin, Mary Jo Pehl, and Bridget Jones) - are self-replicating, bio-engineered organisms that work on the ship, they are microscopic creatures that reside in the Satellite of Love's computer systems. (They are similar to the creatures in *Star Trek: The Next Generation* episode "Evolution", which featured "nanites" taking over the *Enterprise.*) The Nanites made their first appearance in season 8. Based on the concept of nanotechnology, their comical *deus ex machina* activities included such diverse tasks as instant repair and construction, hairstyling, performing a Nanite variation of a flea circus, conducting a microscopic war, and even destroying the Observers' planet after a dangerously vague request from Mike to "take care of [a] little problem". They also ran a microbrewery.

Impact of Nanotechnology

The impact of nanotechnology extends from its medical, ethical, mental, legal and environmental applications, to fields such as engineering, biology, chemistry, computing, materials science, and communications.

Major benefits of nanotechnology include improved manufacturing methods, water purification systems, energy systems, physical enhancement, nanomedicine, better food production methods, nutrition and large-scale infrastructure auto-fabrication. Nanotechnology's reduced size may allow for automation of tasks which were previously inaccessible due to physical restrictions, which in turn may reduce labor, land, or maintenance requirements placed on humans.

Potential risks include environmental, health, and safety issues; transitional effects such as displacement of traditional industries as the products of nanotechnology become dominant, which are of concern to privacy rights advocates. These may be particularly important if potential negative effects of nanoparticles are overlooked.

Whether nanotechnology merits special government regulation is a controversial issue. Regulatory bodies such as the United States Environmental Protection Agency and the Health and Consumer Protection Directorate of the European Commission have started dealing with the potential risks of nanoparticles. The organic food sector has been the first to act with the regulated exclusion of engineered nanoparticles from certified organic produce, firstly in Australia and the UK, and more recently in Canada, as well as for all food certified to Demeter International standards.

Overview

The presence of nanomaterials (materials that contain nanoparticles) is not in itself a threat. It is only certain aspects that can make them risky, in particular their mobility and their increased reactivity. Only if certain properties of certain nanoparticles were harmful to living beings or the environment would we be faced with a genuine hazard. In this case it can be called nanopollution.

In addressing the health and environmental impact of nanomaterials we need to differentiate between two types of nanostructures: (1) Nanocomposites, nanostructured surfaces and nanocomponents (electronic, optical, sensors etc.), where nanoscale particles are incorporated into a substance, material or device ("fixed" nano-particles); and (2) "free" nanoparticles, where at some stage in production or use individual nanoparticles of a substance are present. These free nanoparticles could be nanoscale species of elements, or simple compounds, but also complex compounds where for instance a nanoparticle of a particular element is coated with another substance ("coated" nanoparticle or "core-shell" nanoparticle).

There seems to be consensus that, although one should be aware of materials containing fixed nanoparticles, the immediate concern is with free nanoparticles.

Nanoparticles are very different from their everyday counterparts, so their adverse effects cannot be derived from the known toxicity of the macro-sized material. This poses significant issues for addressing the health and environmental impact of free nanoparticles.

To complicate things further, in talking about nanoparticles it is important that a powder or liquid containing nanoparticles almost never be monodisperse, but contain instead a range of particle sizes. This complicates the experimental analysis as larger nanoparticles might have different properties from smaller ones. Also, nanoparticles show a tendency to aggregate, and such aggregates often behave differently from individual nanoparticles.

Health Impact

The health impacts of nanotechnology are the possible effects that the use of nanotechnological materials and devices will have on human health. As nanotechnology is an emerging field, there is great debate regarding to what extent nanotechnology will benefit or pose risks for human health. Nanotechnology's health impacts can be split into two aspects: the potential for nanotechnological innovations to have medical applications to cure disease, and the potential health hazards posed by exposure to nanomaterials.

Medical Applications

Nanomedicine is the medical application of nanotechnology. The approaches to nanomedicine range from the medical use of nanomaterials, to nanoelectronic biosensors, and even possible future applications of molecular nanotechnology. Nanomedicine seeks to deliver a valuable set of research tools and clinically helpful devices in the near future. The National Nanotechnology Initiative expects new commercial applications in the pharmaceutical industry that may include advanced drug delivery systems, new therapies, and in vivo imaging. Neuro-electronic interfaces and other nanoelectronics-based sensors are another active goal of research. Further down the line, the speculative field of molecular nanotechnology believes that cell repair machines could revolutionize medicine and the medical field.

Nanomedicine research is directly funded, with the US National Institutes of Health in 2005 funding a five-year plan to set up four nanomedicine centers. In April 2006, the journal Nature Materials estimated that 130 nanotech-based drugs and delivery systems were being developed worldwide. Nanomedicine is a large industry, with nanomedicine sales reaching $6.8 billion in 2004. With over 200 companies and 38 products worldwide, a minimum of $3.8 billion in nanotechnology R&D is being invested every year. As the nanomedicine industry continues to grow, it is expected to have a significant impact on the economy.

Health Hazards

Nanotoxicology is the field which studies potential health risks of nanomaterials. The extremely small size of nanomaterials means that they are much more readily taken up by the human body than larger sized particles. How these nanoparticles behave inside the organism is one of the significant issues that needs to be resolved. The behavior of nanoparticles is a function of their size, shape and surface reactivity with the surrounding tissue. Apart from what happens if non-degradable or slowly degradable nanoparticles accumulate in organs, another concern is their potential interaction with biological processes inside the body: because of their large surface, nanoparticles on exposure to tissue and fluids will immediately adsorb onto their surface some of the macromolecules they encounter. The large number of variables influencing toxicity means that it is difficult to generalise about health risks associated with exposure to nanomaterials – each new nanomaterial must be assessed individually and all material properties must be taken into account. Health and environmental issues combine in the workplace of companies engaged in producing or using nanomaterials and in the laboratories engaged in nanoscience and nanotechnology research. It is safe to say that current workplace exposure standards for dusts cannot be applied directly to nanoparticle dusts.

The extremely small size of nanomaterials also means that they are much more readily taken up by the human body than larger sized particles. How these nanoparticles behave inside the body is one of the issues that needs to be resolved. The behavior of nanoparticles is a function of their size, shape and surface reactivity with the surrounding tissue. They could cause overload on phagocytes, cells that ingest and destroy foreign matter, thereby triggering stress reactions that lead to inflammation and weaken the body's defense against other pathogens. Apart from what happens if non-degradable or slowly degradable nanoparticles accumulate in organs, another concern is their potential interaction with biological processes inside the body: because of their large surface, nanoparticles on exposure to tissue and fluids will immediately adsorb onto their surface some of the macromolecules they encounter. This may, for instance, affect the regulatory mechanisms of enzymes and other proteins.

The National Institute for Occupational Safety and Health has conducted initial research on how nanoparticles interact with the body's systems and how workers might be exposed to nano-sized particles in the manufacturing or industrial use of nanomaterials. NIOSH currently offers interim guidelines for working with nanomaterials consistent with the best scientific knowledge. At The National Personal Protective Technology Laboratory of NIOSH, studies investigating the filter penetration of nanoparticles on NIOSH-certified and EU marked respirators, as well as non-certified dust masks have been conducted. These studies found that the most penetrating particle size range was between 30 and 100 nanometers, and leak size was the largest factor in the number of nanoparticles found inside the respirators of the test dummies.

Other properties of nanomaterials that influence toxicity include: chemical composition, shape, surface structure, surface charge, aggregation and solubility, and the presence or absence of functional groups of other chemicals. The large number of variables influencing toxicity means that it is difficult to generalise about health risks associated with exposure to nanomaterials – each new nanomaterial must be assessed individually and all material properties must be taken into account.

Literature reviews have been showing that release of engineered nanoparticles and incurred personal exposure can happen during different work activities. The situation alerts regulatory bodies to necessitate prevention strategies and regulations at nanotechnology workplaces.

Environmental Impact

The environmental impact of nanotechnology is the possible effects that the use of nanotechnological materials and devices will have on the environment. As nanotechnology is an emerging field, there is debate regarding to what extent industrial and commercial use of nanomaterials will affect organisms and ecosystems.

Nanotechnology's environmental impact can be split into two aspects: the potential for nanotechnological innovations to help improve the environment, and the possibly novel type of pollution that nanotechnological materials might cause if released into the environment.

Environmental Applications

Green nanotechnology refers to the use of nanotechnology to enhance the environmental sustainability of processes producing negative externalities. It also refers to the use of the products of nanotechnology to enhance sustainability. It includes making green nano-products and using nano-products in support of sustainability. Green nanotechnology has been described as the development of clean technologies, "to minimize potential environmental and human health risks associated with the manufacture and use of nanotechnology products, and to encourage replacement of existing products with new nano-products that are more environmentally friendly throughout their lifecycle."

Green nanotechnology has two goals: producing nanomaterials and products without harming the environment or human health, and producing nano-products that provide solutions to environmental problems. It uses existing principles of green chemistry and green engineering to make nanomaterials and nano-products without toxic ingredients, at low temperatures using less energy and renewable inputs wherever possible, and using lifecycle thinking in all design and engineering stages.

Pollution

Nanopollution is a generic name for all waste generated by nanodevices or during the nanomaterials manufacturing process. Nanowaste is mainly the group of particles that are released into the environment, or the particles that are thrown away when still on their products.

Social Impact

Beyond the toxicity risks to human health and the environment which are associated with first-generation nanomaterials, nanotechnology has broader societal impact and poses broader social challenges. Social scientists have suggested that nanotechnology's social issues should be understood and assessed not simply as "downstream" risks or impacts. Rather, the challenges should be factored into "upstream" research and decision-making in order to ensure technology development that meets social objectives

Many social scientists and organizations in civil society suggest that technology assessment and governance should also involve public participation

Over 800 nano-related patents were granted in 2003, with numbers increasing to nearly 19,000 internationally by 2012. Corporations are already taking out broad-ranging patents on nanoscale discoveries and inventions. For example, two corporations, NEC and IBM, hold the basic patents on carbon nanotubes, one of the current cornerstones of nanotechnology. Carbon nanotubes have a wide range of uses, and look set to become crucial to several industries from electronics and computers, to strengthened materials to drug delivery and diagnostics. Carbon nanotubes are poised to become a major traded commodity with the potential to replace major conventional raw materials.

Nanotechnologies may provide new solutions for the millions of people in developing countries who lack access to basic services, such as safe water, reliable energy, health care, and education. The 2004 UN Task Force on Science, Technology and Innovation noted that some of the advantages of nanotechnology include production using little labor, land, or maintenance, high productivity, low cost, and modest requirements for materials and energy. However, concerns are frequently raised that the claimed benefits of nanotechnology will not be evenly distributed, and that any benefits (including technical and/or economic) associated with nanotechnology will only reach affluent nations.

Longer-term concerns center on the impact that new technologies will have for society at large, and whether these could possibly lead to either a post-scarcity economy, or alternatively exacerbate the wealth gap between developed and developing nations. The effects of nanotechnology on the society as a whole, on human health and the environment, on trade, on security, on food systems and even on the definition of "human", have not been characterized or politicized.

Regulation

Significant debate exists relating to the question of whether nanotechnology or nanotechnology-based products merit special government regulation. This debate is related to the circumstances in which it is necessary and appropriate to assess new substances prior to their release into the market, community and environment.

Regulatory bodies such as the United States Environmental Protection Agency and the Food and Drug Administration in the U.S. or the Health & Consumer Protection Directorate of the European Commission have started dealing with the potential risks posed by nanoparticles. So far, neither engineered nanoparticles nor the products and materials that contain them are subject to any special regulation regarding production, handling or labelling. The Material Safety Data Sheet that must be issued for some materials often does not differentiate between bulk and nanoscale size of the material in question and even when it does these MSDS are advisory only.

Limited nanotechnology labeling and regulation may exacerbate potential human and environmental health and safety issues associated with nanotechnology. It has been argued that the development of comprehensive regulation of nanotechnology will be vital to ensure that the potential risks associated with the research and commercial application of nanotechnology do not overshadow its potential benefits. Regulation may also be required to meet community expectations about responsible development of nanotechnology, as well as ensuring that public interests are included in shaping the development of nanotechnology.

In "The Consumer Product Safety Commission and Nanotechnology," E. Marla Felcher suggests that the Consumer Product Safety Commission, which is charged with protecting the public against unreasonable risks of injury or death associated with consumer products, is ill-equipped to oversee the safety of complex, high-tech products made using nanotechnology.

References

- Smith, Rebecca (August 19, 2009). "Nanoparticles used in paint could kill, research suggests". London: Telegraph. Retrieved May 19, 2010
- R. V. Lapshin (2004). "Feature-oriented scanning methodology for probe microscopy and nanotechnology" (PDF). Nanotechnology. UK: IOP. 15 (9): 1135–1151. doi:10.1088/0957-4484/15/9/006. ISSN 0957-4484
- Saini, Rajiv; Saini, Santosh; Sharma, Sugandha (2010). "Nanotechnology: The Future Medicine". Journal of Cutaneous and Aesthetic Surgery. 3 (1): 32–33. doi:10.4103/0974-2077.63301. PMC 2890134 . PMID 20606992

- Drexler, K. Eric (1992). Nanosystems: Molecular Machinery, Manufacturing, and Computation. New York: John Wiley & Sons. ISBN 0-471-57547-X
- "Nanoscience and nanotechnologies: opportunities and uncertainties". Royal Society and Royal Academy of Engineering. July 2004. Retrieved 13 May 2011
- Buzea, C.; Pacheco, I. I.; Robbie, K. (2007). "Nanomaterials and nanoparticles: Sources and toxicity". Biointerphases. 2 (4): MR17–MR71. doi:10.1116/1.2815690. PMID 20419892
- Paull, J. & Lyons, K. (2008). "Nanotechnology: The Next Challenge for Organics" (PDF). Journal of Organic Systems. 3: 3–22
- llhoff, Fritz; Lin, Patrick; Moore, Daniel (2010). What is nanotechnology and why does it matter?: from science to ethics. John Wiley and Sons. pp. 3–5. ISBN 1-4051-7545-1
- "Nanotechnology Information Center: Properties, Applications, Research, and Safety Guidelines". American Elements. Retrieved 13 May 2011
- Dams, W. W.; Baughman, R. H. (2005). "RETROSPECTIVE: Richard E. Smalley (1943-2005)". Science. 310 (5756): 1916. doi:10.1126/science.1122120. PMID 16373566
- Bowman D; Fitzharris, M (2007). "Too Small for Concern? Public Health and Nanotechnology". Australian and New Zealand Journal of Public Health. 31 (4): 382–384. doi:10.1111/j.1753-6405.2007.00092.x. PMID 17725022
- Lapshin, R. V. (2011). "Feature-oriented scanning probe microscopy". In H. S. Nalwa. Encyclopedia of Nanoscience and Nanotechnology (PDF). 14. USA: American Scientific Publishers. pp. 105–115. ISBN 1-58883-163-9
- "Analysis: This is the first publicly available on-line inventory of nanotechnology-based consumer products". The Project on Emerging Nanotechnologies. 2008. Retrieved 13 May 2011
- Ghalanbor Z, Marashi SA, Ranjbar B (2005). "Nanotechnology helps medicine: nanoscale swimmers and their future applications". Med Hypotheses. 65 (1): 198–199. doi:10.1016/j.mehy.2005.01.023. PMID 15893147
- Byrne, J. D.; Baugh, J. A. (2008). "The significance of nanoparticles in particle-induced pulmonary fibrosis". McGill journal of medicine : MJM : an international forum for the advancement of medical sciences by students. 11 (1): 43–50. PMC 2322933 . PMID 18523535
- Murray R.G.E. (1993) Advances in Bacterial Paracrystalline Surface Layers. T. J. Beveridge, S. F. Koval (Eds.). Plenum Press. ISBN 978-0-306-44582-8. pp. 3–9
- "CDC – Nanotechnology – NIOSH Workplace Safety and Health Topic". National Institute for Occupational Safety and Health. June 15, 2012. Retrieved 2012-08-24

4

Carbon Nanotube: An Overview

The allotrope of carbon having a cylindrical structure is called a carbon nanotube. Their properties are of extreme significance in nanotechnology. They can be normally classified into single-walled and multi-walled nanotubes. Carbon nanotubes are best understood in confluence with the major topics listed in the following chapter.

Carbon Nanotube

Carbon nanotubes (CNTs) are allotropes of carbon with a cylindrical nanostructure. These cylindrical carbon molecules have unusual properties, which are valuable for nanotechnology, electronics, optics and other fields of materials science and technology. Owing to the material's exceptional strength and stiffness, nanotubes have been constructed with length-to-diameter ratio of up to 132,000,000:1, significantly larger than for any other material.

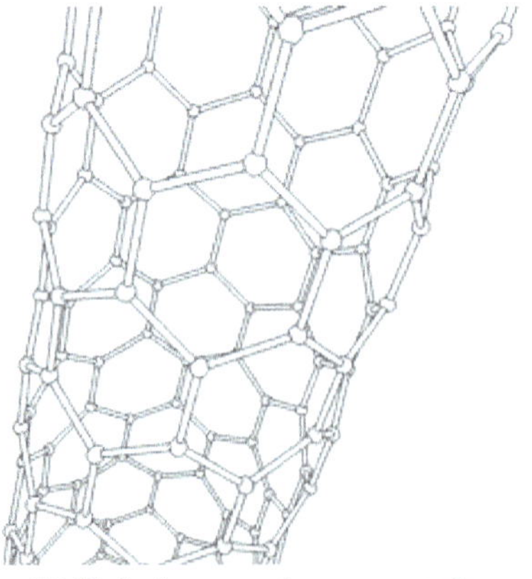

Walled zigzag carbon nanotube

In addition, owing to their extraordinary thermal conductivity, mechanical, and electrical properties, carbon nanotubes find applications as additives to various structural materials. For instance, nanotubes form a tiny portion of the material(s) in some (primarily carbon fiber) baseball bats, golf clubs, car parts or damascus steel.

Nanotubes are members of the fullerene structural family. Their name is derived from their long, hollow structure with the walls formed by one-atom-thick sheets of carbon, called graphene. These sheets are rolled at specific and discrete ("chiral") angles, and the combination of the rolling angle and radius decides the nanotube properties; for example, whether the individual nanotube shell is a metal or semiconductor. Nanotubes are categorized as single-walled nanotubes (SWNTs) and multi-walled nanotubes (MWNTs). Individual nanotubes naturally align themselves into "ropes" held together by van der Waals forces, more specifically, pi-stacking.

Applied quantum chemistry, specifically, orbital hybridization best describes chemical bonding in nanotubes. The chemical bonding of nanotubes is composed entirely of sp^2 bonds, similar to those of graphite. These bonds, which are stronger than the sp^3 bonds found in alkanes and diamond, provide nanotubes with their unique strength.

Types of Carbon Nanotubes and Related Structures

There is no consensus on some terms describing carbon nanotubes in scientific literature: both "-wall" and "-walled" are being used in combination with "single", "double", "triple" or "multi", and the letter C is often omitted in the abbreviation; for example, multi-walled carbon nanotube (MWNT).

Single-walled

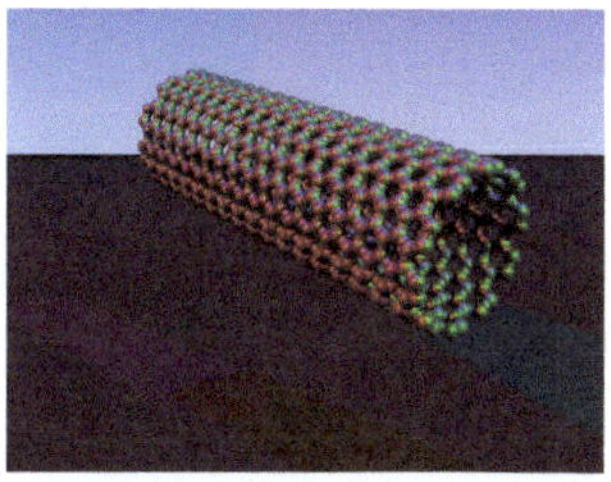
Armchair (*n*,*n*) i.e.: m=n

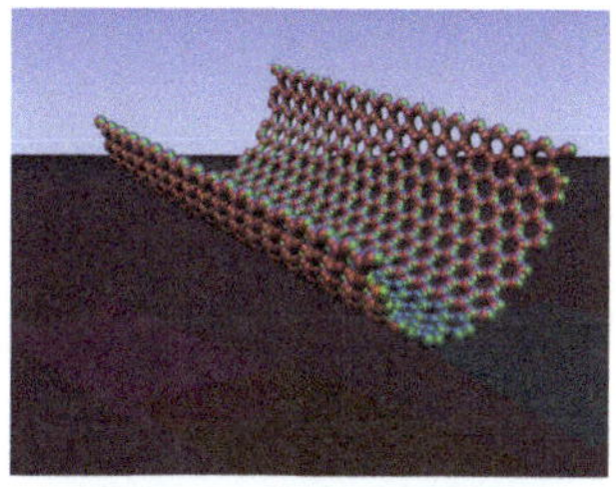
The translation vector is bent, while the chiral vector stays straight

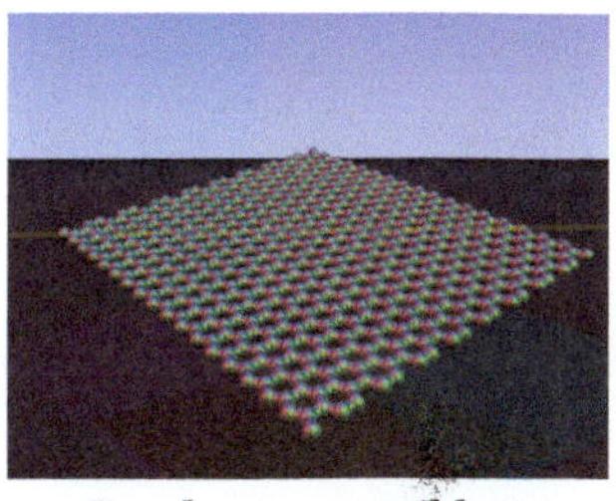
Graphene nanoribbon

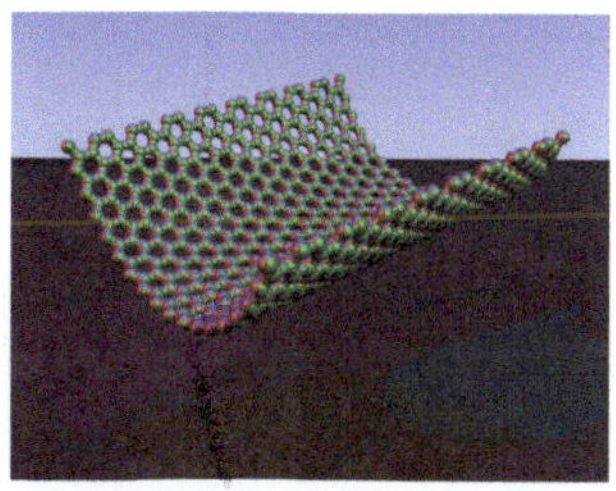
The chiral vector is bent, while the translation vector stays straight

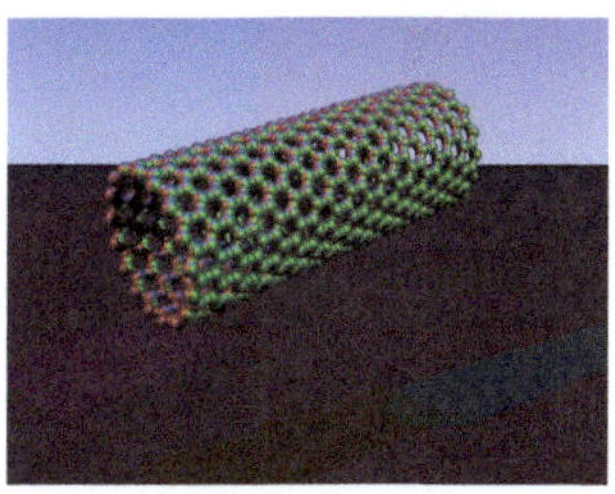
Zigzag (*n*,0)

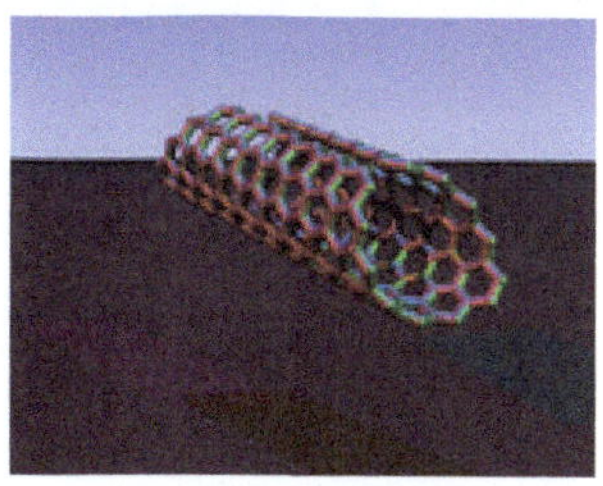
Chiral (*n*,*m*)

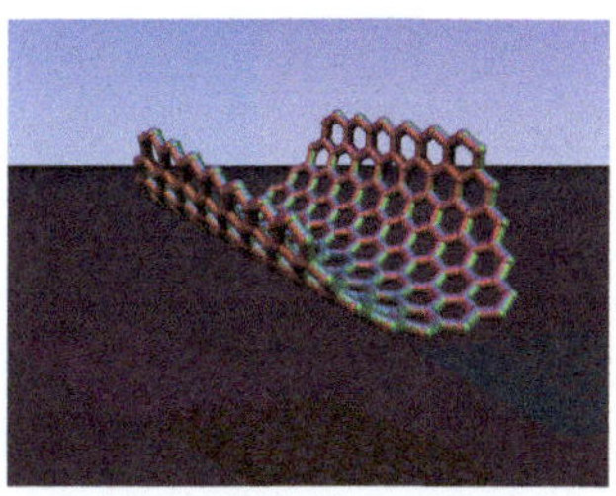
n and *m* can be counted at the end of the tube

Graphene nanoribbon

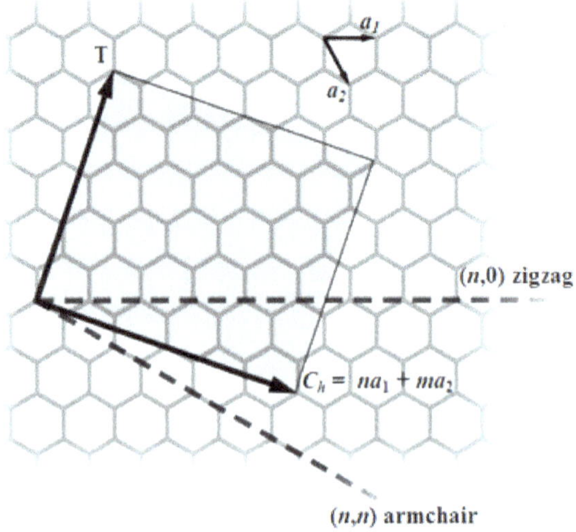

The (n,m) nanotube naming scheme can be thought of as a vector ($\mathbf{C}_h$) in an infinite graphene sheet that describes how to "roll up" the graphene sheet to make the nanotube. **T** denotes the tube axis, and $\mathbf{a}_1$ and $\mathbf{a}_2$ are the unit vectors of graphene in real space.

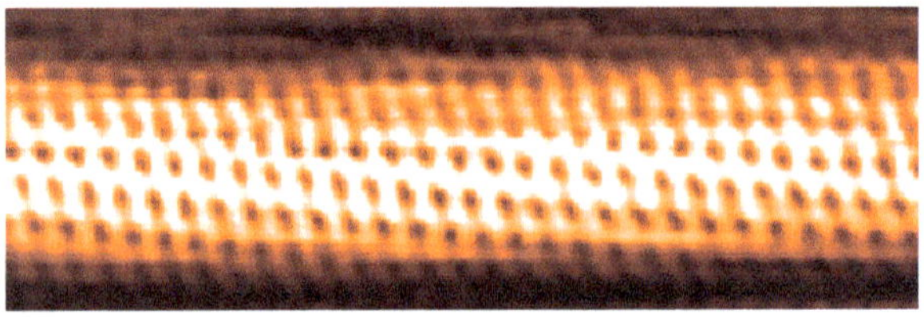

A scanning tunneling microscopy image of single-walled carbon nanotube

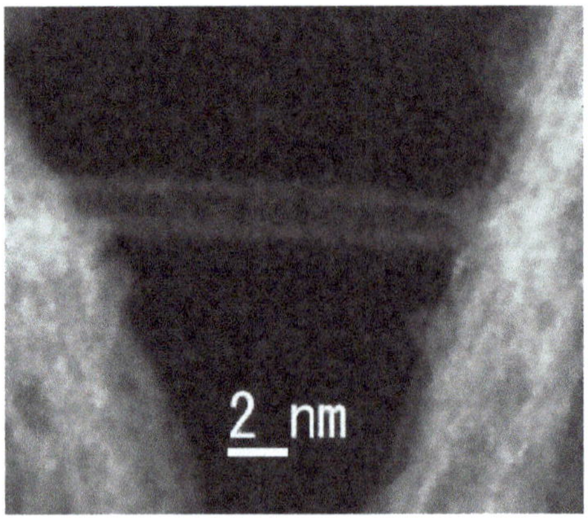

A transmission electron microscopy image of a single-walled carbon nanotube

Most single-walled nanotubes (SWNTs) have a diameter of close to 1 nanometer, and can be many millions of times longer. The structure of a SWNT can be conceptualized by wrapping a one-atom-thick layer of graphite called graphene into a seamless cylinder. The way the graphene sheet is wrapped is represented by a pair of indices (n,m). The integers n and m denote the number of unit vectors along two directions in the honeycomb crystal lattice of graphene. If $m = 0$, the nanotubes are called zigzag nanotubes, and if $n = m$, the nanotubes are called armchair nanotubes. Otherwise, they are called chiral. The diameter of an ideal nanotube can be calculated from its (n,m) indices as follows:

$$d = \frac{a}{\pi}\sqrt{(n^2 + nm + m^2)} = 78.3\sqrt{((n+m)^2 - nm)}\text{pm},$$

where a = 0.246 nm.

SWNTs are an important variety of carbon nanotube because most of their properties change significantly with the (n,m) values, and this dependence is non-monotonic. In particular, their band gap can vary from zero to about 2 eV and their electrical conductivity can show metallic or semiconducting behavior. Single-walled nanotubes are likely candidates for miniaturizing electronics. The most basic building block of these systems is the electric wire, and SWNTs with diameters of an order of a nanometer can be excellent conductors. One useful application of SWNTs is in the development of the first intermolecular field-effect transistors (FET). The first intermolecular logic gate using SWCNT FETs was made in 2001. A logic gate requires both a p-FET and an n-FET. Because SWNTs are p-FETs when exposed to oxygen and n-FETs otherwise, it is possible to expose half of an SWNT to oxygen and protect the other half from it. The resulting SWNT acts as a *not* logic gate with both p and n-type FETs in the same molecule.

Prices for single-walled nanotubes declined from around $1500 per gram as of 2000 to retail prices of around $50 per gram of as-produced 40–60% by weight SWNTs as of March 2010. As of 2016 the retail price of as-produced 75% by weight SWNTs were $2 per gram, cheap enough for widespread use. SWNTs are forecast to make a large impact in electronics applications by 2020 according to the *The Global Market for Carbon Nanotubes* report.

Multi-walled

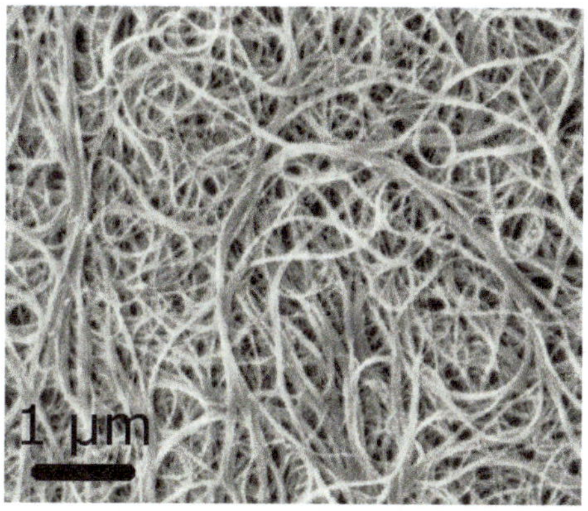

A scanning electron microscopy image of carbon nanotubes bundles

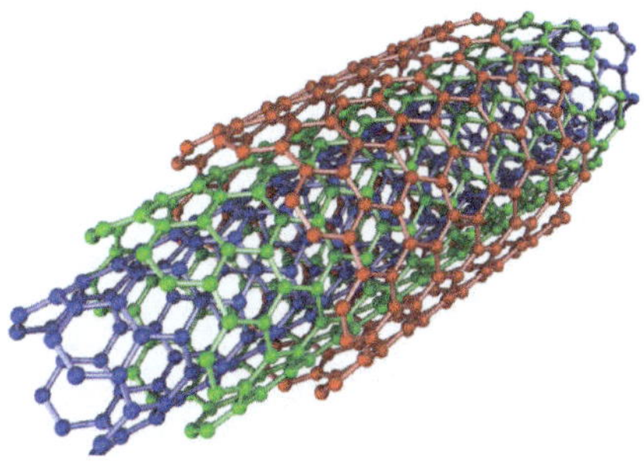

Triple-walled armchair carbon nanotube

Multi-walled nanotubes (MWNTs) consist of multiple rolled layers (concentric tubes) of graphene. There are two models that can be used to describe the structures of multi-walled nanotubes. In the *Russian Doll* model, sheets of graphite are arranged in concentric cylinders, e.g., a (0,8) single-walled nanotube (SWNT) within a larger (0,17) single-walled nanotube. In the *Parchment* model, a single sheet of graphite is rolled in around itself, resembling a scroll of parchment or a rolled newspaper. The interlayer distance in multi-walled nanotubes is close to the distance between graphene layers in graphite, approximately 3.4 Å. The Russian Doll structure is observed more commonly. Its individual shells can be described as SWNTs, which can be metallic or semiconducting. Because of statistical probability and restrictions on the relative diameters of the individual tubes, one of the shells, and thus the whole MWNT, is usually a zero-gap metal.

Double-walled carbon nanotubes (DWNTs) form a special class of nanotubes because their morphology and properties are similar to those of SWNTs but they are more resistant to chemicals. This is especially important when it is necessary to graft chemical functions to the surface of the nanotubes (functionalization) to add properties to the CNT. Covalent functionalization of SWNTs will break some C=C double bonds, leaving "holes" in the structure on the nanotube, and thus modifying both its mechanical and electrical properties. In the case of DWNTs, only the outer wall is modified. DWNT synthesis on the gram-scale was first proposed in 2003 by the CCVD technique, from the selective reduction of oxide solutions in methane and hydrogen.

The telescopic motion ability of inner shells and their unique mechanical properties will permit the use of multi-walled nanotubes as main movable arms in coming nanomechanical devices. Retraction force that occurs to telescopic motion caused by the Lennard-Jones interaction between shells and its value is about 1.5 nN.

Junctions and Crosslinking

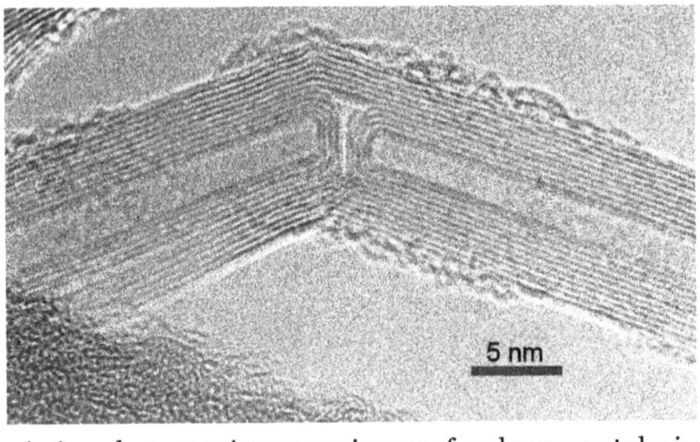

Transmission electron microscope image of carbon nanotube junction

Junctions between 2 or more nanotubes have been widely discussed theoretically. Such junctions are quite frequently observed in samples prepared by arc discharge as well as by chemical vapor deposition. The electronic properties of such junctions were first considered theoretically by Lambin et al., who pointed out that a connection between metallic tube and a semiconducting one would represent a nanoscale heterojunction. Such a junction could therefore form a component of a nanotube-based electronic circuit. The image on the right shows a junction between two multi-walled nanotubes. Junctions between nanotubes and graphene have been considered theoretically, but not widely studied experimentally. Such junctions form the basis of pillared graphene, in which parallel graphene sheets are separated by short nanotubes. Pillared graphene represents a class of three-dimensional carbon nanotube architectures.

3D carbon scaffolds

Recently, several studies have highlighted the prospect of using carbon nanotubes as building blocks to fabricate three-dimensional macroscopic (>100 nm in all three dimensions) all-carbon devices. Lalwani et al. have reported a novel radical initiated thermal crosslinking method to fabricate macroscopic, free-standing, porous, all-carbon scaffolds using single- and multi-walled carbon nanotubes as building blocks. These scaffolds possess macro-, micro-, and nano- structured pores and the porosity can be tailored for specific applications. These 3D all-carbon scaffolds/architectures may be used for the fabrication of the next generation of energy storage, supercapacitors, field emission transistors, high-performance catalysis, photovoltaics, and biomedical devices and implants.

Other Morphologies

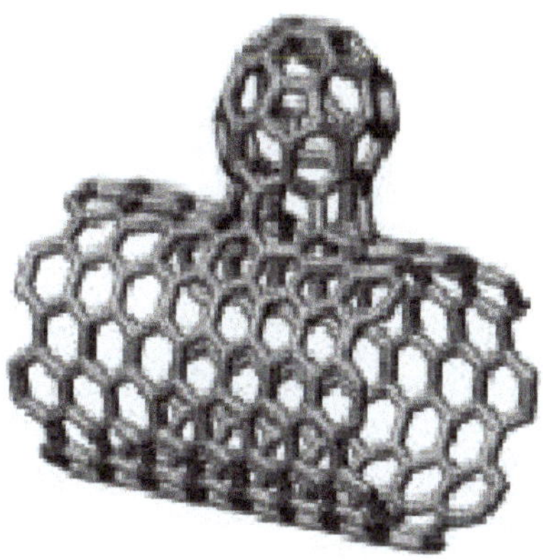

A stable nanobud structure

Carbon nanobuds are a newly created material combining two previously discovered allotropes of carbon: carbon nanotubes and fullerenes. In this new material, fullerene-like "buds" are covalently bonded to the outer sidewalls of the underlying carbon nanotube. This hybrid material has useful properties of both fullerenes and carbon nanotubes. In particular, they have been found to be exceptionally good field emitters. In composite materials, the attached fullerene molecules may function as molecular anchors preventing slipping of the nanotubes, thus improving the composite's mechanical properties.

A carbon peapod is a novel hybrid carbon material which traps fullerene inside a carbon nanotube. It can possess interesting magnetic properties with heating and irradiation. It can also be applied as an oscillator during theoretical investigations and predictions.

In theory, a nanotorus is a carbon nanotube bent into a torus (doughnut shape). Nanotori are predicted to have many unique properties, such as magnetic moments 1000 times larger than previously expected for certain specific radii. Properties such as magnetic moment, thermal stability, etc. vary widely depending on radius of the torus and radius of the tube.

Graphenated carbon nanotubes are a relatively new hybrid that combines graphitic foliates grown along the sidewalls of multiwalled or bamboo style CNTs. The foliate density can vary as a function of deposition conditions (e.g. temperature and time) with their structure ranging from few layers of graphene (< 10) to thicker, more graphite-like. The fundamental advantage of an integrated graphene-CNT structure is the high surface area three-dimensional framework of the CNTs coupled with the high edge density of graphene. Depositing a high density of graphene foliates along the length of aligned CNTs can significantly increase the total charge capacity per unit of nominal area as compared to other carbon nanostructures.

Cup-stacked carbon nanotubes (CSCNTs) differ from other quasi-1D carbon structures, which normally behave as quasi-metallic conductors of electrons. CSCNTs exhibit semiconducting behaviors due to the stacking microstructure of graphene layers.

Extreme Carbon Nanotubes

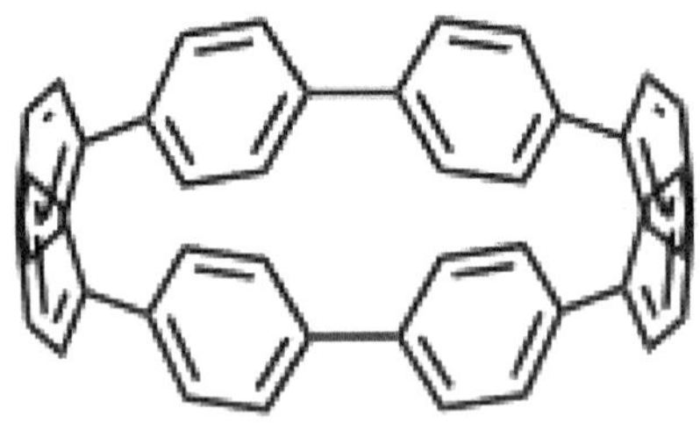

Cycloparaphenylene

The observation of the *longest* carbon nanotubes grown so far are over 1/2 m (550 mm long) was reported in 2013. These nanotubes were grown on Si substrates using an improved chemical vapor deposition (CVD) method and represent electrically uniform arrays of single-walled carbon nanotubes.

The *shortest* carbon nanotube is the organic compound cycloparaphenylene, which was synthesized in 2008.

The *thinnest* carbon nanotube is the armchair (2,2) CNT with a diameter of 0.3 nm. This nanotube was grown inside a multi-walled carbon nanotube. Assigning of carbon nanotube type was done by a combination of high-resolution transmission electron microscopy (HRTEM), Raman spectroscopy and density functional theory (DFT) calculations.

The *thinnest freestanding* single-walled carbon nanotube is about 0.43 nm in diameter. Researchers suggested that it can be either (5,1) or (4,2) SWCNT, but the exact type of carbon nanotube remains questionable. (3,3), (4,3) and (5,1) carbon nanotubes (all about 0.4 nm in diameter) were unambiguously identified using aberration-corrected high-resolution transmission electron microscopy inside double-walled CNTs.

The *highest density* of CNTs was achieved in 2013, grown on a conductive titanium-coated copper surface that was coated with co-catalysts cobalt and molybdenum at lower than typical temperatures of 450 °C. The tubes averaged a height of 380 nm and a mass density of 1.6 g cm^{-3}. The material showed ohmic conductivity (lowest resistance ~22 kΩ).

Properties

Mechanical

Carbon nanotubes are the strongest and stiffest materials yet discovered in terms of tensile strength and elastic modulus respectively. This strength results from the covalent sp^2 bonds formed between the individual carbon atoms. In 2000, a multi-walled carbon nanotube was tested to have a tensile strength of 63 gigapascals (9,100,000 psi). (For illustration, this translates into the ability to endure tension of a weight equivalent to 6,422 kilograms-force (62,980 N; 14,160 lbf) on a cable with cross-section of 1 square millimetre (0.0016 sq in).) Further studies, such as one conducted in 2008, revealed that individual CNT shells have strengths of up to ≈100 gigapascals (15,000,000 psi), which is in agreement with quantum/atomistic models. Since carbon nanotubes have a low density for a solid of 1.3 to 1.4 g/cm^3, its specific strength of up to 48,000 $kN \cdot m \cdot kg^{-1}$ is the best of known materials, compared to high-carbon steel's 154 $kN \cdot m \cdot kg^{-1}$.

Although the strength of individual CNT shells is extremely high, weak shear interactions between adjacent shells and tubes lead to significant reduction in the effective strength of multi-walled carbon nanotubes and carbon nanotube bundles down to only a few GPa. This limitation has been recently addressed by applying high-energy electron irradiation, which crosslinks inner shells and tubes, and effectively increases the strength of these materials to ≈60 GPa for multi-walled carbon nanotubes and ≈17 GPa for double-walled carbon nanotube bundles. CNTs are not nearly as strong under compression. Because of their hollow structure and high aspect ratio, they tend to undergo buckling when placed under compressive, torsional, or bending stress.

On the other hand, there was evidence that in the radial direction they are rather soft. The first transmission electron microscope observation of radial elasticity suggested that even the van der Waals forces can deform two adjacent nanotubes. Later, nanoindentations with atomic force microscope were performed by several groups to quantitatively measure radial elasticity of multi-walled carbon nanotubes and tapping/contact mode atomic force microscopy was also performed on single-walled carbon nanotubes. Young's modulus of on the order of several GPa showed that CNTs are in fact very soft in the radial direction.

Electrical

Unlike graphene, which is a two-dimensional semimetal, carbon nanotubes are either metallic or semiconducting along the tubular axis. For a given (n,m) nanotube, if $n = m$, the nanotube is metallic; if $n - m$ is a multiple of 3, then the nanotube is semiconducting with a very small band gap, otherwise the nanotube is a moderate semiconductor. Thus all armchair ($n = m$) nanotubes are metallic, and nanotubes (6,4), (9,1), etc. are semiconducting. Carbon nanotubes are not semimetallic because the degenerate point (that point where the π [bonding] band meets the π^* [anti-bonding] band, at which the energy goes to zero) is slightly shifted away from the K point in the Brillouin zone due to the curvature of the tube surface, casing hybridization between the σ^* and π^* anti-bonding bands, modifying the band dispersion.

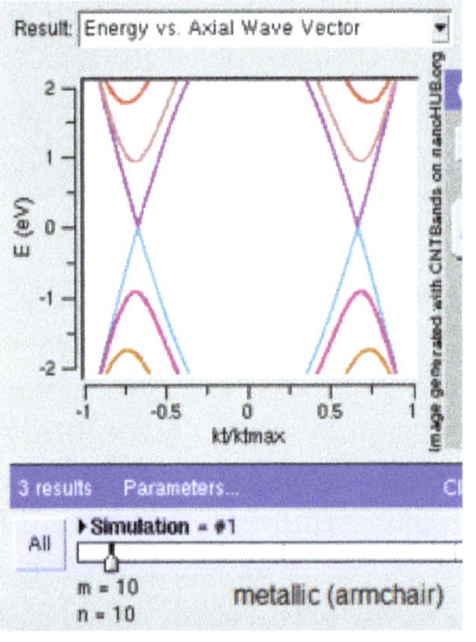

Band structures computed using tight binding approximation for (6,0) CNT (zigzag, metallic), (10,2) CNT (semiconducting) and (10,10) CNT (armchair, metallic).

The rule regarding metallic versus semiconductor behavior has exceptions, because curvature effects in small diameter tubes can strongly influence electrical properties. Thus, a (5,0) SWCNT that should be semiconducting in fact is metallic according to the calculations. Likewise, zigzag and chiral SWCNTs with small diameters that should be metallic have a finite gap (armchair nanotubes remain metallic). In theory, metallic nanotubes can carry an electric current density of 4×10^9 A/cm^2, which is more than 1,000 times greater than those of metals such as copper, where for copper interconnects current densities are limited by electromigration. Carbon nanotubes are thus being explored as interconnects, conductivity enhancing components in composite materials and many groups are attempting to commercialize highly conducting electrical wire assembled from individual carbon nanotubes. There are significant challenges to be overcome, however, such as undesired current saturation under voltage, the much more resistive nanotube-to-nanotube junctions and impurities, all of which lower the electrical conductivity of the macroscopic nanotube wires by orders of magnitude, as compared to the conductivity of the individual nanotubes.

Because of its nanoscale cross-section, electrons propagate only along the tube's axis. As a result, carbon nanotubes are frequently referred to as one-dimensional conductors. The maximum electrical conductance of a single-walled carbon nanotube is $2G_0$, where $G_0 = 2e^2/h$ is the conductance of a single ballistic quantum channel.

Due to the role of the π-electron system in determining the electronic properties of graphene, doping in carbon nanotubes differs from that of bulk crystalline semiconductors from the same group of the periodic table (e.g. silicon). Graphitic substitution of carbon atoms in the nanotube wall by boron or nitrogen dopants leads to p-type and n-type behavior, respectively, as would be expected in silicon. However, some non-substitutional (intercalated or adsorbed) dopants introduced into a carbon nanotube, such as alkali metals as well as electron-rich metallocenes, result in n-type conduction because they donate electrons to the π-electron system of the nanotube. By contrast, π-electron acceptors such as $FeCl_3$ or electron-deficient metallocenes function as p-type dopants since they draw π-electrons away from the top of the valence band.

Intrinsic superconductivity has been reported, although other experiments found no evidence of this, leaving the claim a subject of debate.

Optical

Carbon nanotubes have useful absorption, photoluminescence (fluorescence), and Raman spectroscopy properties. Spectroscopic methods offer the possibility of quick and non-destructive characterization of relatively large amounts of carbon nanotubes. There is a strong demand for such characterization from the industrial point of view: numerous parameters of the nanotube synthesis can be changed, intentionally or unintentionally, to alter the nanotube quality. As shown below, optical absorption, photoluminescence and Raman spectroscopies allow quick and reliable characterization of this "nanotube quality" in terms of non-tubular carbon content, structure (chirality) of the produced nanotubes, and structural defects. Those features determine nearly any other properties such as optical, mechanical, and electrical properties.

Carbon nanotubes are unique "one-dimensional systems" which can be envisioned as rolled single sheets of graphite (or more precisely graphene). This rolling can be done at different angles and curvatures resulting in different nanotube properties. The diameter typically varies in the range 0.4–40 nm (i.e. "only" ~100 times), but the length can vary ~10,000 times, reaching 55.5 cm. The nanotube aspect ratio, or the length-to-diameter ratio, can be as high as 132,000,000:1, which is unequalled by any other material. Consequently, all the properties of the carbon nanotubes relative to those of typical semiconductors are extremely anisotropic (directionally dependent) and tunable.

Whereas mechanical, electrical and electrochemical (supercapacitor) properties of the carbon nanotubes are well established and have immediate applications, the practical use of optical properties is yet unclear. The aforementioned tunability of properties is potentially useful in optics and photonics. In particular, light-emitting diodes (LEDs) and photo-detectors based on a single nanotube have been produced in the lab. Their unique feature is not the efficiency, which is yet relatively low, but the narrow selectivity in the wavelength of emission and detection of light and the possibility of its fine tuning through the nanotube structure. In addition, bolometer and optoelectronic memory devices have been realised on ensembles of single-walled carbon nanotubes.

Crystallographic defects also affect the tube's electrical properties. A common result is lowered conductivity through the defective region of the tube. A defect in armchair-type tubes (which can conduct electricity) can cause the surrounding region to become semiconducting, and single monatomic vacancies induce magnetic properties.

Thermal

All nanotubes are expected to be very good thermal conductors along the tube, exhibiting a property known as "ballistic conduction", but good insulators lateral to the tube axis. Measurements show that an individual SWNT has a room-temperature thermal conductivity along its axis of about 3500 $W·m^{-1}·K^{-1}$; compare this to copper, a metal well known for its good thermal conductivity, which transmits 385 $W·m^{-1}·K^{-1}$. An individual SWNT has a room-temperature thermal conductivity across its axis (in the radial direction) of about 1.52 $W·m^{-1}·K^{-1}$, which is about as thermally conductive as soil. Macroscopic assemblies of nanotubes such as films or fibres have reached up to 1500 $W \cdot m^{-1} \cdot K^{-1}$ so far. The temperature stability of carbon nanotubes is estimated to be up to 2800 °C in vacuum and about 750 °C in air.

Crystallographic defects strongly affect the tube's thermal properties. Such defects lead to phonon scattering, which in turn increases the relaxation rate of the phonons. This reduces the mean free path and reduces the thermal conductivity of nanotube structures. Phonon transport simulations indicate that substitutional defects such as nitrogen or boron will primarily lead to scattering of high-frequency optical phonons. However, larger-scale defects such as Stone Wales defects cause phonon scattering over a wide range of frequencies, leading to a greater reduction in thermal conductivity.

Synthesis

Techniques have been developed to produce nanotubes in sizable quantities, including arc discharge, laser ablation, high-pressure carbon monoxide disproportionation, and chemical vapor deposition (CVD). Most of these processes take place in a vacuum or with process gases. The CVD growth method is popular, as it yields high purity and has a high degree of control over diameter, length and morphology. Using particulate catalysts, large quantities of nanotubes can be synthesized by these methods; advances in catalysis and continuous growth are making CNTs more commercially viable.

Vertically aligned carbon nanotube arrays are also grown by thermal chemical vapor deposition. A substrate (quartz, silicon, stainless steel, etc.) is coated with a catalytic metal (Fe, Co, Ni) layer. Typically that layer is iron, and is deposited via sputtering to a thickness of 1–5 nm. A 10–50 nm underlayer of alumina is often also put down on the substrate first. This imparts controllable wetting and good interfacial properties. When the substrate is heated to the growth temperature (~700 °C), the continuous iron film breaks up into small islands... each island then nucleates a carbon nanotube. The sputtered thickness controls the island size, and this in turn determines the nanotube diameter. Thinner iron layers drive down the diameter of the islands, and they drive down the diameter of the nanotubes grown. The amount of time that the metal island can sit at the growth temperature is limited, as they are mobile, and can merge into larger (but fewer) islands. Annealing at the growth temperature reduces the site density (number of CNT/mm^2) while increasing the catalyst diameter.

Chemical Modification

Carbon nanotubes can be functionalized to attain desired properties that can be used in a wide variety of applications. The two main methods of carbon nanotube functionalization are covalent and non-covalent modifications. Because of their hydrophobic nature, carbon nanotubes tend to agglomerate hindering their dispersion in solvents or viscous polymer melts. The resulting nanotube bundles or aggregates reduce the mechanical performance of the final composite. The surface of the carbon nanotubes can be modified to reduce the hydrophobicity and improve interfacial adhesion to a bulk polymer through chemical attachment.

Applications

Current

Current use and application of nanotubes has mostly been limited to the use of bulk nanotubes, which is a mass of rather unorganized fragments of nanotubes. Bulk nanotube materials may never achieve a tensile strength similar to that of individual tubes, but such composites may, nevertheless, yield strengths sufficient for many applications. Bulk carbon nanotubes have already been used as composite fibers in polymers to improve the mechanical, thermal and electrical properties of the bulk product.

- Easton-Bell Sports, Inc. have been in partnership with Zyvex Performance Materials, using CNT technology in a number of their bicycle components—including flat and riser handlebars, cranks, forks, seatposts, stems and aero bars.
- Zyvex Technologies has also built a 54' maritime vessel, the Piranha Unmanned Surface Vessel, as a technology demonstrator for what is possible using CNT technology. CNTs help improve the structural performance of the vessel, resulting in a lightweight 8,000 lb boat that can carry a payload of 15,000 lb over a range of 2,500 miles.
- Amroy Europe Oy manufactures Hybtonite carbon nanoepoxy resins where carbon nanotubes have been chemically activated to bond to epoxy, resulting in a composite material that is 20% to 30% stronger than other composite materials. It has been used for wind turbines, marine paints and a variety of sports gear such as skis, ice hockey sticks, baseball bats, hunting arrows, and surfboards.

Other current applications include:

- tips for atomic force microscope probes
- in tissue engineering, carbon nanotubes can act as scaffolding for bone growth

There is also ongoing research in using carbon nanotubes as a scaffold for diverse microfabrication techniques.

Potential

The strength and flexibility of carbon nanotubes makes them of potential use in controlling other nanoscale structures, which suggests they will have an important role in nanotechnology engineering. The highest tensile strength of an individual multi-walled carbon nanotube has been tested to be 63 GPa. Carbon nanotubes were found in Damascus steel from the 17th century, possibly helping to account for the legendary strength of the swords made of it. Recently, several studies have highlighted the prospect of using carbon nanotubes as building blocks to fabricate three-dimensional macroscopic (>1mm in all three dimensions) all-carbon devices. Lalwani et al. have reported a novel radical initiated thermal crosslinking method to fabricated macroscopic, free-standing, porous, all-carbon scaffolds using single- and multi-walled carbon nanotubes as building blocks. These scaffolds possess macro-, micro-, and nano- structured pores and the porosity can be tailored for specific applications. These 3D all-carbon scaffolds/architectures may be used for the fabrication of the next generation of energy storage, supercapacitors, field emission transistors, high-performance catalysis, photovoltaics, and biomedical devices and implants.

Large quantities of pure CNTs can be made into a fiber (a.k.a. filament) by wet spinning. The fiber is either directly spun from the synthesis pot or spun from pre-made dissolved CNTs. Individual fibers can be turned into a yarn. Apart from its strength and flexibility, the main advantage is making an electrically conducting yarn. The electronic properties of individual CNT fibers (i.e. bundle of individual CNT) is governed by the two-dimensional structure of CNTs. At 300 K, CNT fibers have a resistivity one order of magnitude higher than the best electrical conductors.

CNT-based yarns are suitable for applications in energy and electrochemical water treatment when coated with an ion-exchange membrane. Also, CNT-based yarns could replace copper as a winding material. Pyrhönen et al. (2015) have built a motor using CNT winding.

Safety and Health

The National Institute for Occupational Safety and Health (NIOSH) is the leading United States federal agency conducting research and providing guidance on the occupational safety and health implications and applications of nanotechnology. Early scientific studies have indicated that some of these nanoscale particles may pose a greater health risk than the larger bulk form of these materials. In 2013, NIOSH published a Current Intelligence Bulletin detailing the potential hazards and recommended exposure limit for carbon nanotubes and fibers.

As of October 2016, single wall carbon nanotubes have been registered through the European Union's Registration, Evaluation, Authorization and Restriction of Chemicals (REACH) regulations, based on evaluation of the potentially hazardous properties of SWCNT. Based on this registration, SWCNT commercialization is allowed in the EU up to 10 metric tons. Currently, the type of SWCNT registered through REACH is limited to the specific type of single wall carbon nanotubes manufactured by OCSiAl, which submitted the application.

History

The true identity of the discoverers of carbon nanotubes is a subject of some controversy. A 2006 editorial written by Marc Monthioux and Vladimir Kuznetsov in the journal *Carbon* described the interesting and often-misstated origin of the carbon nanotube. A large percentage of academic and popular literature attributes the discovery of hollow, nanometer-size tubes composed of graphitic carbon to Sumio Iijima of NEC in 1991. He published a paper describing his discovery which initiated a flurry of excitement and could be credited by inspiring the many scientists now studying applications of carbon nanotubes. Though Iijima has been given much of the credit for discovering carbon nanotubes, it turns out that the timeline of carbon nanotubes goes back much further than 1991.

In 1952, L. V. Radushkevich and V. M. Lukyanovich published clear images of 50 nanometer diameter tubes made of carbon in the Soviet *Journal of Physical Chemistry*. This discovery was largely unnoticed, as the article was published in Russian, and Western scientists' access to Soviet press was limited during the Cold War. Monthioux and Kuznetsov mentioned in their *Carbon* editorial.

The fact is, Radushkevich and Lukyanovich [..] should be credited for the discovery that carbon filaments could be hollow and have a nanometer- size diameter, that is to say for the discovery of carbon nanotubes.

In 1976, Morinobu Endo of CNRS observed hollow tubes of rolled up graphite sheets synthesised by a chemical vapour-growth technique. The first specimens observed would later come to be known as single-walled carbon nanotubes (SWNTs). Endo, in his early review of vapor-phase-grown carbon fibers (VPCF), also reminded us that he had observed a hollow tube, linearly extended with parallel carbon layer faces near the fiber core. This appears to be the observation of multi-walled carbon nanotubes at the center of the fiber. The mass-produced MWCNTs today are strongly related to the VPGCF developed by Endo. In fact, they call it the "Endo-process", out of respect for his early work and patents.

In 1979, John Abrahamson presented evidence of carbon nanotubes at the 14th Biennial Conference of Carbon at Pennsylvania State University. The conference paper described carbon nanotubes as carbon fibers that were produced on carbon anodes during arc discharge. A characterization of these fibers was given as well as hypotheses for their growth in a nitrogen atmosphere at low pressures.

In 1981, a group of Soviet scientists published the results of chemical and structural characterization of carbon nanoparticles produced by a thermocatalytical disproportionation of carbon monoxide. Using TEM images and XRD patterns, the authors suggested that their "carbon multi-layer tubular crystals" were formed by rolling graphene layers into cylinders. They speculated that by rolling graphene layers into a cylinder, many different arrangements of graphene hexagonal nets are possible. They suggested two possibilities of such arrangements: circular arrangement (armchair nanotube) and a spiral, helical arrangement (chiral tube).

In 1987, Howard G. Tennent of Hyperion Catalysis was issued a U.S. patent for the production of "cylindrical discrete carbon fibrils" with a "constant diameter between about 3.5 and about 70 nanometers..., length 10^2 times the diameter, and an outer region of multiple essentially continuous layers of ordered carbon atoms and a distinct inner core...."

Iijima's discovery of multi-walled carbon nanotubes in the insoluble material of arc-burned graphite rods in 1991 and Mintmire, Dunlap, and White's independent prediction that if single-walled carbon nanotubes could be made, then they would exhibit remarkable conducting properties helped create the initial buzz that is now associated with carbon nanotubes. Nanotube research accelerated greatly following the independent discoveries by Bethune at IBM and Iijima at NEC of *single-walled* carbon nanotubes and methods to specifically produce them by adding transition-metal catalysts to the carbon in an arc discharge. The arc discharge technique was well-known to produce the famed Buckminster fullerene on a preparative scale, and these results appeared to extend the run of accidental discoveries relating to fullerenes. The discovery of nanotubes remains a contentious issue. Many believe that Iijima's report in 1991 is of particular importance because it brought carbon nanotubes into the awareness of the scientific community as a whole.

Synthesis of Carbon Nanotubes

Arc discharge synthesis Laser ablation synthesis Thermal synthesis

- Chemical vapor deposition(CVD)
- High-pressure carbon monoxide synthesis (HiPco)

PECVD synthesis

Arc discharge synthesis

* Arc discharge was the first method for producing both SWNTs and MWNTs
* Arc discharge synthesis uses a low-voltage (~12 to 25 V), high-current (50 to 120 amps) and power supply (an arc welder can be used).
* An arc is produced across a 1-mm gap between two graphite electrodes 5 to 20 mm in diameter. An inert gas such as He or Ar is used as the atmosphere for the reaction, at a pressure of 100 to 1000 torr.

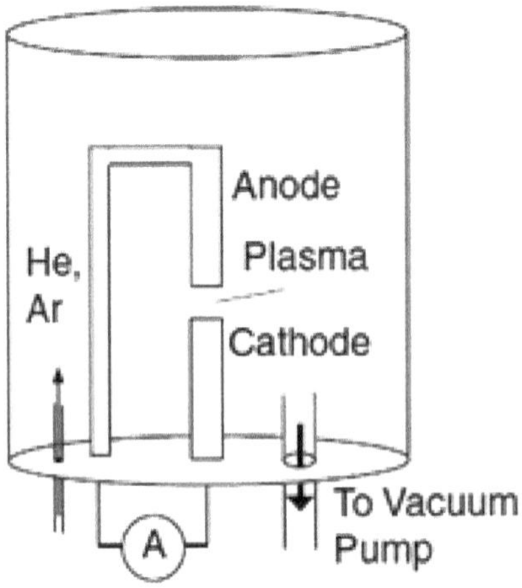

Advantages

Control Over Growth

By tailoring the Ar:He gas ratio, the diameter of the SWNTs formed can be controlled, with greater Ar yielding smaller diameters. The anode–cathode distance can be changed to vary the strength of the plasma formed in between. Several metal catalyst compositions can be used ,currently Y:Ni mixture is used which shows yield up to 90% SWNT with an average diameter of 1.2 to 1.4 nm

b. Both SWNTs and MWNTs made from this process and the process is relatively cheaper.

Disadvantages

Many side products are formed along with the SWNTs and MWNTs The product requires extensive purification

Laser Ablation Synthesis

The laser ablation technique uses a 1.2 % of Co/Ni with 98.8 % of graphite composite target. It is placed in a 1200°C quartz tube furnace with an inert atmosphere of 500 Torr of Ar or He. It is vaporized with a laser pulse. A pulsed- or continuous-wave laser can be used.

Nanometer-size metal catalyst particles are formed in the plume of vaporized graphite. The metal particles catalyze the growth of SWCNTs in the plasma plume.

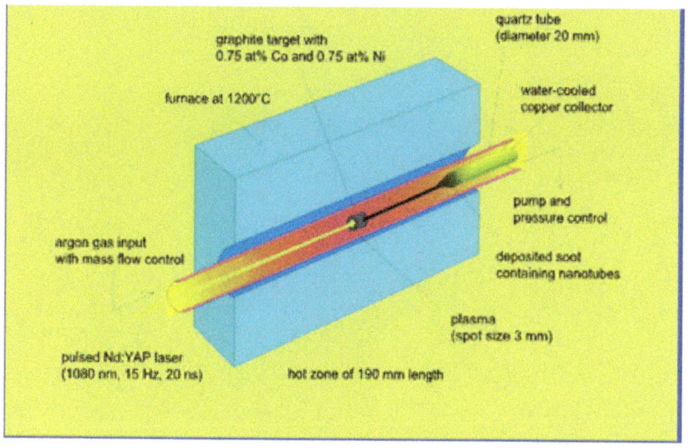

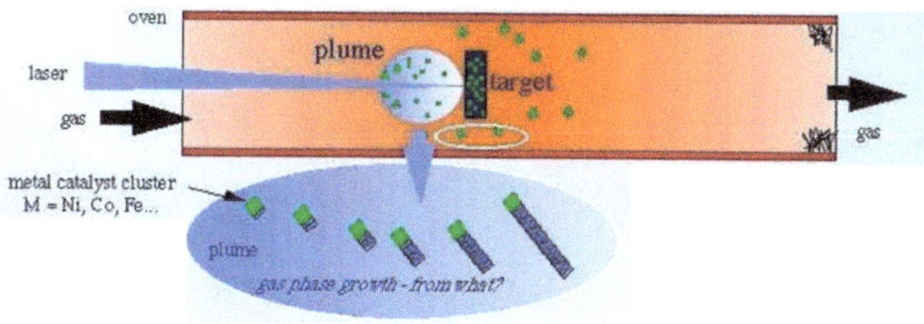

- Chemical Vapour Deposition
  - ◇ Chemical vapour deposition (CVD) synthesis is achieved by putting a carbon source in the gas phase and using an energy source, such as a plasma or a resistively heated coil, to transfer energy to a gaseous carbon molecule.
  - ◇ The energy source is used to "crack" the gaseous molecule CH_4 molecule into reactive atomic carbon.
  - ◇ Then, the carbon diffuses towards the substrate, which is heated and coated with a catalyst such as Ni, Fe or Co where it will bind to form nanotubes.
- High-pressure Carbon Monoxide Synthesis

The metal catalyst is formed in situ when $Fe(CO)_5$ or $Ni(CO)_4$ is injected into the reactor along with a stream of carbon monoxide (CO) gas at 900 to 1100°C and at a pressure of 30 to 50 atm. The reaction to make SWNTs is the disproportionation of CO by nanometer-size metal catalyst particles. Yields are claimed to be 97% pure.

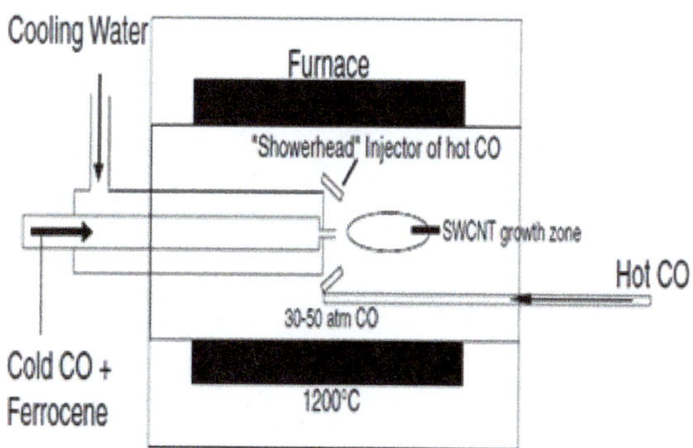

Fullerene

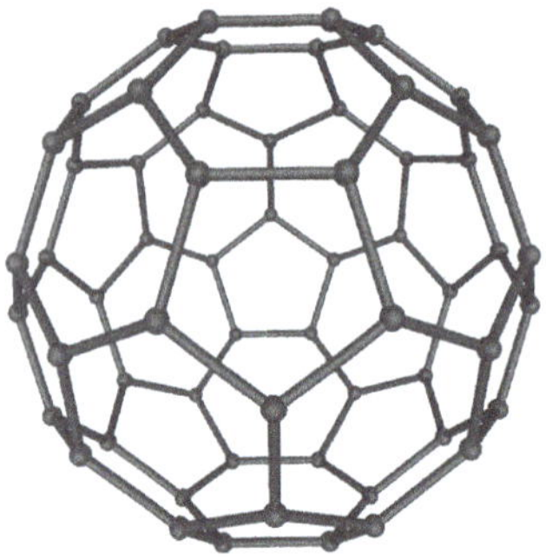

Buckminsterfullerene C_{60} (left) and carbon nanotubes (right) are two examples of structures in the fullerene family.

A fullerene is a molecule of carbon in the form of a hollow sphere, ellipsoid, tube, and many other shapes. Spherical fullerenes, also referred to as Buckminsterfullerenes (buckyballs), resemble the balls used in football (soccer). Cylindrical fullerenes are also called carbon nanotubes (buckytubes). Fullerenes are similar in structure to graphite, which is composed of stacked graphene sheets of linked hexagonal rings; unless they are cylindrical, they must also contain pentagonal (or sometimes heptagonal) rings.

The first fullerene molecule to be discovered, and the family's namesake, buckminsterfullerene (C_{60}), was manufactured in 1985 by Richard Smalley, Robert Curl, James Heath, Sean O'Brien, and Harold Kroto at Rice University. The name was an homage to Buckminster Fuller, whose geodesic domes it resembles. The structure was also identified some five years earlier by Sumio Iijima, from an electron microscope image, where it formed the core of a "bucky onion". Fullerenes have since been found to occur in nature. More recently, fullerenes have been detected in outer space. According to astronomer Letizia Stanghellini, "It's possible that buckyballs from outer space provided seeds for life on Earth."

The discovery of fullerenes greatly expanded the number of known carbon allotropes, which until recently were limited to graphite, graphene, diamond, and amorphous carbon such as soot and charcoal. Buckyballs and buckytubes have been the subject of intense research, both for their unique chemistry and for their technological applications, especially in materials science, electronics, and nanotechnology.

History

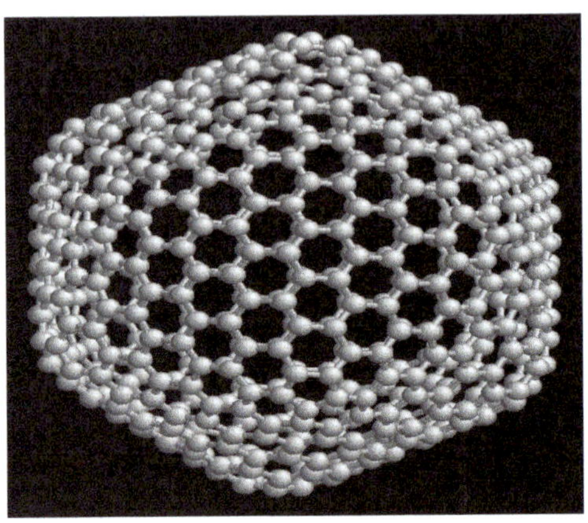

The icosahedral fullerene C_{540}, another member of the family of fullerenes.

The icosahedral $C_{60}H_{60}$ cage was mentioned in 1965 as a possible topological structure. Eiji Osawa of Toyohashi University of Technology predicted the existence of C_{60} in 1970. He noticed that the structure of a corannulene molecule was a subset of an Association football shape, and he hypothesised that a full ball shape could also exist. Japanese scientific journals reported his idea, but neither it nor any translations of it reached Europe or the Americas.

Also in 1970, R. W. Henson (then of the Atomic Energy Research Establishment) proposed the structure and made a model of C_{60}. Unfortunately, the evidence for this new form of carbon was very weak and was not accepted, even by his colleagues. The results were never published but were acknowledged in *Carbon* in 1999.

In 1973 independently from Henson, a group of scientists from the USSR, directed by Prof. Bochvar, made a quantum-chemical analysis of the stability of C_{60} and calculated its electronic structure. As in the previous cases, the scientific community did not accept the theoretical prediction. The paper was published in 1973 in *Proceedings of the USSR Academy of Sciences* (in Russian).

In mass spectrometry discrete peaks appeared corresponding to molecules with the exact mass of sixty or seventy or more carbon atoms. In 1985 Harold Kroto of the University of Sussex, James R. Heath, Sean O'Brien, Robert Curl and Richard Smalley from Rice University, discovered C_{60}, and shortly thereafter came to discover the fullerenes. Kroto, Curl, and Smalley were awarded the 1996 Nobel Prize in Chemistry for their roles in the discovery of this class of molecules. C_{60} and other fullerenes were later noticed occurring outside the laboratory (for example, in normal candle-soot). By 1990 it was relatively easy to produce gram-sized samples of fullerene powder using the techniques of Donald Huffman, Wolfgang Krätschmer, Lowell D. Lamb, and Konstantinos Fostiropoulos. Fullerene purification remains a challenge to chemists and to a large extent determines fullerene prices. So-called endohedral fullerenes have ions or small molecules incorporated inside the cage atoms. Fullerene is an unusual reactant in many organic reactions such as the Bingel reaction discovered in 1993. Carbon nanotubes were first discovered and synthesized in 1991.

Minute quantities of the fullerenes, in the form of C_{60}, C_{70}, C_{76}, C_{82} and C_{84} molecules, are produced in nature, hidden in soot and formed by lightning discharges in the atmosphere. In 1992, fullerenes were found in a family of minerals known as Shungites in Karelia, Russia.

In 2010, fullerenes (C_{60}) have been discovered in a cloud of cosmic dust surrounding a distant star 6500 light years away. Using NASA's Spitzer infrared telescope the scientists spotted the molecules' unmistakable infrared signature. Sir Harry Kroto, who shared the 1996 Nobel Prize in Chemistry for the discovery of buckyballs commented: "This most exciting breakthrough provides convincing evidence that the buckyball has, as I long suspected, existed since time immemorial in the dark recesses of our galaxy."

Naming

The discoverers of the Buckminsterfullerene (C_{60}) allotrope of carbon named it after Richard Buckminster Fuller, a noted architectural modeler who popularized the geodesic dome. Since buckminsterfullerenes have a similar shape to those of such domes, they thought the name appropriate. As the discovery of the fullerene family came *after* buckminsterfullerene, the shortened name 'fullerene' is used to refer to the family of fullerenes. The suffix "-ene" indicates that each C atom is covalently bonded to three others (instead of the maximum of four), a situation that classically would correspond to the existence of bonds involving two pairs of electrons ("double bonds").

Types of Fullerene

Since the discovery of fullerenes in 1985, structural variations on fullerenes have evolved well beyond the individual clusters themselves. Examples include:

- Buckyball clusters: smallest member is C_{20} (unsaturated version of dodecahedrane) and the most common is C_{60};
- Nanotubes: hollow tubes of very small dimensions, having single or multiple walls; potential applications in electronics industry;
- Megatubes: larger in diameter than nanotubes and prepared with walls of different thickness; potentially used for the transport of a variety of molecules of different sizes;
- polymers: chain, two-dimensional and three-dimensional polymers are formed under high-pressure high-temperature conditions; single-strand polymers are formed using the Atom Transfer Radical Addition Polymerization (ATRAP) route;
- nano"onions": spherical particles based on multiple carbon layers surrounding a buckyball core; proposed for lubricants;
- linked "ball-and-chain" dimers: two buckyballs linked by a carbon chain;
- fullerene rings.

Buckyballs

Buckminsterfullerene

Buckminsterfullerene is the smallest fullerene molecule containing pentagonal and hexagonal rings in which no two pentagons share an edge (which can be destabilizing, as in pentalene). It is also most common in terms of natural occurrence, as it can often be found in soot.

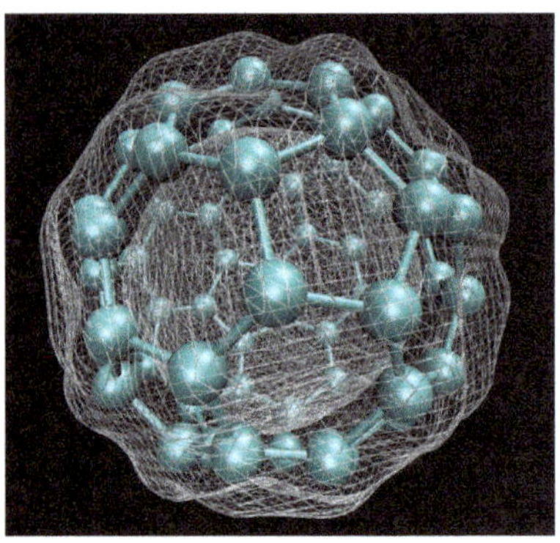

C_{60} with isosurface of ground state electron density as calculated with DFT

The structure of C_{60} is a truncated icosahedron, which resembles an association football ball of the type made of twenty hexagons and twelve pentagons, with a carbon atom at the vertices of each polygon and a bond along each polygon edge.

The van der Waals diameter of a C_{60} molecule is about 1.1 nanometers (nm). The nucleus to nucleus diameter of a C_{60} molecule is about 0.71 nm.

The C_{60} molecule has two bond lengths. The 6:6 ring bonds (between two hexagons) can be considered "double bonds" and are shorter than the 6:5 bonds (between a hexagon and a pentagon). Its average bond length is 1.4 angstroms.

Silicon buckyballs have been created around metal ions.

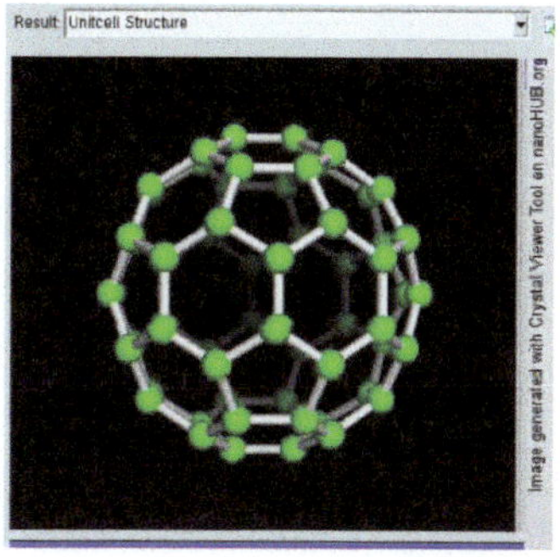

view of C_{60}, one kind of fullerene

Boron Buckyball

A type of buckyball which uses boron atoms, instead of the usual carbon, was predicted and described in 2007. The B_{80} structure, with each atom forming 5 or 6 bonds, is predicted to be more stable than the C_{60} buckyball. One reason for this given by the researchers is that B_{80} is actually more like the original geodesic dome structure popularized by Buckminster Fuller, which uses triangles rather than hexagons. However, this work has been subject to much criticism by quantum chemists as it was concluded that the predicted I_h symmetric structure was vibrationally unstable and the resulting cage undergoes a spontaneous symmetry break, yielding a puckered cage with rare T_h symmetry (symmetry of a volleyball). The number of six-member rings in this molecule is 20 and number of five-member rings is 12. There is an additional atom in the center of each six-member ring, bonded to each atom surrounding it. By employing a systematic global search algorithm, later it was found that the previously proposed B80 fullerene is not global minimum for 80 atom boron clusters and hence can not be found in nature. In the same paper by Sandip De et al., it was concluded that boron's energy landscape is significantly different from other fullerenes already found in nature hence pure boron fullerenes are unlikely to exist in nature.

Other Buckyballs

Another fairly common fullerene is C_{70}, but fullerenes with 72, 76, 84 and even up to 100 carbon atoms are commonly obtained.

In mathematical terms, the structure of a fullerene is a trivalent convex polyhedron with pentagonal and hexagonal faces. In graph theory, the term fullerene refers to any 3-regular, planar graph with all faces of size 5 or 6 (including the external face). It follows from Euler's polyhedron formula, $V - E + F = 2$ (where V, E, F are the numbers of vertices, edges, and faces), that there are exactly 12 pentagons in a fullerene and $V/2 - 10$ hexagons.

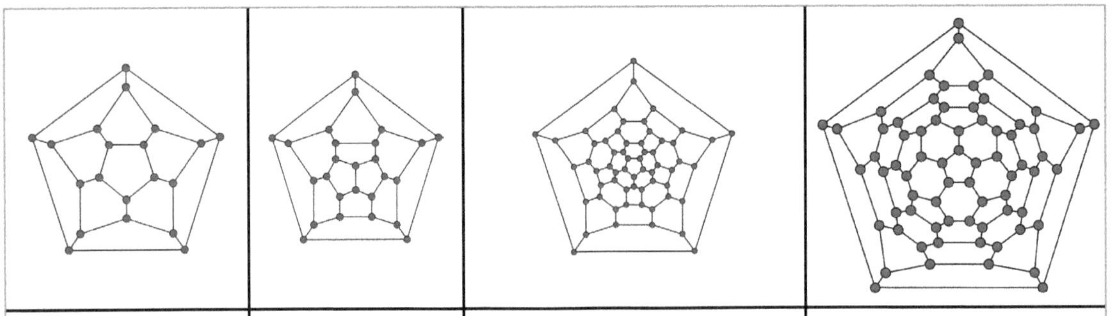

20-fullerene (dodecahedral graph)	26-fullerene graph	60-fullerene (truncated icosahedral graph)	70-fullerene graph

The smallest fullerene is the dodecahedral C_{20}. There are no fullerenes with 22 vertices. The number of fullerenes C_{2n} grows with increasing $n = 12, 13, 14, \ldots$, roughly in proportion to n^9 (sequence A007894 in the OEIS). For instance, there are 1812 non-isomorphic fullerenes C_{60}. Note that only one form of C_{60}, the buckminsterfullerene alias truncated icosahedron, has no pair of adjacent pentagons (the smallest such fullerene). To further illustrate the growth, there are 214,127,713 non-isomorphic fullerenes C_{200}, 15,655,672 of which have no adjacent pentagons.

Heterofullerenes have heteroatoms substituting carbons in cage or tube-shaped structures. They were discovered in 1993 and greatly expand the overall fullerene class of compounds. Notable examples include boron, nitrogen (azafullerene), oxygen, and phosphorus derivatives.

Trimetasphere carbon nanomaterials were discovered by researchers at Virginia Tech and licensed exclusively to Luna Innovations. This class of novel molecules comprises 80 carbon atoms (C_{80}) forming a sphere which encloses a complex of three metal atoms and one nitrogen atom. These fullerenes encapsulate metals which puts them in the subset referred to as metallofullerenes. Trimetaspheres have the potential for use in diagnostics (as safe imaging agents), therapeutics and in organic solar cells.

Carbon Nanotubes

Nanotubes are cylindrical fullerenes. These tubes of carbon are usually only a few nanometres wide, but they can range from less than a micrometer to several millimeters in length. They often have closed ends, but can be open-ended as well. There are also cases in which the tube reduces in diameter before closing off. Their unique molecular structure results in extraordinary macroscopic properties, including high tensile strength, high electrical conductivity, high ductility, high heat conductivity, and relative chemical inactivity (as it is cylindrical and "planar" — that is, it has no "exposed" atoms that can be easily displaced). One proposed use of carbon nanotubes is in paper batteries, developed in 2007 by researchers at Rensselaer Polytechnic Institute. Another highly speculative proposed use in the field of space technologies is to produce high-tensile carbon cables required by a space elevator.

Carbon Nanobuds

Nanobuds have been obtained by adding buckminsterfullerenes to carbon nanotubes.

Fullerite

The C_{60} fullerene in crystalline form

Fullerites are the solid-state manifestation of fullerenes and related compounds and materials.

"Ultrahard fullerite" is a coined term frequently used to describe material produced by high-pressure high-temperature (HPHT) processing of fullerite. Such treatment converts fullerite into a nanocrystalline form of diamond which has been reported to exhibit remarkable mechanical properties.

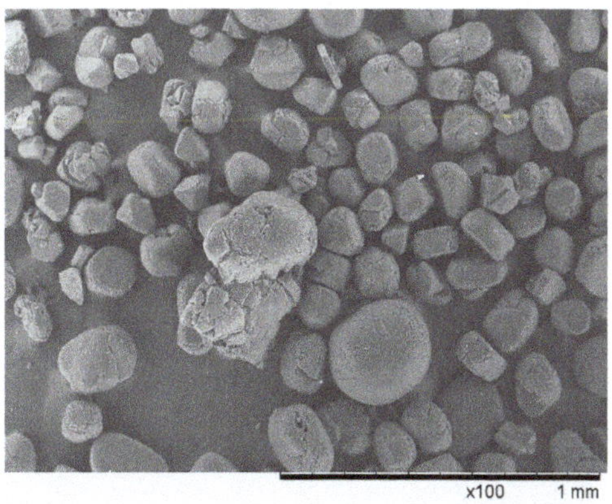

Fullerite (scanning electron microscope image)

Inorganic Fullerenes

Materials with fullerene-like molecular structures but lacking carbon include MoS_2, WS_2, TiS_2 and NbS_2. Under isostatic pressure, these new materials were found to be stable up to at least 350 tons/ cm^2.

Properties

For the past decade, the chemical and physical properties of fullerenes have been a hot topic in the field of research and development, and are likely to continue to be for a long time. *Popular Science* has discussed possible uses of fullerenes (graphene) in armor. In April 2003, fullerenes were under study for potential medicinal use: binding specific antibiotics to the structure to target resistant bacteria and even target certain cancer cells such as melanoma. The October 2005 issue of *Chemistry & Biology* contains an article describing the use of fullerenes as light-activated antimicrobial agents.

In the field of nanotechnology, heat resistance and superconductivity are some of the more heavily studied properties.

A common method used to produce fullerenes is to send a large current between two nearby graphite electrodes in an inert atmosphere. The resulting carbon plasma arc between the electrodes cools into sooty residue from which many fullerenes can be isolated.

There are many calculations that have been done using ab-initio quantum methods applied to fullerenes. By DFT and TD-DFT methods one can obtain IR, Raman and UV spectra. Results of such calculations can be compared with experimental results.

Aromaticity

Researchers have been able to increase the reactivity of fullerenes by attaching active groups to their surfaces. Buckminsterfullerene does not exhibit "superaromaticity": that is, the electrons in the hexagonal rings do not delocalize over the whole molecule.

A spherical fullerene of n carbon atoms has n pi-bonding electrons, free to delocalize. These should try to delocalize over the whole molecule. The quantum mechanics of such an arrangement should be like one shell only of the well-known quantum mechanical structure of a single atom, with a stable filled shell for $n = 2, 8, 18, 32, 50, 72, 98, 128$, etc.; i.e. twice a perfect square number; but this series does not include 60. This $2(N + 1)^2$ rule (with N integer) for spherical aromaticity is the three-dimensional analogue of Hückel's rule. The 10+ cation would satisfy this rule, and should be aromatic. This has been shown to be the case using quantum chemical modelling, which showed the existence of strong diamagnetic sphere currents in the cation.

As a result, C_{60} in water tends to pick up two more electrons and become an anion. The nC_{60} described below may be the result of C_{60} trying to form a loose metallic bond.

Chemistry

Fullerenes are stable, but not totally unreactive. The sp^2-hybridized carbon atoms, which are at their energy minimum in planar graphite, must be bent to form the closed sphere or tube, which produces angle strain. The characteristic reaction of fullerenes is electrophilic addition at 6,6-double bonds, which reduces angle strain by changing sp^2-hybridized carbons into sp^3-hybridized ones. The change in hybridized orbitals causes the bond angles to decrease from about 120° in the sp^2 orbitals to about 109.5° in the sp^3 orbitals. This decrease in bond angles allows for the bonds to bend less when closing the sphere or tube, and thus, the molecule becomes more stable.

Other atoms can be trapped inside fullerenes to form inclusion compounds known as endohedral fullerenes. An unusual example is the egg-shaped fullerene $Tb_3N@C_{84}$, which violates the isolated pentagon rule. Recent evidence for a meteor impact at the end of the Permian period was found by analyzing noble gases so preserved. Metallofullerene-based inoculates using the rhonditic steel process are beginning production as one of the first commercially viable uses of buckyballs.

Solubility

C_{60} in solution

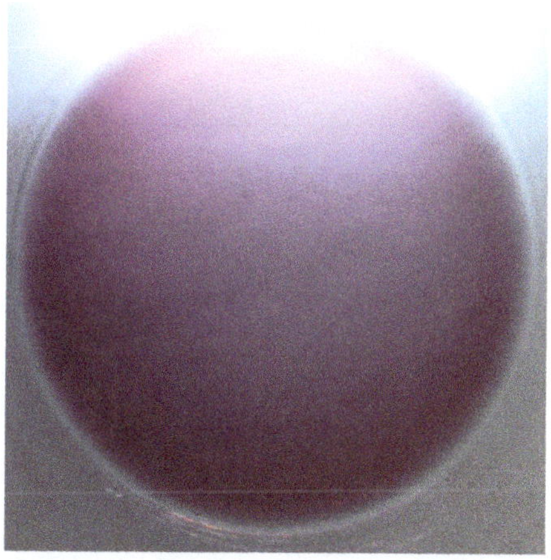

C_{60} in extra virgin olive oil showing the characteristic purple color of pristine C_{60} solutions.

Fullerenes are sparingly soluble in many solvents. Common solvents for the fullerenes include aromatics, such as toluene, and others like carbon disulfide. Solutions of pure buckminsterfullerene have a deep purple color. Solutions of C_{70} are a reddish brown. The higher fullerenes C_{76} to C_{84} have a variety of colors. C_{76} has two optical forms, while other higher fullerenes have several structural isomers. Fullerenes are the only known allotrope of carbon that can be dissolved in common solvents at room temperature.

Solvent	C_{60} mg/mL	C_{70} mg/mL
1-chloronaphthalene	51	ND
1-methylnaphthalene	33	ND
1,2-dichlorobenzene	24	36.2
1,2,4-trimethylbenzene	18	ND
tetrahydronaphthalene	16	ND
carbon disulfide	8	9.875
1,2,3-tribromopropane	8	ND
chlorobenzene	7	ND

p-xylene	5	3.985
bromoform	5	ND
cumene	4	ND
toluene	3	1.406
benzene	1.5	1.3
carbon tetrachloride	0.447	0.121
chloroform	0.25	ND
n-hexane	0.046	0.013
cyclohexane	0.035	0.08
tetrahydrofuran	0.006	ND
acetonitrile	0.004	ND
methanol	4.0×10^{-5}	ND
water	1.3×10^{-11}	ND
pentane	0.004	0.002
heptane	ND	0.047
octane	0.025	0.042
isooctane	0.026	ND
decane	0.070	0.053
dodecane	0.091	0.098
tetradecane	0.126	ND
acetone	ND	0.0019
isopropanol	ND	0.0021
dioxane	0.0041	ND
mesitylene	0.997	1.472
dichloromethane	0.254	0.080
ND, not determined		

Some fullerene structures are not soluble because they have a small band gap between the ground and excited states. These include the small fullerenes C_{28}, C_{36} and C_{50}. The C_{72} structure is also in this class, but the endohedral version with a trapped lanthanide-group atom is soluble due to the interaction of the metal atom and the electronic states of the fullerene. Researchers had originally been puzzled by C_{72} being absent in fullerene plasma-generated soot extract, but found in endohedral samples. Small band gap fullerenes are highly reactive and bind to other fullerenes or to soot particles.

Solvents that are able to dissolve buckminsterfullerene (C_{60} and C_{70}) are listed at left in order from highest solubility. The solubility value given is the approximate saturated concentration.

Solubility of C_{60} in some solvents shows unusual behaviour due to existence of solvate phases (analogues of crystallohydrates). For example, solubility of C_{60} in benzene solution shows maximum at about 313 K. Crystallization from benzene solution at temperatures below maximum results in formation of triclinic solid solvate with four benzene molecules $C_{60}{\cdot}4C_6H_6$ which is rather unstable in air. Out of solution, this structure decomposes into usual face-centered cubic (fcc) C_{60} in few minutes' time. At temperatures above solubility maximum the solvate is not stable even when immersed in saturated solution and melts with formation of fcc C_{60}. Crystallization at temperatures above the solubility maximum results in formation of pure fcc C_{60}. Millimeter-sized crystals of C_{60} and C_{70} can be grown from solution both for solvates and for pure fullerenes.

Quantum Mechanics

In 1999, researchers from the University of Vienna demonstrated that wave-particle duality applied to molecules such as fullerene.

Chirality

Some fullerenes (e.g. C_{76}, C_{78}, C_{80}, and C_{84}) are inherently chiral because they are D_2-symmetric, and have been successfully resolved. Research efforts are ongoing to develop specific sensors for their enantiomers.

Construction

Two theories have been proposed to describe the molecular mechanisms that make fullerenes. The older, "bottom-up" theory proposes that they are built atom-by-atom. The alternative "top-down" approach claims that fullerenes form when much larger structures break into constituent parts.

In 2013 researchers discovered that asymmetrical fullerenes formed from larger structures settle into stable fullerenes. The synthesized substance was a particular metallofullerene consisting of 84 carbon atoms with two additional carbon atoms and two yttrium atoms inside the cage. The process produced approximately 100 micrograms.

However, they found that the asymmetrical molecule could theoretically collapse to form nearly every known fullerene and metallofullerene. Minor perturbations involving the breaking of a few molecular bonds cause the cage to become highly symmetrical and stable. This insight supports the theory that fullerenes can be formed from graphene when the appropriate molecular bonds are severed.

Production Technology

Fullerene production processes comprise the following five subprocesses: (i) synthesis of fullerenes or fullerene-containing soot; (ii) extraction; (iii) separation (purification) for each fullerene molecule, yielding pure fullerenes such as C_{60}; (iv) synthesis of derivatives (mostly using the techniques of organic synthesis); (v) other post-processing such as dispersion into a matrix. The two synthesis methods used in practice are the arc method, and the combustion method. The latter, discovered at the Massachusetts Institute of Technology, is preferred for large scale industrial production.

Applications

Fullerenes have been extensively used for several biomedical applications including the design of high-performance MRI contrast agents, X-Ray imaging contrast agents, photodynamic therapy and drug and gene delivery, summarized in several comprehensive reviews.

Tumor Research

While past cancer research has involved radiation therapy, photodynamic therapy is important to study because breakthroughs in treatments for tumor cells will give more options to patients with different conditions. Recent experiments using HeLa cells in cancer research involves the development of new photosensitizers with increased ability to be absorbed by cancer cells and still trigger cell death. It is also important that a new photosensitizer does not stay in the body for a long time to prevent unwanted cell damage.

Fullerenes can be made to be absorbed by HeLa cells. The C_{60} derivatives can be delivered to the cells by using the functional groups L-phenylalanine, folic acid, and L-arginine among others. Functionalizing the fullerenes aims to increase the solubility of the molecule by the cancer cells. Cancer cells take up these molecules at an increased rate because of an upregulation of transporters in the cancer cell, in this case amino acid transporters will bring in the L-arginine and L-phenylalanine functional groups of the fullerenes.

Once absorbed by the cells, the C_{60} derivatives would react to light radiation by turning molecular oxygen into reactive oxygen which triggers apoptosis in the HeLa cells and other cancer cells that can absorb the fullerene molecule. This research shows that a reactive substance can target cancer cells and then be triggered by light radiation, minimizing damage to surrounding tissues while undergoing treatment.

When absorbed by cancer cells and exposed to light radiation, the reaction that creates reactive oxygen damages the DNA, proteins, and lipids that make up the cancer cell. This cellular damage forces the cancerous cell to go through apoptosis, which can lead to the reduction in size of a tumor. Once the light radiation treatment is finished the fullerene will reabsorb the free radicals to prevent damage of other tissues. Since this treatment focuses on cancer cells, it is a good option for patients whose cancer cells are within reach of light radiation. As this research continues, the treatment may penetrate deeper into the body and be absorbed by cancer cells more effectively.

Safety and Toxicity

A comprehensive and recent review on fullerene toxicity is given by Lalwani et al. These authors review the works on fullerene toxicity beginning in the early 1990s to present, and conclude that very little evidence gathered since the discovery of fullerenes indicate that C_{60} is toxic. The toxicity of these carbon nanoparticles is not only dose and time-dependent, but also depends on a number of other factors such as: type (e.g., C_{60}, C_{70}, $M@C_{60}$, $M@C_{82}$, functional groups used to water solubilize these nanoparticles (e.g., OH, COOH), and method of administration (e.g., intravenous, intraperitoneal). The authors therefore recommend that pharmacology of every new fullerene- or metallofullerene-based complex must be assessed individually as a different compound.

Popular Culture

Examples of fullerenes in popular culture are numerous. Fullerenes appeared in fiction well before scientists took serious interest in them. In a humorously speculative 1966 column for *New Scientist*, David Jones suggested that it may be possible to create giant hollow carbon molecules by distorting a plane hexagonal net by the addition of impurity atoms.

On 4 September 2010, Google used an interactively rotatable fullerene C_{60} as the second 'o' in their logo to celebrate the 25th anniversary of the discovery of the fullerenes.

Carbon Nanobud

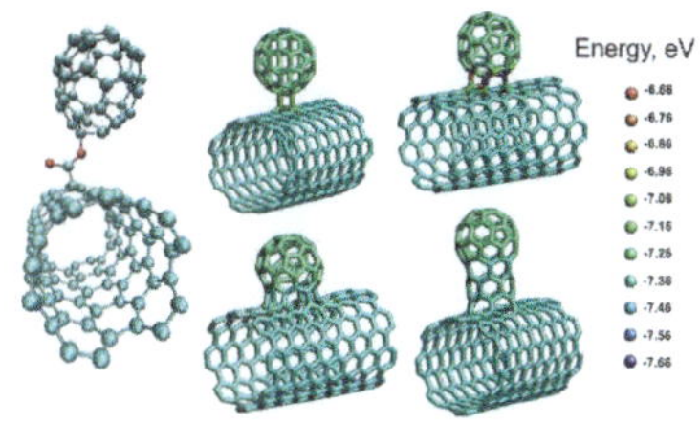

Computer models of several stable nanobud structures

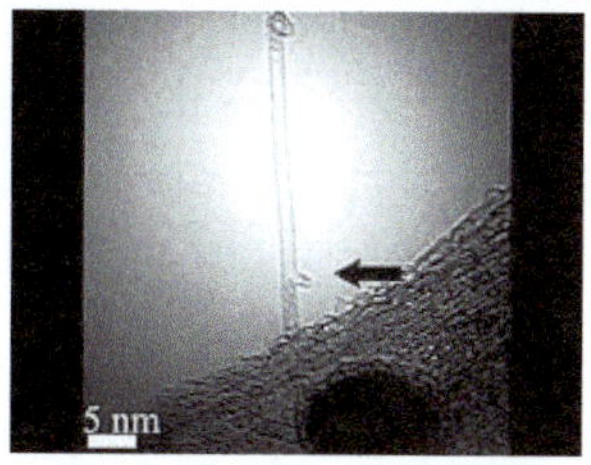

Capture of an additional fullerene molecule by a nanobud

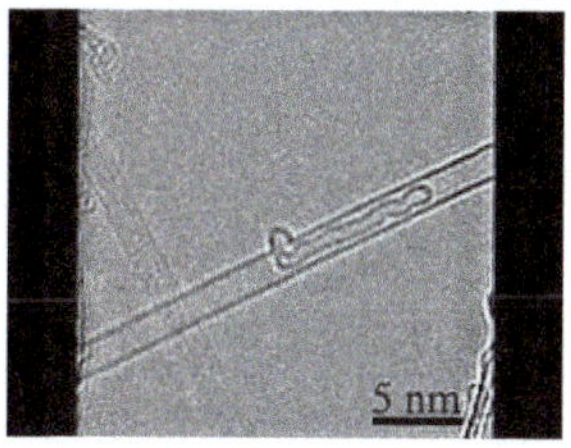

In situ observation of a carbon nanobud by transmission electron microscopy

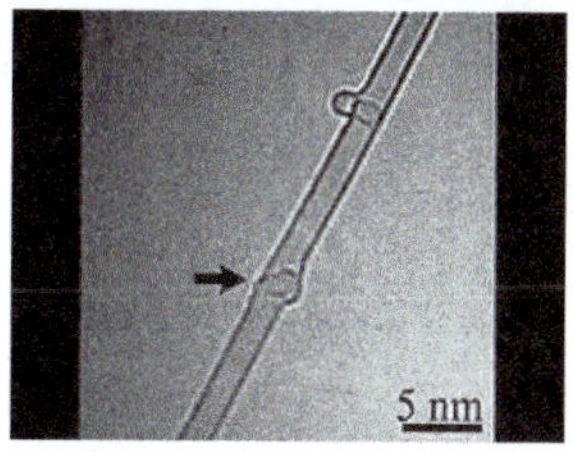

Generation of fullerene molecules (carbon peapod) inside a nanobud

In nanotechnology, a carbon nanobud is a material that combines carbon nanotubes and spheroidal fullerenes, both allotropes of carbon, in the same structure, forming "buds" attached to the tubes. Carbon nanobuds were discovered and synthesized in 2006.

In this new material, fullerenes are covalently bonded to the outer sidewalls of the underlying nanotube. Consequently, nanobuds exhibit properties of both carbon nanotubes and fullerenes. For instance, the mechanical properties and the electrical conductivity of the nanobuds are similar to those of corresponding carbon nanotubes. However, because of the higher reactivity of the attached fullerene molecules, the hybrid material can be further functionalized through known fullerene chemistry. Additionally, the attached fullerene molecules can be used as molecular anchors to prevent slipping of the nanotubes in various composite materials, thus modifying the composite's mechanical properties.

Owing to the large number of highly curved fullerene surfaces acting as electron emission sites on conductive carbon nanotubes, nanobuds possess advantageous field electron emission characteristics. Randomly oriented nanobuds have already been demonstrated to have an extremely low work function for field electron emission. Reported test measurements show (macroscopic) field thresholds of about 0.65 V/μm, (non-functionalized single-walled carbon nanotubes have a macroscopic field threshold for field electron emission ~2 V/μm) and a much higher current density as compared with that of the corresponding pure single-walled carbon nanotubes. The electron transport properties of certain nanobud classes have been treated theoretically. The study shows that electrons indeed pass to the neck and bud region of the nanobud system.

Canatu Oy, a Finnish company, claims the intellectual property rights for nanobud material, its synthesis processes, and several applications.

Applications

Properties such as chemical reactivity, good dispersion and variable band gap electronic structure suggest wide applicability of nanobuds. As the production processes are scalable, the nanobud applications may have industrial importance. Several theoretical works also mentioned the magnetism in nanobuds. Canatu has demonstrated curved and highly foldable touch screens made with nanobuds.

Applications of Carbon Nano Tubes

- Field Emission (LED, etc)
- Conductive plastics
- Conductive adhesives & Connectors
- Molecular electronics
- Energy storage
- Thermal materials (conduct or insulate)
- Structural composites (Boeing 787,buildings,etc)
- Catalytic & biomedical supports
- Others

Fullerene Purification

Fullerene purification is the process of obtaining a fullerene compound free of contamination. In fullerene production mixtures of C_{60}., C_{70} and higher homologues are always formed. Fullerene purification is key to fullerene science and determines fullerene prices and the success of practical applications of fullerenes. The first available purification method for C_{60} fullerene was by HPLC from which small amounts could be generated at large expense.

A practical laboratory-scale method for purification of soot enriched in C_{60} and C_{70} starts with extraction in toluene followed by filtration with a paper filter. The solvent is evaporated and the residue (the toluene-soluble soot fraction) re-dissolved in toluene and subjected to column chromatography. C_{60} elutes first with a purple color and C_{70} is next displaying a reddish-brown color.

Chemical Properties of Fullerenes:

Fullerenes tend to react as electrophiles. An additional driving force is relief of strain when double bonds become saturated. Key in this type of reaction is the level of functionalization i.e. monoaddition or multiple additions and in case of multiple additions their topological relationships (new substituents huddled together or evenly spaced). In conformity with IUPAC rules, the terms methanofullerene are used to indicate the ring-closed (cyclopropane) fullerene derivatives, and fulleroid to ring-open (methanoannulene) structures.

Fullerenes react as electrophiles with a host of nucleophiles in nucleophilic additions. The intermediary formed carbanion is captured by another electrophile. Examples of nucleophiles are Grignard reagents and organolithium reagents. For example the reaction of C_{60} with methylmagnesium chloride stops quantitatively at the penta-adduct with the methyl groups centered around a cyclopentadienyl anion which is subsequently protonated. Another nucleophilic reaction is the Bingel reaction. Fullerene reacts with chlorobenzene and aluminum chloride in a Friedel-Crafts alkylation type reaction. In this hydroarylation the reaction product is the 1,2-addition adduct (Ar-CC-H).

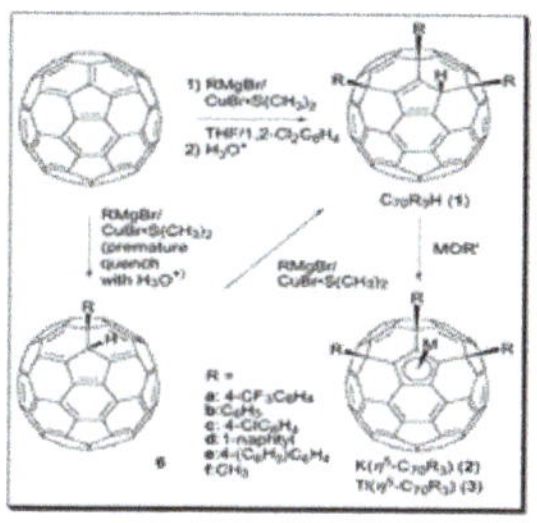

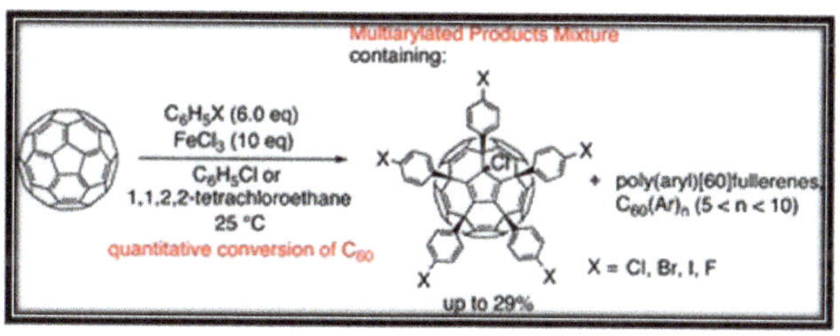

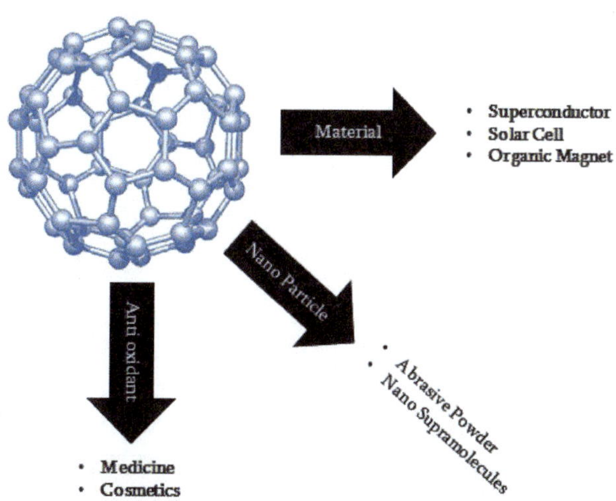

Derivatives of Fullerenes are one of the best know electron accepto materials till date in organic based solar cells.

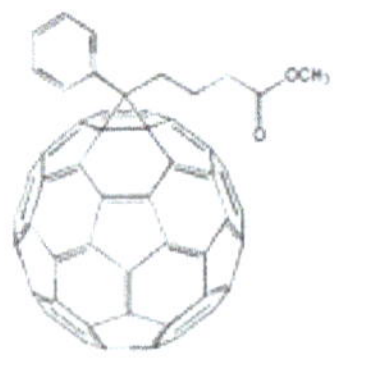

PCBM

PCBM: [6,6]phenyl-C_{61}-butyric acid methyl ester, acceptor small molecule

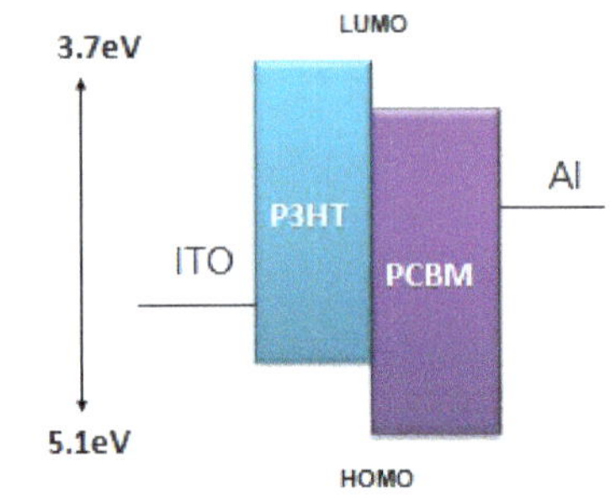

When exposed to light, fullerene can produce singlet oxygen in high quantum yields. This action, together with direct electron transfer from excited state of fullerene and DNA bases, can be used to cleave DNA.

Fullerenes (C_{60}) and their derivatives have potential antiviral activity, which has strong implications on the treatment of HIV-infection.

Cationic, anionic and amino acid type fullerene derivatives inhibit HIV-reverse transcriptase and hepatitis C virus replication.

Fullerenes are one of the ingredients used in the manufacturing of anti-aging cosmetic creams.

Fullerene Chemistry

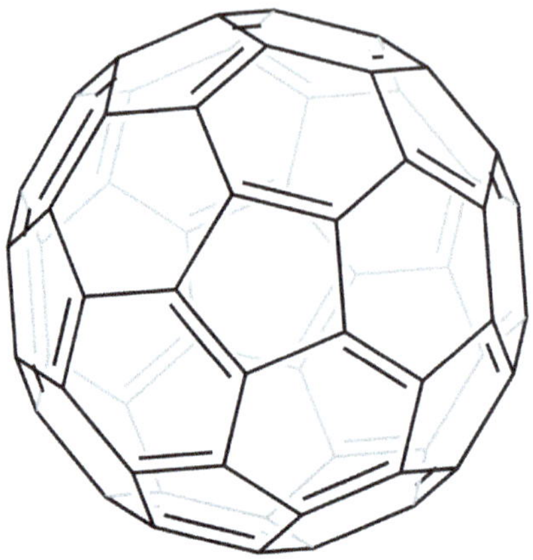

Fullerene chemistry is a field of organic chemistry devoted to the chemical properties of fullerenes. Research in this field is driven by the need to functionalize fullerenes and tune their properties. For example, fullerene is notoriously insoluble and adding a suitable group can enhance solubility. By adding a polymerizable group, a fullerene polymer can be obtained. Functionalized fullerenes are divided into two classes: exohedral fullerenes with substituents outside the cage and endohedral fullerenes with trapped molecules inside the cage.

This article covers the chemistry of buckyballs, while the chemistry of carbon nanotubes is covered in Carbon nanotube chemistry.

Chemical Properties of Fullerenes

Fullerene or C_{60} is soccer-ball-shaped or I_h with 12 pentagons and 20 hexagons. According to Euler's theorem these 12 pentagons are required for closure of the carbon network consisting of n hexagons and C_{60} is the first stable fullerene because it is the smallest possible to obey this rule. In this structure none of the pentagons make contact with each other. Both C_{60} and its relative C_{70} obey this so-called isolated pentagon rule (IPR). The next homologue C_{84} has 24 IPR isomers of which several are isolated and another 51,568 non-IPR isomers. Non-IPR fullerenes have thus far only been isolated as endohedral fullerenes such as $Tb_3N@C_{84}$ with two fused pentagons at the apex of an egg-shaped cage. or as fullerenes with exohedral stabilization such as $C_{50}Cl_{10}$ and reportedly $C_{60}H_8$. Lower fullerenes do not obey isolated pentagon rule (IPR).

Because of the molecule's spherical shape the carbon atoms are highly pyramidalized, which has far-reaching consequences for reactivity. It is estimated that strain energy constitutes 80% of the heat of formation. The conjugated carbon atoms respond to deviation from planarity by orbital rehybridization of the sp^2 orbitals and pi orbitals to a $sp^{2.27}$ orbital with a gain in p-character. The p lobes extend further outside the surface than they do into the interior of the sphere and this is one of the reasons a fullerene is electronegative. The other reason is that the empty low-lying pi* orbitals also have a high s character.

The double bonds in fullerene are not all the same. Two groups can be identified: 30 so-called double bonds connect two hexagons and 60 bonds connect a hexagon and a pentagon. Of the two the bonds are shorter with more double-bond character and therefore a hexagon is often represented as a cyclohexatriene and a pentagon as a pentalene or radialene. In other words, although the carbon atoms in fullerene are all conjugated the superstructure is not a super aromatic compound. The X-ray diffraction bond length values are 139.1 pm for the bond and 145.5 pm for the bond.

C_{60} fullerene has 60 pi electrons but a closed shell configuration requires 72 electrons. The fullerene is able to acquire the missing electrons by reaction with potassium to form first the $K_6C_{60}^{6-}$ salt and then the $K_{12}C_{60}^{12-}$ In this compound the bond length alternation observed in the parent molecule has vanished.

Fullerene Reactions

Fullerenes tend to react as electrophiles. An additional driving force is relief of strain when double bonds become saturated. Key in this type of reaction is the level of functionalization i.e. monoaddition or multiple additions and in case of multiple additions their topological relationships (new substituents huddled together or evenly spaced). In conformity with IUPAC rules, the terms methanofullerene are used to indicate the ring-closed (cyclopropane) fullerene derivatives, and fulleroid to ring-open (methanoannulene) structures.

Nucleophilic Addition

Fullerenes react as electrophiles with a host of nucleophiles in nucleophilic additions. The intermediary formed carbanion is captured by another electrophile. Examples of nucleophiles are Grignard reagents and organolithium reagents. For example, the reaction of C_{60} with methylmagnesium chloride stops quantitatively at the penta-adduct with the methyl groups centered around a cyclopentadienyl anion which is subsequently protonated. Another nucleophilic reaction is the Bingel reaction. Fullerene reacts with chlorobenzene and aluminium chloride in a Friedel-Crafts alkylation type reaction. In this hydroarylation the reaction product is the 1,2-addition adduct (Ar-CC-H).

Pericyclic Reactions

The bonds of fullerenes react as dienes or dienophiles in cycloadditions for instance Diels-Alder reactions. 4-membered rings can be obtained by [2+2]cycloadditions for instance with benzyne. An example of a 1,3-dipolar cycloaddition to a 5-membered ring is the Prato reaction.

Hydrogenation

Fullerenes are easily hydrogenated by several methods. Examples of hydrofullerenes are $C_{60}H_{18}$ and $C_{60}H_{36}$. However, completely hydrogenated $C_{60}H_{60}$ is only hypothetical because of large strain. Highly hydrogenated fullerenes are not stable, as prolonged hydrogenation of fullerenes by direct reaction with hydrogen gas at high temperature conditions results in cage fragmentation. At the final reaction stage this causes collapse of cage structure with formation of polycyclic aromatic hydrocarbons.

Oxidation

Although more difficult than reduction, oxidation of fullerene is possible for instance with oxygen and osmium tetraoxide.

Hydroxylation

Fullerenes can be hydroxylated to fullerenols or fullerols. Water solubility depends on the total number of hydroxyl groups that can be attached. One method is fullerene reaction in diluted sulfuric acid and potassium nitrate to $C_{60}(OH)_{15}$. Another method is reaction in diluted sodium hydroxide catalysed by TBAH adding 24 to 26 hydroxyl groups. Hydroxylation has also been reported using solvent-free NaOH / hydrogen peroxide. $C_{60}(OH)_8$ was prepared using a multistep procedure starting from a mixed peroxide fullerene. The maximum number of hydroxyl groups that can be attached (hydrogen peroxide method) stands at 36–40.

Electrophilic Addition

Fullerenes react in electrophilic additions as well. The reaction with bromine can add up to 24 bromine atoms to the sphere. The record holder for fluorine addition is $C_{60}F_{48}$. According to in silico predictions the as yet elusive $C_{60}F_{60}$ may have some of the fluorine atoms in endo positions (pointing inwards) and may resemble a tube more than it does a sphere.

Retro Additions

Protocols have been investigated for removing substituents via retroadditions after they have served their purpose. Examples are the retro-Bingel reaction and the retro-Prato reaction.

Carbene Additions

Fullerenes react with carbenes to methanofullerenes. The reaction of fullerene with dichlorocarbene (created by sodium trichloroacetate pyrolysis) was first reported in 1993. A single addition takes place along a bond.

Radical Additions

Fullerenes can be considered radical scavengers. With a simple hydrocarbon radical such as the t-butyl radical obtained by thermolysis or photolysis from a suitable precursor the tBuC60 radical is formed that can be studied. The unpaired electron does not delocalize over the entire sphere but takes up positions in the vicinity of the tBu substituent.

Fullerenes as Ligands

Fullerene is a ligand in organometallic chemistry. The double bond is electron-deficient and usually forms metallic bonds with $\eta = 2$ hapticity. Bonding modes such as $\eta = 5$ or $\eta = 6$ can be induced by modification of the coordination sphere.

- C_{60} fullerene reacts with tungsten hexacarbonyl $W(CO)_6$ to the $(\eta^2\text{-}C_{60})W(CO)_5$ complex in a hexane solution in direct sunlight.

Variants

Open-cage Fullerenes

A part of fullerene research is devoted to so-called open-cage fullerenes whereby one or more bonds are removed chemically exposing an orifice. In this way it is possible to insert into it small molecules such as hydrogen, helium or lithium. The first such open-cage fullerene was reported in 1995. In endohedral hydrogen fullerenes the opening, hydrogen insertion and closing back up has already been demonstrated.

Heterofullerenes

In heterofullerenes at least one carbon atom is replaced by another element. Based on spectroscopy, substitutions have been reported with boron (borafullerenes), nitrogen (azafullerenes), oxygen, arsenic, germanium, phosphorus, silicon, iron, copper, nickel, rhodium and iridium. Reports on isolated heterofullerenes are limited to those based on nitrogen and oxygen.

Fullerene Dimers

The C_{60} fullerene dimerizes in a formal [2+2] cycloaddition to a C_{120} bucky dumbbell in the solid state by mechanochemistry (high-speed vibration milling) with potassium cyanide as a catalyst. The trimer has also been reported using 4-aminopyridine as catalyst (4% yield) and observed with scanning tunneling microscopy as a monolayer.

Synthesis

Multistep fullerene Synthesis

Although the procedure for the synthesis of the C_{60} fullerene is well established (generation of a large current between two nearby graphite electrodes in an inert atmosphere) a 2002 study described an organic synthesis of the compound starting from simple organic compounds.

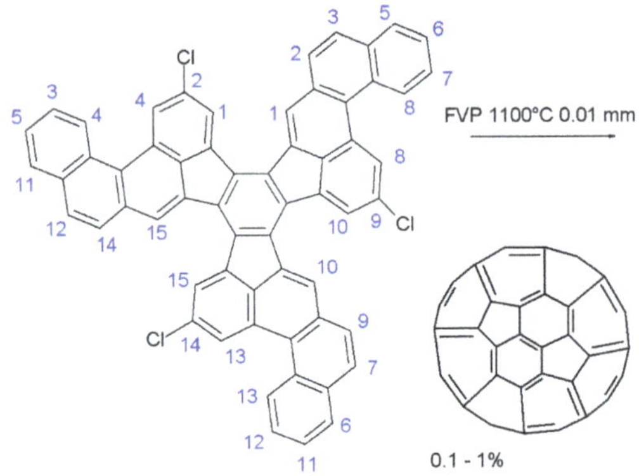

In the final step a large polycyclic aromatic hydrocarbon consisting of 13 hexagons and three pentagons was submitted to flash vacuum pyrolysis at 1100 °C and 0.01 Torr. The three carbon chlorine bonds served as free radical incubators and the ball was stitched up in a no-doubt complex series of radical reactions. The chemical yield was low: 0.1 to 1%. A small percentage of fullerenes is formed in any process which involves burning of hydrocarbons, e.g. in candle burning. The yield through a combustion method is often above 1%. The method proposed above does not provide any advantage for synthesis of fullerenes compared to the usual combustion method, and therefore, the organic synthesis of fullerenes remains a challenge for chemistry.

A similar exercise aimed at construction of a C78 cage in 2008 (but leaving out the precursor's halogens) did not result in a sufficient yield but at least the introduction of Stone Wales defects could be ruled out. C60 synthesis through a fluorinated fullerene precursor was reported in 2013

Purification

Fullerene purification is the process of obtaining a fullerene compound free of contamination. In fullerene production mixtures of C_{60}, C_{70} and higher homologues are always formed. Fullerene purification is key to fullerene science and determines fullerene prices and the success of practical applications of fullerenes. The first available purification method for C_{60} fullerene was by HPLC from which small amounts could be generated at large expense.

A practical laboratory-scale method for purification of soot enriched in C_{60} and C_{70} starts with extraction in toluene followed by filtration with a paper filter. The solvent is evaporated and the residue (the toluene-soluble soot fraction) redissolved in toluene and subjected to column chromatography. C_{60} elutes first with a purple color and C_{70} is next displaying a reddish-brown color.

In nanotube processing the established purification method for removing amorphous carbon and metals is by competitive oxidation (often a sulfuric acid / nitric acid mixture). It is assumed that this oxidation creates oxygen containing groups (hydroxyl, carbonyl, carboxyl) on the nanotube surface which electrostatically stabilize them in water and which can later be utilized in chemical functionalization. One report reveals that the oxygen containing groups in actuality combine with carbon contaminations absorbed to the nanotube wall that can be removed by a simple base wash. Cleaned nanotubes are reported to have reduced D/G ratio indicative of less functionalization, and the absence of oxygen is also apparent from IR spectroscopy and X-ray photoelectron spectroscopy.

Experimental Purification Strategies

A recent kilogram-scale fullerene purification strategy was demonstrated by Nagata *et al.* In this method C_{60} was separated from a mixture of C_{60}, C_{70} and higher fullerene compounds by first adding the amidine compound DBU to a solution of the mixture in 1,2,3-trimethylbenzene. DBU as it turns out only reacts to C_{70} fullerenes and higher which reaction products separate out and can be removed by filtration. C_{60} fullerenes do not have any affinity for DBU and are subsequently isolated. Other diamine compounds like DABCO do not share this selectivity.

C_{60} but not C_{70} forms a 1:2 inclusion compound with cyclodextrin (CD). A separation method for both fullerenes based on this principle is made possible by anchoring cyclodextrin to colloidal gold particles through a sulfur-sulfur bridge. The Au/CD compound is very stable and soluble in water and selectively extracts C_{60} from the insoluble mixture after refluxing for several days. The C_{70} fullerene component is then removed by simple filtration. C_{60} is driven out from the Au/CD compound by adding adamantol which has a higher affinity for the cyclodextrin cavity. Au/CD is completely recycled when adamantol in turn is driven out by adding ethanol and ethanol removed by evaporation; 50 mg of Au/CD captures 5 mg of C_{60} fullerene.

Buckminsterfullerene

Buckminsterfullerene (or bucky-ball) is a spherical fullerene molecule with the formula C_{60}. It has a cage-like fused-ring structure (truncated icosahedron) which resembles a soccer ball (football), made of twenty hexagons and twelve pentagons, with a carbon atom at each vertex of each polygon and a bond along each polygon edge.

It was first generated in 1985 by Harold Kroto, James R. Heath, Sean O'Brien, Robert Curl, and Richard Smalley at Rice University. Kroto, Curl and Smalley were awarded the 1996 Nobel Prize in Chemistry for their roles in the discovery of buckminsterfullerene and the related class of molecules, the fullerenes. The name is a reference to Buckminster Fuller, as C_{60} resembles his trademark geodesic domes. Buckminsterfullerene is the most common naturally occurring fullerene molecule, as it can be found in small quantities in soot. Solid and gaseous forms of the molecule have been detected in deep space.

Buckminsterfullerene is one of the largest objects to have been shown to exhibit wave–particle duality; as stated in the theory every object exhibits this behavior. Its discovery led to the exploration of a new field of chemistry involving the study of fullerenes.

Etymology

Buckminsterfullerene derives from the name of the noted futurist and inventor Buckminster Fuller. One of his designs of a geodesic dome structure bears great resemblance to C_{60}; as a result, the discoverers of the allotrope named the newfound molecule after him. The general public, however, sometimes refers to buckminsterfullerene, and even Fuller's dome structure, as buckyballs.

History

The structure associated with fullerenes was described by Leonardo da Vinci. Albrecht Dürer also reproduced a similar icosahedron containing 12 pentagonal and 20 hexagonal faces but there are no clear documentations of this.

Discovery

Theoretical predictions of buckyball molecules appeared in the late 1960s – early 1970s, but they went largely unnoticed. In the early 1970s, the chemistry of unsaturated carbon configurations was studied by a group at the University of Sussex, led by Harry Kroto and David Walton. In the 1980s a technique was developed by Richard Smalley and Robert Curl at Rice University, Texas to isolate these substances. They used laser vaporization of a suitable target to produce clusters of atoms. Kroto realized that by using a graphite target, any carbon chains formed could be studied. Another interesting fact is that, at the same time, astrophysicists were working along with spectroscopists to study infrared emissions from giant red carbon stars. Smalley and team were able to use a laser vaporization technique to create carbon clusters which could potentially emit infrared at the same wavelength as had been emitted by the red carbon star. Hence, the inspiration came to Smalley and team to use the laser technique on graphite to create the first fullerene molecule.

Harold Kroto

Richard Smalley

C_{60} was discovered in 1985 by Robert Curl, Harold Kroto and Richard Smalley. Using laser evaporation of graphite they found C_n clusters (where n>20 and even) of which the most common were

C_{60} and C_{70}. A solid rotating graphite disk was used as the surface from which carbon was vaporized using a laser beam creating hot plasma that was then passed through a stream of high-density helium gas. The carbon species were subsequently cooled and ionized resulting in the formation of clusters. Clusters ranged in molecular masses but Kroto and Smalley found predominance in a C_{60} cluster that could be enhanced further by letting the plasma react longer.They also discovered that the C_{60} molecule formed a cage-like structure, a regular truncated icosahedron.

Many soccer balls have the same shape as buckminsterfullerene, C_{60}.

For this discovery they were awarded the 1996 Nobel Prize in Chemistry. The discovery of buckyballs was surprising, as the scientists aimed the experiment at producing carbon plasmas to replicate and characterize unidentified interstellar matter. Mass spectrometry analysis of the product indicated the formation of spheroidal carbon molecules.

The experimental evidence, a strong peak at 720 atomic mass units, indicated that a carbon molecule with 60 carbon atoms was forming, but provided no structural information. The research group concluded after reactivity experiments, that the most likely structure was a spheroidal molecule. The idea was quickly rationalized as the basis of an icosahedral symmetry closed cage structure. Kroto mentioned geodesic dome structures of the noted futurist and inventor Buckminster Fuller as influences in the naming of this particular substance as buckminsterfullerene.

Synthesis

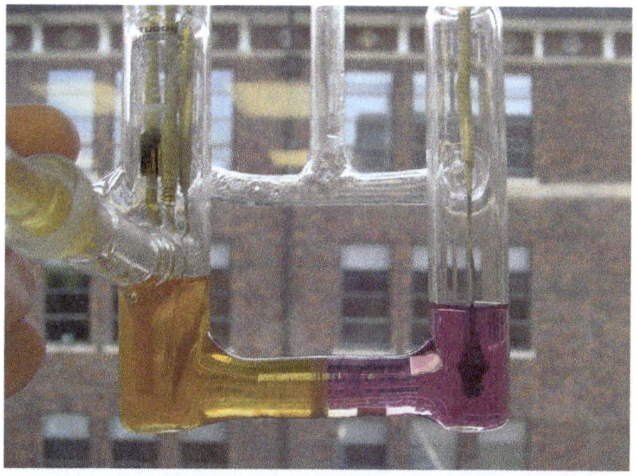

High-vacuum electrolysis of a C_{60}-fullerene derivative. Slow diffusion into the anode (right side) yields the characteristic purple color of pure C_{60}.

In 1989 physicists Wolfgang Krätschmer, Konstantinos Fostiropoulos, and Donald R. Huffman observed unusual optical absorptions in thin films of carbon dust (soot). The soot had been generated by an arc-process between two graphite electrodes in a helium atmosphere where the electrode material evaporates and condenses forming soot in the quenching atmosphere. Among other features, the IR spectra of the soot showed four discrete bands in close agreement to those proposed for C_{60}. Another paper on the characterization and verification of the molecular structure followed on in the same year (1990) from their thin film experiments, and detailed also the extraction of an evaporable as well as benzene soluble material from the arc-generated soot. This extract had TEM and X-ray crystal analysis consistent with arrays of spherical C_{60} molecules, approximately 1.0 nm in van der Waals diameter as well as the expected molecular mass of 720 u for C_{60} (and 840 u for C_{70}) in their mass spectra. The method was simple and efficient to prepare the material in gram amounts per day (1990) which has boosted the fullerene research and is even today applied for the industrial production of fullerenes.

Alternatively, soot is produced by laser ablation of graphite or pyrolysis of aromatic hydrocarbons. Fullerenes are extracted from the soot using a multistep procedure. First, the soot is dissolved in appropriate organic solvents. This step yields a solution containing up to 75% of C_{60}, as well as other fullerenes. These fractions are separated using chromatography. Generally, the fullerenes are dissolved in hydrocarbon or halogenated hydrocarbon and separated using alumina columns.

Further Developments

The versatility of fullerene molecules has led to a large amount of research exploring their properties. One interesting property is the relatively large volume of the internal space of the molecule. Atoms of different elements may be placed inside the molecular cage formed by the carbon atoms.

Beam-experiments conducted between 1985 and 1990 provided more evidence for the stability of C_{60} while supporting the closed-cage structural theory and predicting some of the bulk properties such a molecule would have. Around this time, intense theoretical group theory activity also predicted that C_{60} should have only four IR-active vibrational bands, on account of its icosahedral symmetry.

In 2012 a toxicity study by Tarek Baati and Fathi Moussa from the University of Paris, showed that C_{60} dissolved in olive oil was not toxic to rodents. In a video interview with Professor Fathi Moussa regarding the study, further information was provided regarding the toxicity study, and the method of action whereby the lifespan of the rodents was increased by 90% relative to controls when the animals were dosed with C_{60} olive oil.

Properties

Molecule

The structure of a buckminsterfullerene is a truncated icosahedron with 60 vertices and 32 faces (20 hexagons and 12 pentagons where no pentagons share a vertex) with a carbon atom at the vertices of each polygon and a bond along each polygon edge. The van der Waals diameter of a

C_{60} molecule is about 1.01 nanometers (nm). The nucleus to nucleus diameter of a C_{60} molecule is about 0.71 nm. The C_{60} molecule has two bond lengths. The 6:6 ring bonds (between two hexagons) can be considered "double bonds" and are shorter than the 6:5 bonds (between a hexagon and a pentagon). Its average bond length is 0.14 nm. Each carbon atom in the structure is bonded covalently with 3 others.

The C_{60} molecule is extremely stable, withstanding high temperatures and high pressures. The exposed surface of the structure can selectively react with other species while maintaining the spherical geometry. Atoms and small molecules can be trapped within the molecule without reacting.

C_{60} undergoes six reversible, one-electron reductions to C_{60}^{6-}, but oxidation is irreversible. The first reduction needs ≈1.0 V (Fc/Fc^{+}), showing that C_{60} is a moderately effective electron acceptor. C_{60} tends to avoid having double bonds in the pentagonal rings, which makes electron delocalization poor, and results in C_{60} not being "superaromatic". C_{60} behaves very much like an electron deficient alkene and readily reacts with electron rich species.

A carbon atom in the C_{60} molecule can be substituted by a nitrogen or boron atom yielding a $C_{59}N$ or $C_{59}B$ respectively.

Orthogonal projections					
Centered by	**Vertex**	**Edge 5–6**	**Edge 6–6**	**Face Hexagon**	**Face Pentagon**
Image	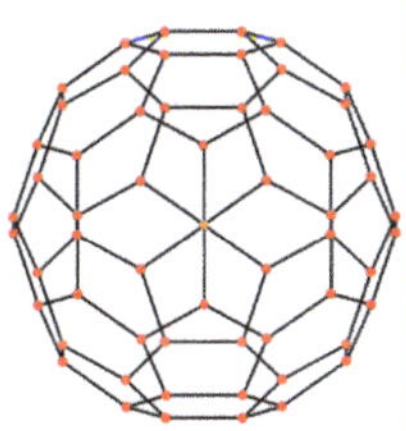	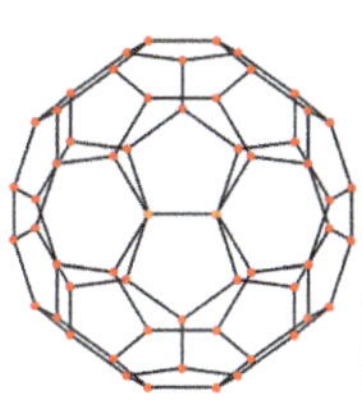	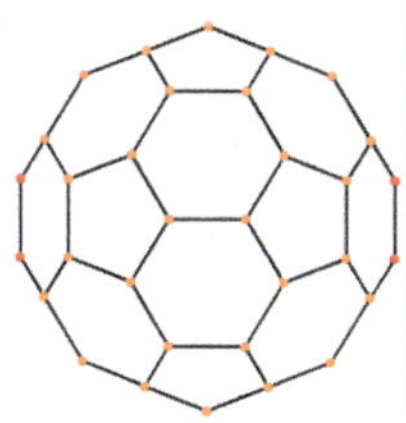	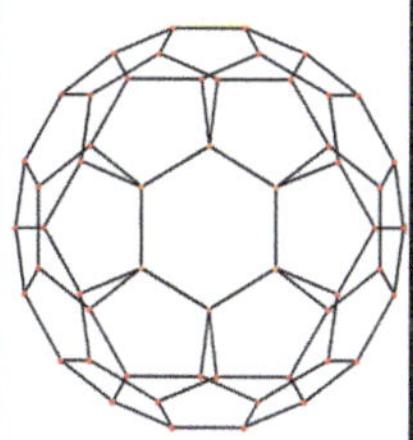	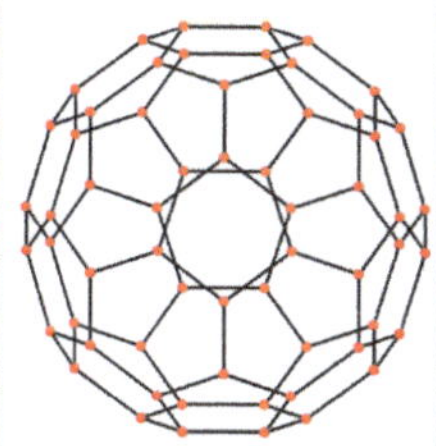
Projective symmetry	[2]	[2]	[2]	[6]	[10]

Solution

C_{60} solution

Solubility of C_{60}	
Solvent	**Solubility (g/L)**
1-chloronaphthalene	51
1-methylnaphthalene	33
1,2-dichlorobenzene	24
1,2,4-trimethylbenzene	18
tetrahydronaphthalene	16
carbon disulfide	8
1,2,3-tribromopropane	8
xylene	5
bromoform	5
cumene	4
toluene	3
benzene	1.5
carbon tetrachloride	0.447
chloroform	0.25
n-hexane	0.046
cyclohexane	0.035
tetrahydrofuran	0.006
acetonitrile	0.004
methanol	0.00004
water	1.3×10^{-11}
pentane	0.004
octane	0.025
isooctane	0.026
decane	0.070
dodecane	0.091
tetradecane	0.126
dioxane	0.0041
mesitylene	0.997
dichloromethane	0.254

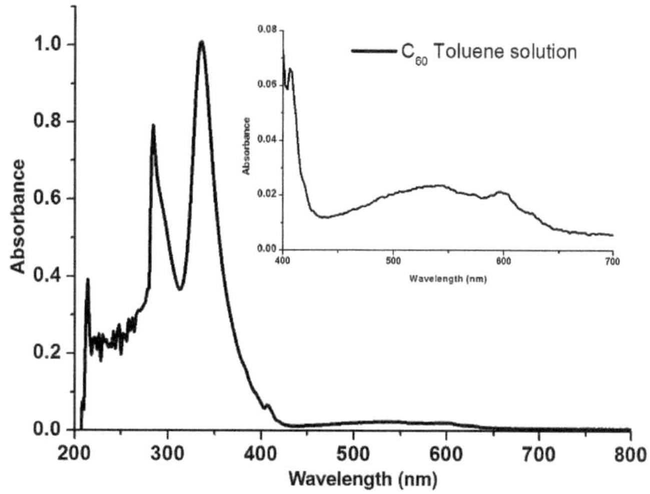

Optical absorption spectrum of C_{60} solution, showing reduced absorption for the blue (~450 nm) and red (~700 nm) light that results in the purple color.

Fullerenes are sparingly soluble in aromatic solvents such as toluene and carbon disulfide, but insoluble in water. Solutions of pure C_{60} have a deep purple color which leaves a brown residue upon evaporation. The reason for this color change is the relatively narrow energy width of the band of molecular levels responsible for green light absorption by individual C_{60} molecules. Thus individual molecules transmit some blue and red light resulting in a purple color. Upon drying, intermolecular interaction results in the overlap and broadening of the energy bands, thereby eliminating the blue light transmittance and causing the purple to brown color change.

C_{60} crystallises with some solvents in the lattice ("solvates"). For example, crystallization of C_{60} in benzene solution yields triclinic crystals with the formula $C_{60}{\cdot}4C_6H_6$. Like other solvates, this one readily releases benzene to give the usual fcc C_{60}. Millimeter-sized crystals of C_{60} and C_{70} can be grown from solution both for solvates and for pure fullerenes.

Solid

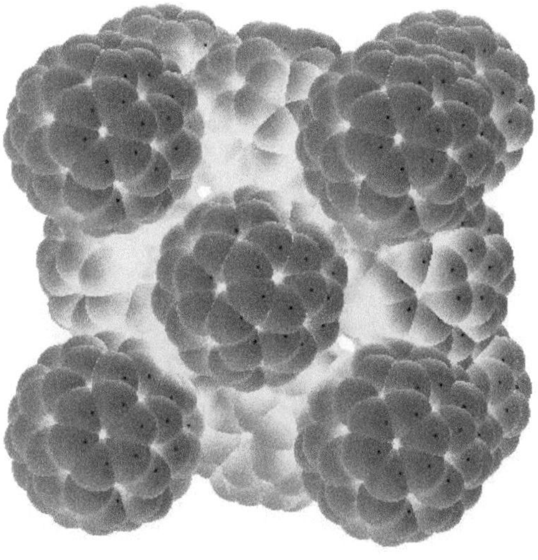

C_{60} crystal structure

In solid buckminsterfullerene, the molecules C_{60} stick together via the van der Waals forces in the fcc (face-centered cubic) motif. At low temperatures the individual molecules are locked against rotation. Upon heating, they start rotating at about −20 °C. This results in a first-order phase transition to a fcc structure and a small, yet abrupt increase in the lattice constant from 1.411 to 1.4154 nm.

C_{60} solid is as soft as graphite, but when compressed to less than 70% of its volume it transforms into a superhard form of diamond. C_{60} films and solution have strong non-linear optical properties; in particular, their optical absorption increases with light intensity (saturable absorption).

C_{60} forms a brownish solid with an optical absorption threshold at ≈1.6 eV. It is an n-type semiconductor with a low activation energy of 0.1–0.3 eV; this conductivity is attributed to intrinsic or oxygen-related defects. Fcc C_{60} contains voids at its octahedral and tetrahedral sites which are sufficient large (0.6 and 0.2 nm respectively) to accommodate impurity atoms. When alkali metals are doped into these voids, C_{60} converts from a semiconductor into a conductor or even superconductor.

Band Structure and Superconductivity

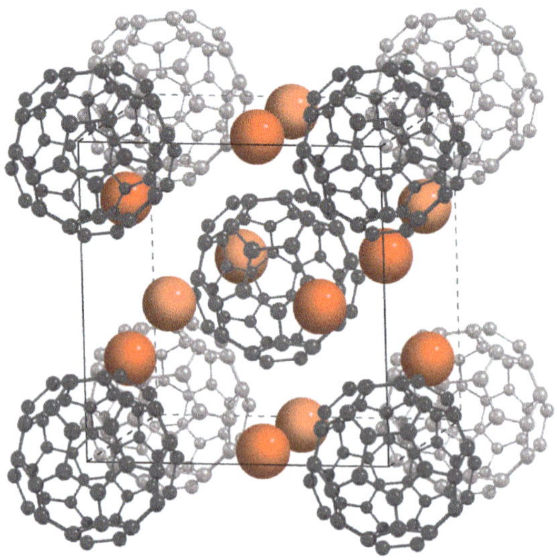

Cs_3C_{60} crystal structure

In 1991, Haddon *et al.* found that intercalation of alkali-metal atoms in solid C_{60} leads to metallic behavior. In 1991, it was revealed that potassium-doped C_{60} becomes superconducting at 18 K. This was the highest transition temperature for a molecular superconductor. Since then, superconductivity has been reported in fullerene doped with various other alkali metals. It has been shown that the superconducting transition temperature in alkaline-metal-doped fullerene increases with the unit-cell volume V. As caesium forms the largest alkali ion, caesium-doped fullerene is an important material in this family. Recently, superconductivity at 38 K has been reported in bulk Cs_3C_{60}, but only under applied pressure. The highest superconducting transition temperature of 33 K at ambient pressure is reported for Cs_2RbC_{60}.

The increase of transition temperature with the unit-cell volume had been believed to be evidence for the BCS mechanism of C_{60} solid superconductivity, because inter C_{60} separation can be related to an increase in the density of states on the Fermi level, $N(\varepsilon_F)$. Therefore, there have been many efforts to increase the interfullerene separation, in particular, intercalating neutral molecules into the A_3C_{60} lattice to increase the interfullerene spacing while the valence of C_{60} is kept unchanged. However, this ammoniation technique has revealed a new aspect of fullerene intercalation compounds: the Mott transition and the correlation between the orientation/orbital order of C_{60} molecules and the magnetic structure.

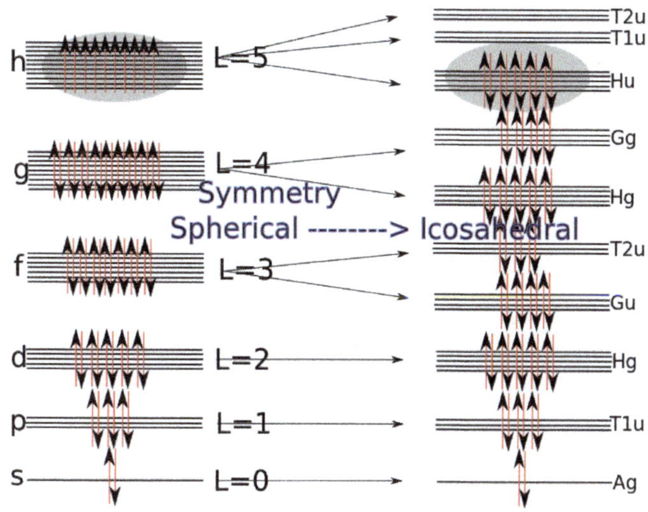

Electronic structure of C_{60} under "ideal" spherical (left) and "real" icosahedral symmetry (right).

The C_{60} molecules compose a solid of weakly bound molecules. The fullerites are therefore molecular solids, in which the molecular properties still survive. The discrete levels of a free C_{60} molecule are only weakly broadened in the solid, which leads to a set of essentially nonoverlapping bands with a narrow width of about 0.5 eV. For an undoped C_{60} solid, the 5-fold h_u band is the HOMO level, and the 3-fold t_{1u} band is the empty LUMO level, and this system is a band insulator. But when the C_{60} solid is doped with metal atoms, the metal atoms give electrons to the t_{1u} band or the upper 3-fold t_{1g} band. This partial electron occupation of the band may lead to metallic behavior. However, A_4C_{60} is an insulator, although the t_{1u} band is only partially filled and it should be a metal according to band theory. This unpredicted behavior may be explained by the Jahn–Teller effect, where spontaneous deformations of high-symmetry molecules induce the splitting of degenerate levels to gain the electronic energy. The Jahn–Teller type electron-phonon interaction is strong enough in C_{60} solids to destroy the band picture for particular valence states.

A narrow band or strongly correlated electronic system and degenerated ground states are important points to understand in explaining superconductivity in fullerene solids. When the inter-electron repulsion U is greater than the bandwidth, an insulating localized electron ground state is produced in the simple Mott–Hubbard model. This explains the absence of superconductivity at ambient pressure in caesium-doped C_{60} solids. Electron-correlation-driven localization of the t_{1u} electrons exceeds the critical value, leading to the Mott insulator. The application of high pressure decreases the interfullerene spacing, therefore caesium-doped C_{60} solids turn to metallic and superconducting.

A fully developed theory of C_{60} solids superconductivity is still lacking, but it has been widely accepted that strong electronic correlations and the Jahn–Teller electron–phonon coupling produce local electron-pairings that show a high transition temperature close to the insulator–metal transition.

Chemical Reactions and Properties

Hydrogenation

C_{60} exhibits a small degree of aromatic character, but it still reflects localized double and single C–C bond characters. Therefore, C_{60} can undergo addition with hydrogen to give polyhydrofullerenes. C_{60} also undergoes Birch reduction. For example, C_{60} reacts with lithium in liquid ammonia, followed by *tert*-butanol to give a mixture of polyhydrofullerenes such as $C_{60}H_{18}$, $C_{60}H_{32}$, $C_{60}H_{36}$, with $C_{60}H_{32}$ being the dominating product. This mixture of polyhydrofullerenes can be re-oxidized by 2,3-dichloro-5,6-dicyano-1,4-benzoquinone to give C_{60} again.

Selective hydrogenation method exists. Reaction of C_{60} with 9,9′,10,10′-dihydroanthracene under the same conditions, depending on the time of reaction, gives $C_{60}H_{32}$ and $C_{60}H_{18}$ respectively and selectively.

C_{60} can be hydrogenated, suggesting that a modified buckminsterfullerene called organometallic buckyballs (OBBs) could become a vehicle for "high density, room temperature, ambient pressure storage of hydrogen". These OBBs are created by binding atoms of a transition metal (TM) to C_{60} or $C_{48}B_{12}$ and then binding many hydrogen atoms to this TM atom, dispersing them evenly throughout the inside of the organometallic buckyball. The study found that the theoretical amount of H_2 that can be retrieved from the OBB at ambient pressure approaches 9 wt %, a mass fraction that has been designated as optimal for hydrogen fuel by the U.S. Department of Energy.

Halogenation

Addition of fluorine, chlorine, and bromine occurs for C_{60}.

Fluorine atoms are small enough for a 1,2-addition, while Cl_2 and Br_2 add to remote C atoms due to steric factors. For example, in $C_{60}Br_8$ and $C_{60}Br_{24}$, the Br atoms are in 1,3- or 1,4-positions with respect to each other.

Under various conditions a vast number of halogenated derivatives of C_{60} can be produced, some with extraordinary selectivity on one or two isomers over the other possible ones.

Addition of fluorine and chlorine usually results in a flattening of the C_{60} framework into a drum-shaped molecule.

Addition of Oxygen Atoms

Solutions of C_{60} can be oxygenated to the epoxide $C_{60}O$. Ozonation of C_{60} in 1,2-xylene at 257K gives an intermediate ozonide $C_{60}O_3$, which can be decomposed into 2 forms of $C_{60}O$. Decomposition of $C_{60}O_3$ at 296 K gives the epoxide, but photolysis gives a product in which the O atom bridges a 5,6-edge.

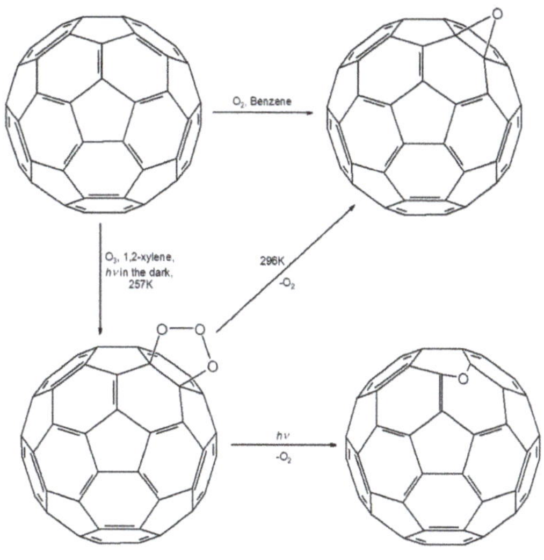

Cycloadditions

The Diels–Alder reaction is commonly employed to functionalize C_{60}. Reaction of C_{60} with appropriate substituted diene gives the corresponding adduct.

The Diels–Alder reaction between C_{60} and 3,6-diaryl-1,2,4,5-tetrazines affords C_{62}. The C_{62} has the structure in which a four-membered ring is surrounded by four six-membered rings.

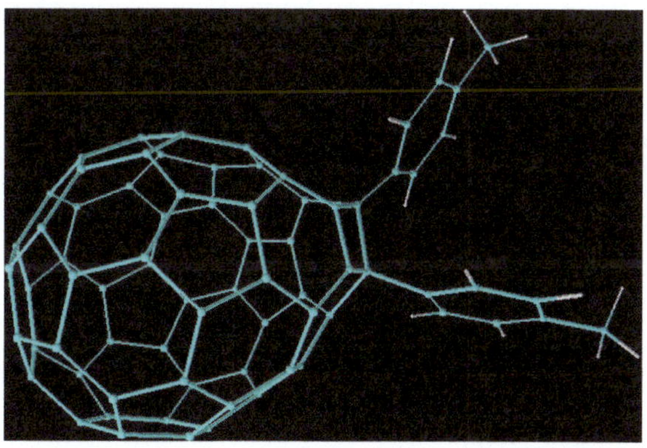

A C_{62} derivative [$C_{62}(C_6H_4\text{-4-Me})_2$] synthesized from C_{60} and 3,6-bis(4-methylphenyl)-3,6-dihydro-1,2,4,5-tetrazine

The C_{60} molecules can also be coupled through a [2+2] cycloaddition, giving the dumbbell-shaped compound C_{120}. The coupling is achieved by high-speed vibrating milling of C_{60} with a catalytic amount of KCN. The reaction is reversible as C_{120} dissociates back to two C_{60} molecules when heated at 450 K (177 °C; 350 °F). Under high pressure and temperature, repeated [2+2] cycloaddition between C_{60} results in a polymerized fullerene chains and networks. These polymers remain stable at ambient pressure and temperature once formed, and have remarkably interesting electronic and magnetic properties, such as being ferromagnetic above room temperature.

Free Radical Reactions

Reactions of C_{60} with free radicals readily occur. When C_{60} is mixed with a disulfide RSSR, the radical $C_{60}SR\bullet$ forms spontaneously upon irradiation of the mixture.

Stability of the radical species $C_{60}Y^{\cdot}$ depends largely on steric factors of Y. When *tert*-butyl halide is photolyzed and allowed to react with C_{60}, a reversible inter-cage C–C bond is formed:

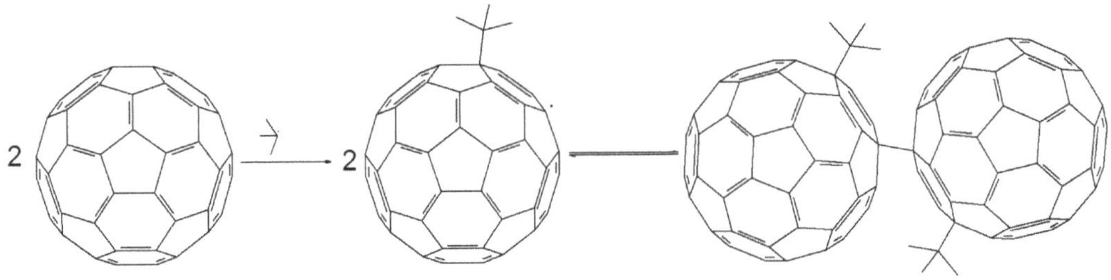

Cyclopropanation (Bingel Reaction)

Cyclopropanation (the Bingel reaction) is another common method for functionalizing C_{60}. Cyclopropanation of C_{60} mostly occurs at the junction of 2 hexagons due to steric factors.

The first cyclopropanation was carried out by treating the β-bromomalonate with C_{60} in the presence of a base. Cyclopropanation also occur readily with diazomethanes. For example, diphenyldiazomethane reacts readily with C_{60} to give the compound $C_{61}Ph_2$. Phenyl-C_{61}-butyric acid methyl ester derivative prepared through cyclopropanation has been studied for use in organic solar cells.

Redox reactions – C_{60} anions and Cations

The LUMO in C_{60} is triply degenerate, with the HOMO–LUMO separation relatively small. This small gap suggests that reduction of C_{60} should occur at mild potentials leading to fulleride anions, $[C_{60}]^{n-}$ (n = 1–6). The midpoint potentials of 1-electron reduction of buckminsterfullerene and its anions is given in the table below:

Reduction potential of C_{60} at 213 K	
Half-reaction	***E*° (V)**
$C_{60} + e^- \rightleftharpoons C_{60}^-$	−0.169
$C_{60}^- + e^- \rightleftharpoons C_{60}^{2-}$	−0.599
$C_{60}^{2-} + e^- \rightleftharpoons C_{60}^{3-}$	−1.129
$C_{60}^{3-} + e^- \rightleftharpoons C_{60}^{4-}$	−1.579
$C_{60}^{4-} + e^- \rightleftharpoons C_{60}^{5-}$	−2.069
$C_{60}^{5-} + e^- \rightleftharpoons C_{60}^{6-}$	−2.479

C_{60} forms a variety of charge-transfer complexes, for example with tetrakis(dimethylamino)ethylene:

$$C_{60} + C_2(NMe_2)_4 \rightarrow [C_2(NMe_2)_4]^+[C_{60}]^-$$

This salt exhibits ferromagnetism at 16 K.

The paramagnetic fulleride ion $[C_{60}]^{2-}$ has been isolated as the $[K(crypt\text{-}222)]^+$ salt. It is synthesized by treating C_{60} with metallic potassium in the presence of 2.2.2-cryptand. The most common fulleride ion, however, is $[C_{60}]^{3-}$. Alkali metal salts of this trianion are superconducting. In M_3C_{60} (M = Na, K, Rb), the M^+ ions occupy the interstitial holes in a lattice composed of ccp lattice composed of nearly spherical C_{60} anions. In Cs_3C_{60}, the cages are arranged in a bcc lattice.

Critical temperatures (T_c) of the fulleride salts M_3C_{60}	
Salt	**T_c (K)**
Na_3C_{60}	(non-superconducting)
K_3C_{60}	18
Rb_3C_{60}	28
Cs_3C_{60}	40

C_{60} oxidizes with difficulty. Three reversible oxidation processes have been observed by using cyclic voltammetry with ultra-dry methylene chloride and a supporting electrolyte with extremely high oxidation resistance and low nucleophilicity, such as $[^nBu_4N]$ $[AsF_6]$.

Reduction potentials of C_{60} oxidation at low temperatures	
Half-reaction	**E^o (V)**
$C_{60} \rightleftharpoons C_{60}^{+}$	+1.27
$C_{60}^{+} \rightleftharpoons C_{60}^{2+}$	+1.71
$C_{60}^{2+} \rightleftharpoons C_{60}^{3+}$	+2.14

Which the $[C_{60}]^{2+}$ ion is very unstable, and the third process can be studied only at low temperatures.

The redox potentials of C_{60} can be modified supramolecularly. A dibenzo-18-crown-6 derivative of C_{60} has been made, featuring a voltage sensor device, with the reversible binding of K^+ ion causing an anodic shift of 90mV of the first C_{60} reduction.

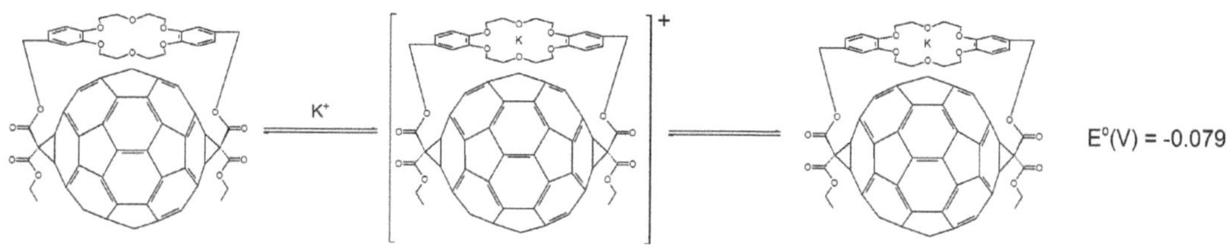

Metal Complexes

C_{60} forms complexes akin to the more common alkenes. Complexes have been reported molybdenum, tungsten, platinum, palladium, iridium, and titanium. The pentacarbonyl species are produced by photochemical reactions.

$$M(CO)_6 + C_{60} \rightarrow M(\eta^2\text{-}C_{60})(CO)_5 + CO \ (M = Mo, W)$$

In the case of platinum complex, the labile ethylene ligand is the leaving group in a thermal reaction:

$$Pt(\eta^2\text{-}C_2H_4)(PPh_3)_2 + C_{60} \rightarrow Pt(\eta^2\text{-}C_{60})(PPh_3)_2 + C_2H_4$$

Titanocene complexes have also been reported:

$$(\eta^5\text{-Cp})_2\text{Ti}(\eta^2\text{-}(CH_3)_3SiC{\equiv}CSi(CH_3)_3) + C_{60} \rightarrow (\eta^5\text{-Cp})_2\text{Ti}(\eta^2\text{-}C_{60}) + (CH_3)_3SiC{\equiv}CSi(CH_3)_3$$

Coordinatively unsaturated precursors, such as Vaska's complex, for adducts with C_{60}:

$$\textit{trans}\text{-Ir(CO)Cl}(PPh_3)_2 + C_{60} \rightarrow \text{Ir(CO)Cl}(\eta^2\text{-}C_{60})(PPh_3)_2$$

One such iridium complex, $[Ir(\eta^2\text{-}C_{60})(CO)Cl(Ph_2CH_2C_6H_4OCH_2Ph)_2]$ has been prepared where the metal center projects two electron-rich 'arms' that embrace the C_{60} guest.

Endohedral Fullerenes

Metal atoms or certain small molecules such as H_2 and noble gas can be encapsulated inside the C_{60} cage. These endohedral fullerenes are usually synthesized by doping in the metal atoms in an arc reactor or by laser evaporation. These methods gives low yields of endohedral fullerenes, and a better method involves the opening of the cage, packing in the atoms or molecules, and closing the opening using certain organic reactions. This method, however, is still immature and only a few species have been synthesized this way.

Endohedral fullerenes show distinct and intriguing chemical properties that can be completely different from the encapsulated atom or molecule, as well as the fullerene itself. The encapsulated atoms have been shown to perform circular motions inside the C_{60} cage, and its motion has been followed by using NMR spectroscopy.

Applications

No application of C_{60} has been commercialized. In the medical field, elements such as helium (that can be detected in minute quantities) can be used as chemical tracers in impregnated buckyballs.

Water-soluble derivatives of C_{60} were discovered to exert an inhibition on the three isoforms of nitric oxide synthase, with slightly different potencies.

The optical absorption properties of C_{60} match solar spectrum in a way that suggests that $C_{60\text{-}}$ based films could be useful for photovoltaic applications. Because of its high electronic affinity it is one of the most common electron acceptors used in donor/acceptor based solar cells. Conversion efficiencies up to 5.7% have been reported in C_{60}– polymer cells.

References

- Ma, K.L. (2011). "Electronic transport properties of junctions between carbon nanotubes and graphene nanoribbons". European Physical Journal B. 83: 487–492. doi:10.1140/epjb/e2011-20313-9
- Liu, L.; Guo, G.; Jayanthi, C.; Wu, S. (2002). "Colossal Paramagnetic Moments in Metallic Carbon Nanotori". Phys. Rev. Lett. 88 (21): 217206. Bibcode:2002PhRvL..88u7206L. PMID 12059501. doi:10.1103/PhysRevLett.88.217206
- Pacios Pujadó, Mercè (2012). Carbon Nanotubes as Platforms for Biosensors with Electrochemical and Electronic Transduction. Springer Heidelberg. pp. XX,208. ISBN 978-3-642-31421-6. doi:10.1007/978-3-642-31421-6
- "CDC – NIOSH Publications and Products – Current Intelligence Bulletin 65: Occupational Exposure to Carbon Nanotubes and Nanofibers (2013–145)". www.cdc.gov. Retrieved 2017-02-01

- Osawa, Eiji (2002). Perspectives of Fullerene Nanotechnology. Springer Science & Business Media. pp. 29–. ISBN 978-0-7923-7174-8
- Chernozatonskii, L.A. (1992). "Carbon nanotube connectors and planar jungle gyms". Physics Letters A. 172: 173–176. doi:10.1016/0375-9601(92)90978-u
- Sinha, Saion; Barjami, Saimir; Iannacchione, Germano; Schwab, Alexander; Muench, George (5 June 2005). "Off-axis thermal properties of carbon nanotube films". Journal of Nanoparticle Research. 7 (6): 651–657. doi:10.1007/s11051-005-8382-9
- Lambin, P. (1996). "Atomic structure and electronic properties of bent carbon nanotubes". Synth. Met. 77: 249–1254. doi:10.1016/0379-6779(96)80097-x
- Takeuchi, K.; Hayashi, T.; Kim, Y. A.; Fujisawa, K. and Endo, M. (February 2014) "The state-of-the-art science and applications of carbon nanotubes", nanojournal.ifmo.ru. Volume 5, Issue 1, pp 15
- Menon, M. (1997). "Carbon Nanotube "T Junctions": Nanoscale Metal-Semiconductor-Metal Contact Devices". Physical Review Letters. 79: 4453– 4456. doi:10.1103/physrevlett.79.4453
- Oberlin, A.; Endo, M.; Koyama, T. (1976). "Filamentous growth of carbon through benzene decomposition" (PDF). Journal of Crystal Growth. 32 (3): 335–349. Bibcode:1976JCrGr..32..335O. doi:10.1016/0022-0248(76)90115-9
- Iijima, S (1980). "Direct observation of the tetrahedral bonding in graphitized carbon black by high resolution electron microscopy". Journal of Crystal Growth. 50 (3): 675–683. Bibcode:1980JCrGr..50..675I. doi:10.1016/0022-0248(80)90013-5
- Zavalniuk, V.; Marchenko, S. (2011). "Theoretical analysis of telescopic oscillations in multi-walled carbon nanotubes". Low Temperature Physics. 37 (4): 337. Bibcode:2011LTP....37..337Z. arXiv:0903.2461 . doi:10.1063/1.3592692
- "REACH Registration Completed for Single-Wall Carbon Nanotubes". pcimag.com. PCI Mag. 16 October 2016. Retrieved 24 November 2016
- Schultz, H.P. (1965). "Topological Organic Chemistry. Polyhedranes and Prismanes". Journal of Organic Chemistry. 30 (5): 1361–1364. doi:10.1021/jo01016a005
- Ruoff, R.S.; Tse, Doris S.; Malhotra, Ripudaman; Lorents, Donald C. (1993). "Solubility of fullerene (C_{60}) in a variety of solvents" (PDF). Journal of Physical Chemistry. 97 (13): 3379–3383. doi:10.1021/j100115a049
- Semenov, K. N.; Charykov, N. A.; Keskinov, V. A.; Piartman, A. K.; Blokhin, A. A.; Kopyrin, A. A. (2010). "Solubility of Light Fullerenes in Organic Solvents". Journal of Chemical & Engineering Data. 55: 13–36. doi:10.1021/je900296s
- Talyzin, A.V. (1997). "Phase Transition C_{60}–C_{60}*$4C_6H_6$ in Liquid Benzene". Journal of Physical Chemistry B. 101 (47): 9679–9681. doi:10.1021/jp9720303

5

Quantum Dot: An Integrated Study

Quantum dots are semiconductors that are measurable in nanoparticles. They emit light of a particular frequency when brought in contact with electricity or magnetic force. These frequencies can be manipulated to change the dot's properties. The topics discussed in the chapter are of great importance to broaden the existing knowledge on quantum dots.

Quantum Dot

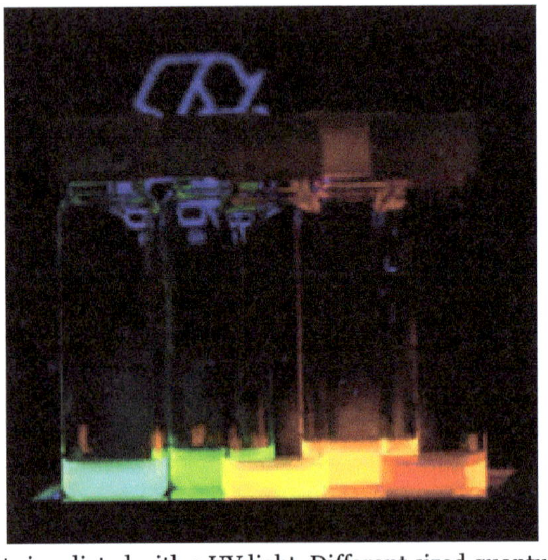

Colloidal quantum dots irradiated with a UV light. Different sized quantum dots emit different color light due to quantum confinement.

Quantum dots (QD) are very small semiconductor particles, only several nanometres in size, so small that their optical and electronic properties differ from those of larger particles. They are a central theme in nanotechnology. Many types of quantum dot will emit light of specific frequencies if electricity or light is applied to them, and these frequencies can be precisely tuned by changing the dots' size, shape and material, giving rise to many applications.

In the language of materials science, nanoscale semiconductor materials tightly confine either electrons or electron holes. Quantum dots are also sometimes referred to as artificial atoms, a term that emphasizes that a quantum dot is a single object with bound, discrete electronic states, as is the case with naturally occurring atoms or molecules.

Quantum dots exhibit properties that are intermediate between those of bulk semiconductors and those of discrete molecules. Their optoelectronic properties change as a function of both size and shape. Larger QDs (radius of 5–6 nm, for example) emit longer wavelengths resulting in emission colors such as orange or red. Smaller QDs (radius of 2–3 nm, for example) emit shorter wavelengths resulting in colors like blue and green, although the specific colors and sizes vary depending on the exact composition of the QD.

Because of their highly tunable properties, QDs are of wide interest. Potential applications include transistors, solar cells, LEDs, diode lasers and second-harmonic generation, quantum computing, and medical imaging. Additionally, their small size allows for QDs to be suspended in solution which leads to possible uses in inkjet printing and spin-coating. These processing techniques result in less-expensive and less time consuming methods of semiconductor fabrication.

Production

Quantum Dots with gradually stepping emission from violet to deep red are being produced in a kg scale at PlasmaChem GmbH.

There are several ways to prepare quantum dots, the principal ones involving colloids.

Colloidal Synthesis

Colloidal semiconductor nanocrystals are synthesized from solutions, much like traditional chemical processes. The main difference is the product neither precipitates as a bulk solid nor remains dissolved. Heating the solution at high temperature, the precursors decompose forming monomers which then nucleate and generate nanocrystals. Temperature is a critical factor in determining optimal conditions for the nanocrystal growth. It must be high enough to allow for rearrangement and annealing of atoms during the synthesis process while being low enough to promote crystal growth. The concentration of monomers is another critical factor that has to be stringently controlled during nanocrystal growth. The growth process of nanocrystals can occur in two different regimes, "focusing" and "defocusing". At high monomer concentrations, the critical size (the size where nanocrystals neither grow nor shrink) is relatively small, resulting in growth of nearly all particles. In this regime, smaller particles grow faster than large ones (since larger crystals need more atoms to grow than small crystals) resulting in "focusing" of the size distribution to yield nearly monodisperse particles. The size focusing is optimal when the monomer concentration is kept such that the average nanocrystal size present is always slightly larger than the critical size. Over time, the monomer concentration diminishes, the critical size becomes larger than the average size present, and the distribution "defocuses".

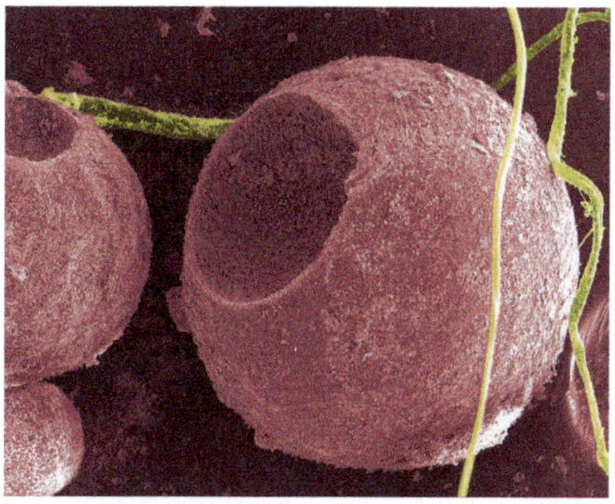

Cadmium sulfide quantum dots on cells

There are colloidal methods to produce many different semiconductors. Typical dots are made of binary compounds such as lead sulfide, lead selenide, cadmium selenide, cadmium sulfide, indium arsenide, and indium phosphide. Dots may also be made from ternary compounds such as cadmium selenide sulfide. These quantum dots can contain as few as 100 to 100,000 atoms within the quantum dot volume, with a diameter of ~ 10 to 50 atoms. This corresponds to about 2 to 10 nanometers, and at 10 nm in diameter, nearly 3 million quantum dots could be lined up end to end and fit within the width of a human thumb.

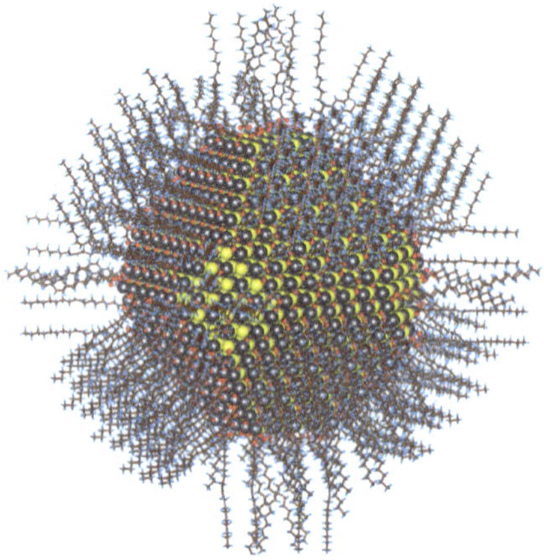

Ideallized image of colloidal nanoparticle of lead sulfide (selenide) with complete passivation by oleic acid, oleyl amine and hydroxyl ligands (size ~5nm)

Large batches of quantum dots may be synthesized via colloidal synthesis. Due to this scalability and the convenience of benchtop conditions, colloidal synthetic methods are promising for commercial applications. It is acknowledged to be the least toxic of all the different forms of synthesis.

Plasma Synthesis

Plasma synthesis has evolved to be one of the most popular gas-phase approaches for the production of quantum dots, especially those with covalent bonds. For example, silicon (Si) and germanium (Ge) quantum dots have been synthesized by using nonthermal plasma. The size, shape, surface and composition of quantum dots can all be controlled in nonthermal plasma. Doping that seems quite challenging for quantum dots has also been realized in plasma synthesis. Quantum dots synthesized by plasma are usually in the form of powder, for which surface modification may be carried out. This can lead to excellent dispersion of quantum dots in either organic solvents or water (i. e., colloidal quantum dots).

Fabrication

- Self-assembled quantum dots are typically between 5 and 50 nm in size. Quantum dots defined by lithographically patterned gate electrodes, or by etching on two-dimensional electron gasses in semiconductor heterostructures can have lateral dimensions between 20 and 100 nm.
- Some quantum dots are small regions of one material buried in another with a larger band gap. These can be so-called core–shell structures, e.g., with CdSe in the core and ZnS in the shell or from special forms of silica called ormosil.

- Quantum dots sometimes occur spontaneously in quantum well structures due to monolayer fluctuations in the well's thickness.
- Self-assembled quantum dots nucleate spontaneously under certain conditions during molecular beam epitaxy (MBE) and metallorganic vapor phase epitaxy (MOVPE), when a material is grown on a substrate to which it is not lattice matched. The resulting strain produces coherently strained islands on top of a two-dimensional wetting layer. This growth mode is known as Stranski–Krastanov growth. The islands can be subsequently buried to form the quantum dot. This fabrication method has potential for applications in quantum cryptography (i.e. single photon sources) and quantum computation. The main limitations of this method are the cost of fabrication and the lack of control over positioning of individual dots.
- Individual quantum dots can be created from two-dimensional electron or hole gases present in remotely doped quantum wells or semiconductor heterostructures called lateral quantum dots. The sample surface is coated with a thin layer of resist. A lateral pattern is then defined in the resist by electron beam lithography. This pattern can then be transferred to the electron or hole gas by etching, or by depositing metal electrodes (lift-off process) that allow the application of external voltages between the electron gas and the electrodes. Such quantum dots are mainly of interest for experiments and applications involving electron or hole transport, i.e., an electrical current.
- The energy spectrum of a quantum dot can be engineered by controlling the geometrical size, shape, and the strength of the confinement potential. Also, in contrast to atoms, it is relatively easy to connect quantum dots by tunnel barriers to conducting leads, which allows the application of the techniques of tunneling spectroscopy for their investigation.

The quantum dot absorption features correspond to transitions between discrete,three-dimensional particle in a box states of the electron and the hole, both confined to the same nanometer-size box.These discrete transitions are reminiscent of atomic spectra and have resulted in quantum dots also being called *artificial atoms*.

- Confinement in quantum dots can also arise from electrostatic potentials (generated by external electrodes, doping, strain, or impurities).
- Complementary metal-oxide-semiconductor (CMOS) technology can be employed to fabricate silicon quantum dots. Ultra small (L=20 nm, W=20 nm) CMOS transistors behave as single electron quantum dots when operated at cryogenic temperature over a range of −269 °C (4 K) to about −258 °C (15 K). The transistor displays Coulomb blockade due to progressive charging of electrons one by one. The number of electrons confined in the channel is driven by the gate voltage, starting from an occupation of zero electrons, and it can be set to 1 or many.

Viral Assembly

Genetically engineered M13 bacteriophage viruses allow preparation of quantum dot biocomposite structures. It had previously been shown that genetically engineered viruses can recognize specific semiconductor surfaces through the method of selection by combinatorial phage display.

Additionally, it is known that liquid crystalline structures of wild-type viruses (Fd, M13, and TMV) are adjustable by controlling the solution concentrations, solution ionic strength, and the external magnetic field applied to the solutions. Consequently, the specific recognition properties of the virus can be used to organize inorganic nanocrystals, forming ordered arrays over the length scale defined by liquid crystal formation. Using this information, Lee et al. (2000) were able to create self-assembled, highly oriented, self-supporting films from a phage and ZnS precursor solution. This system allowed them to vary both the length of bacteriophage and the type of inorganic material through genetic modification and selection.

Electrochemical Assembly

Highly ordered arrays of quantum dots may also be self-assembled by electrochemical techniques. A template is created by causing an ionic reaction at an electrolyte-metal interface which results in the spontaneous assembly of nanostructures, including quantum dots, onto the metal which is then used as a mask for mesa-etching these nanostructures on a chosen substrate.

Bulk-manufacture

Quantum dot manufacturing relies on a process called "high temperature dual injection" which has been scaled by multiple companies for commercial applications that require large quantities (hundreds of kilograms to tonnes) of quantum dots. This reproducible production method can be applied to a wide range of quantum dot sizes and compositions.

The bonding in certain cadmium-free quantum dots, such as III-V-based quantum dots, is more covalent than that in II-VI materials, therefore it is more difficult to separate nanoparticle nucleation and growth via a high temperature dual injection synthesis. An alternative method of quantum dot synthesis, the "molecular seeding" process, provides a reproducible route to the production of high quality quantum dots in large volumes. The process utilises identical molecules of a molecular cluster compound as the nucleation sites for nanoparticle growth, thus avoiding the need for a high temperature injection step. Particle growth is maintained by the periodic addition of precursors at moderate temperatures until the desired particle size is reached. The molecular seeding process is not limited to the production of cadmium-free quantum dots; for example, the process can be used to synthesise kilogram batches of high quality II-VI quantum dots in just a few hours.

Another approach for the mass production of colloidal quantum dots can be seen in the transfer of the well-known hot-injection methodology for the synthesis to a technical continuous flow system. The batch-to-batch variations arising from the needs during the mentioned methodology can be overcome by utilizing technical components for mixing and growth as well as transport and temperature adjustments. For the production of CdSe based semiconductor nanoparticles this method has been investigated and tuned to production amounts of kg per month. Since the use of technical components allows for easy interchange in regards of maximum through-put and size, it can be further enhanced to tens or even hundreds of kilograms.

In 2011 a consortium of U.S. and Dutch companies reported a "milestone" in high volume quantum dot manufacturing by applying the traditional high temperature dual injection method to a flow system.

On January 23, 2013 Dow entered into an exclusive licensing agreement with UK-based Nanoco for the use of their low-temperature molecular seeding method for bulk manufacture of cadmium-free quantum dots for electronic displays, and on September 24, 2014 Dow commenced work on the production facility in South Korea capable of producing sufficient quantum dots for "millions of cadmium-free televisions and other devices, such as tablets". Mass production is due to commence in mid-2015. On 24 March 2015 Dow announced a partnership deal with LG Electronics to develop the use of cadmium free quantum dots in displays.

Heavy Metal-free Quantum Dots

In many regions of the world there is now a restriction or ban on the use of heavy metals in many household goods, which means that most cadmium based quantum dots are unusable for consumer-goods applications.

For commercial viability, a range of restricted, heavy metal-free quantum dots has been developed showing bright emissions in the visible and near infra-red region of the spectrum and have similar optical properties to those of CdSe quantum dots. Among these systems are InP/ZnS and CuInS/ZnS, for example.

Peptides are being researched as potential quantum dot material. Since peptides occur naturally in all organisms, such dots would likely be nontoxic and easily biodegraded.

Safety

Toxicity

Some quantum dots pose risks to human health and the environment under certain conditions. Notably, the studies on quantum dot toxicity are focused on cadmium containing particles and has yet to be demonstrated in animal models after physiologically relevant dosing. In vitro studies, based on cell cultures, on quantum dots (QD) toxicity suggests that their toxicity may derive from multiple factors including its physicochemical characteristics (size, shape, composition, surface functional groups, and surface charges) and environment. Assessing their potential toxicity is complex as these factors include properties such as QD size, charge, concentration, chemical composition, capping ligands, and also on their oxidative, mechanical and photolytic stability.

Many studies have focused on the mechanism of QD cytotoxicity using model cell cultures. It has been demonstrated that after exposure to ultraviolet radiation or oxidized by air CdSe QDs release free cadmium ions causing cell death. Group II-VI QDs also have been reported to induce the formation of reactive oxygen species after exposure to light, which in turn can damage cellular components such as proteins, lipids and DNA. Some studies have also demonstrated that addition of a ZnS shell inhibit the process of reactive oxygen species in CdSe QDs. Another aspect of QD toxicity is the process of their size dependent intracellular pathways that concentrate these particles in cellular organelles that are inaccessible by metal ions, which may result in unique patterns of cytotoxicity compared to their constituent metal ions. The reports of QD localization in the cell nucleus present additional modes of toxicity because they may induce DNA mutation, which in turn will propagate through future generation of cells causing diseases.

Although concentration of QDs in certain organelles have been reported in in vivo studies using animal models, interestingly, no alterations in animal behavior, weight, hematological markers or organ damage has been found through either histological or biochemical analysis. These finding have led scientists to believe that intracellular dose is the most important deterring factor for QD toxicity. Therefore, factors determining the QD endocytosis that determine the effective intracellular concentration, such as QD size, shape and surface chemistry determine their toxicity. Excretion of QDs through urine in animal models also have demonstrated via injecting radio-labeled ZnS capped CdSe QDs where the ligand shell was labelled with Tc. Though multiple other studies have concluded retention of QDs in cellular levels, exocytosis of QDs is still poorly studied in the literature.

While significant research efforts have broadened the understanding of toxicity of QDs, there are large discrepancies in the literature and questions still remains to be answered. Diversity of this class material as compared to normal chemical substances makes the assessment of their toxicity very challenging. As their toxicity may also be dynamic depending on the environmental factors such as pH level, light exposure and cell type, traditional methods of assessing toxicity of chemicals such as LD_{50} are not applicable for QDs. Therefore, researchers are focusing on introducing novel approaches and adapting existing methods to include this unique class of materials. Furthermore, novel strategies to engineer safer QDs are still under exploration by the scientific community. A recent novelty in the field is the discovery of carbon quantum dots, a new generation of optically-active nanoparticles potentially capable of replacing semiconductor QDs, but with the advantage of much lower toxicity.

Optical Properties

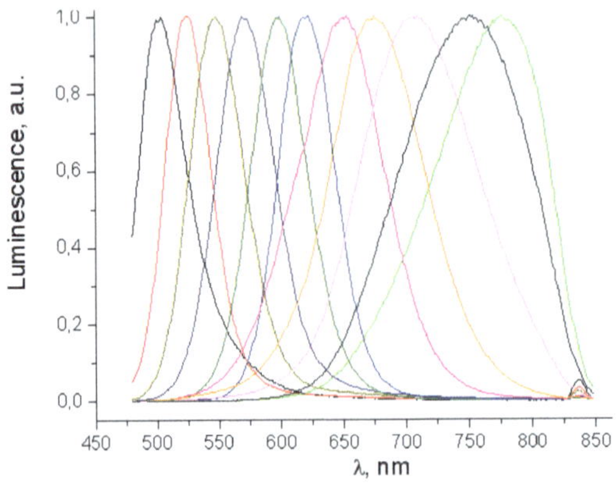

Fluorescence spectra of CdTe quantum dots of various sizes. Different sized quantum dots emit different color light due to quantum confinement.

In semiconductors, light absorption generally leads to an electron being excited from the valence to the conduction band, leaving behind a hole. The electron and the hole can bind to each other to form an exciton. When this exciton recombines (i.e. the electron resumes its ground state), the exciton's energy can be emitted as light. This is called fluorescence. In a simplified model, the energy of the emitted photon can be understood as the sum of the band gap energy between the highest occupied level and the lowest unoccupied energy level, the confinement energies of the hole and the excited electron, and the bound energy of the exciton (the electron-hole pair):

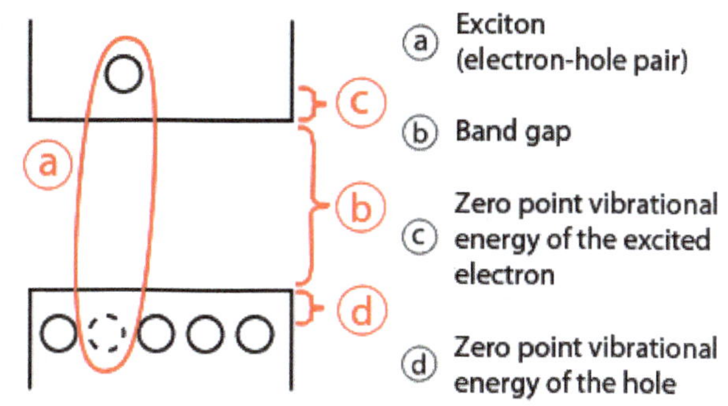

As the confinement energy depends on the quantum dot's size, both absorption onset and fluorescence emission can be tuned by changing the size of the quantum dot during its synthesis. The larger the dot, the redder (lower energy) its absorption onset and fluorescence spectrum. Conversely, smaller dots absorb and emit bluer (higher energy) light. Recent articles in *Nanotechnology* and in other journals have begun to suggest that the shape of the quantum dot may be a factor in the coloration as well, but as yet not enough information is available. Furthermore, it was shown that the lifetime of fluorescence is determined by the size of the quantum dot. Larger dots have more closely spaced energy levels in which the electron-hole pair can be trapped. Therefore, electron-hole pairs in larger dots live longer causing larger dots to show a longer lifetime.

To improve fluorescence quantum yield, quantum dots can be made with "shells" of a larger band-gap semiconductor material around them. The improvement is suggested to be due to the reduced access of electron and hole to non-radiative surface recombination pathways in some cases, but also due to reduced auger recombination in others.

Potential Applications

Quantum dots are particularly promising for optical applications due to their high extinction coefficient. They operate like a single electron transistor and show the Coulomb blockade effect. Quantum dots have also been suggested as implementations of qubits for quantum information processing.

Tuning the size of quantum dots is attractive for many potential applications. For instance, larger quantum dots have a greater spectrum-shift towards red compared to smaller dots, and exhibit less pronounced quantum properties. Conversely, the smaller particles allow one to take advantage of more subtle quantum effects.

Being zero-dimensional, quantum dots have a sharper density of states than higher-dimensional structures. As a result, they have superior transport and optical properties. They have potential uses in diode lasers, amplifiers, and biological sensors. Quantum dots may be excited within a locally enhanced electromagnetic field produced by gold nanoparticles, which can then be observed from the surface plasmon resonance in the photoluminescent excitation spectrum of (CdSe)ZnS nanocrystals. High-quality quantum dots are well suited for optical encoding and multiplexing applications due to their broad excitation profiles and narrow/symmetric emission spectra. The new generations of quantum dots have far-reaching potential for the study of intracellular processes at the single-molecule level, high-resolution cellular imaging, long-term in vivo observation of cell trafficking, tumor targeting, and diagnostics.

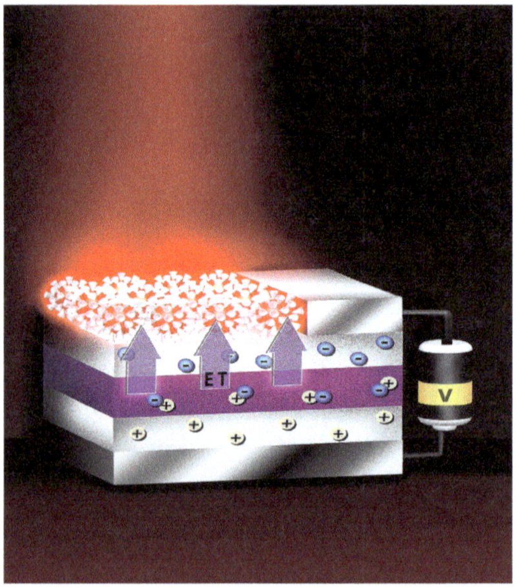

A device that produces visible light, through energy transfer from thin layers of quantum wells to crystals above the layers.

CdSe nanocrystals are efficient triplet photosensitizers. Laser excitation of small CdSe nanoparticles enables the extraction of the excited state energy from the Quantum Dots into bulk solution, thus opening the door to a wide range of potential applications such as photodynamic therapy, photovoltaic devices, molecular electronics, and catalysis.

Computing

Quantum dot technology is potentially relevant to solid-state quantum computation. By applying small voltages to the leads, current through the quantum dot can be controlled and thereby precise measurements of the spin and other properties therein can be made. With several entangled quantum dots, or qubits, plus a way of performing operations, quantum calculations and the computers that would perform them might be possible.

Biology

In modern biological analysis, various kinds of organic dyes are used. However, as technology advances, greater flexibility in these dyes is sought. To this end, quantum dots have quickly filled in the role, being found to be superior to traditional organic dyes on several counts, one of the most immediately obvious being brightness (owing to the high extinction coefficient combined with a comparable quantum yield to fluorescent dyes) as well as their stability (allowing much less photobleaching). It has been estimated that quantum dots are 20 times brighter and 100 times more stable than traditional fluorescent reporters. For single-particle tracking, the irregular blinking of quantum dots is a minor drawback. However, there have been groups which have developed quantum dots which are essentially nonblinking and demonstrated their utility in single molecule tracking experiments.

The use of quantum dots for highly sensitive cellular imaging has seen major advances. The improved photostability of quantum dots, for example, allows the acquisition of many consecutive focal-plane images that can be reconstructed into a high-resolution three-dimensional image. Another application that takes advantage of the extraordinary photostability of quantum dot probes is the real-time tracking of molecules and cells over extended periods of time. Antibodies, streptavidin, peptides, DNA, nucleic acid aptamers, or small-molecule ligands can be used to target quantum dots to specific proteins on cells. Researchers were able to observe quantum dots in lymph nodes of mice for more than 4 months.

Semiconductor quantum dots have also been employed for in vitro imaging of pre-labeled cells. The ability to image single-cell migration in real time is expected to be important to several research areas such as embryogenesis, cancer metastasis, stem cell therapeutics, and lymphocyte immunology.

One application of quantum dots in biology is as donor fluorophores in Förster resonance energy transfer, where the large extinction coefficient and spectral purity of these fluorophores make them superior to molecular fluorophores It is also worth noting that the broad absorbance of QDs allows selective excitation of the QD donor and a minimum excitation of a dye acceptor in FRET-based studies. The applicability of the FRET model, which assumes that the Quantum Dot can be approximated as a point dipole, has recently been demonstrated

The use of quantum dots for tumor targeting under in vivo conditions employ two targeting schemes: active targeting and passive targeting. In the case of active targeting, quantum dots are functionalized with tumor-specific binding sites to selectively bind to tumor cells. Passive targeting uses the enhanced permeation and retention of tumor cells for the delivery of quantum dot probes. Fast-growing tumor cells typically have more permeable membranes than healthy cells, allowing the leakage of small nanoparticles into the cell body. Moreover, tumor cells lack an effective lymphatic drainage system, which leads to subsequent nanoparticle-accumulation.

Quantum dot probes exhibit in vivo toxicity. For example, CdSe nanocrystals are highly toxic to cultured cells under UV illumination, because the particles dissolve, in a process known as photolysis, to release toxic cadmium ions into the culture medium. In the absence of UV irradiation, however, quantum dots with a stable polymer coating have been found to be essentially nontoxic. Hydrogel encapsulation of quantum dots allows for quantum dots to be introduced into a stable aqueous solution, reducing the possibility of cadmium leakage.Then again, only little is known about the excretion process of quantum dots from living organisms.

In another potential application, quantum dots are being investigated as the inorganic fluorophore for intra-operative detection of tumors using fluorescence spectroscopy.

Delivery of undamaged quantum dots to the cell cytoplasm has been a challenge with existing techniques. Vector-based methods have resulted in aggregation and endosomal sequestration of quantum dots while electroporation can damage the semi-conducting particles and aggregate delivered dots in the cytosol. Via cell squeezing, quantum dots can be efficiently delivered without inducing aggregation, trapping material in endosomes, or significant loss of cell viability. Moreover, it has shown that individual quantum dots delivered by this approach are detectable in the cell cytosol, thus illustrating the potential of this technique for single molecule tracking studies.

Photovoltaic Devices

The tunable absorption spectrum and high extinction coefficients of quantum dots make them attractive for light harvesting technologies such as photovoltaics. Quantum dots may be able to increase the efficiency and reduce the cost of today's typical silicon photovoltaic cells. According to an experimental proof from 2004, quantum dots of lead selenide can produce more than one exciton from one high energy photon via the process of carrier multiplication or multiple exciton generation (MEG). This compares favorably to today's photovoltaic cells which can only manage one exciton per high-energy photon, with high kinetic energy carriers losing their energy as heat. Quantum dot photovoltaics would theoretically be cheaper to manufacture, as they can be made "using simple chemical reactions."

Quantum Dot Only Solar Cells

Aromatic self-assembled monolayers (SAMs) (e.g. 4-nitrobenzoic acid) can be used to improve the band alignment at electrodes for better efficiencies. This technique has provided a record power conversion efficiency (PCE) of 10.7%. The SAM is positioned between ZnO-PbS colloidal quantum dot (CQD) film junction to modify band alignment via the dipole moment of the constituent SAM molecule, and the band tuning may be modified via the density, dipole and the orientation of the SAM molecule.

Quantum Dot in Hybrid Solar Cells

Colloidal quantum dots are also used in inorganic/organic hybrid solar cells. These solar cells are attractive because of potentially their low-cost fabrication and relatively high efficiency. Incorporation of metal oxides, such as ZnO, TiO_2, and Nb_2O_5 nanomaterials into organic photovoltaics have been commercialized using full roll-to-roll processing. A 13.2% power conversion efficiency is claimed in Si nanowire/PEDOT:PSS hybrid solar cells.

Quantum Dot with Nanowire in Solar Cells

Another potential use involves capped single-crystal ZnO nanowires with CdSe quantum dots, immersed in mercaptopropionic acid as hole transport medium in order to obtain a QD-sensitized solar cell. The morphology of the nanowires allowed the electrons to have a direct pathway to the photoanode. This form of solar cell exhibits 50-60% internal quantum efficiencies.

Nanowires with quantum dot coatings on silicon nanowires (SiNW) and carbon quantum dots. The use of SiNWs instead of planar silicon enhances the antiflection properties of Si. The SiNW exhibits a light-trapping effect due to light trapping in the SiNW. This use of SiNWs in conjunction with carbon quantum dots resulted in a solar cell that reached 9.10% PCE.

Graphene quantum dots have also been blended with organic electronic materials to improve efficiency and lower cost in photovoltaic devices and organic light emitting diodes (OLEDs) in compared to graphene sheets. These graphene quantum dots were functionalized with organic ligands that experience photoluminescence from UV-Vis absorption.

Light Emitting Devices

Several methods are proposed for using quantum dots to improve existing light-emitting diode

(LED) design, including "Quantum Dot Light Emitting Diode" (QD-LED) displays and "Quantum Dot White Light Emitting Diode" (QD-WLED) displays. Because Quantum dots naturally produce monochromatic light, they can be more efficient than light sources which must be color filtered. QD-LEDs can be fabricated on a silicon substrate, which allows them to be integrated onto standard silicon-based integrated circuits or microelectromechanical systems. Quantum dots are valued for displays, because they emit light in very specific gaussian distributions. This can result in a display with visibly more accurate colors. A conventional color liquid crystal display (LCD) is usually backlit by fluorescent lamps (CCFLs) or conventional white LEDs that are color filtered to produce red, green, and blue pixels. An improvement is using conventional blue-emitting LEDs as the light sources and converting part of the emitted light into *pure* green and red light by the appropriate quantum dots placed in front of the blue LED or using a quantum dot infused diffuser sheet in the backlight optical stack. This type of white light as the backlight of an LCD panel allows for the best color gamut at lower cost than a RGB LED combination using three LEDs.

The ability of QDs to precisely convert and tune a spectrum makes them attractive for LCD displays. Previous LCD displays can waste energy converting red-green poor, blue-yellow rich white light into a more balanced lighting. By using QDs, only the necessary colors for ideal images are contained in the screen. The result is a screen that is brighter, clearer, and more energy-efficient. The first commercial application of quantum dots was the Sony XBR X900A series of flat panel televisions released in 2013.

In June 2006, QD Vision announced technical success in making a proof-of-concept quantum dot display and show a bright emission in the visible and near infra-red region of the spectrum. A QD-LED integrated at a scanning microscopy tip was used to demonstrate fluorescence near-field scanning optical microscopy (NSOM) imaging.

Photodetector Devices

Quantum dot photodetectors (QDPs) can be fabricated either via solution-processing, or from conventional single-crystalline semiconductors. Conventional single-crystalline semiconductor QDPs are precluded from integration with flexible organic electronics due to the incompatibility of their growth conditions with the process windows required by organic semiconductors. On the other hand, solution-processed QDPs can be readily integrated with an almost infinite variety of substrates, and also postprocessed atop other integrated circuits. Such colloidal QDPs have potential applications in surveillance, machine vision, industrial inspection, spectroscopy, and fluorescent biomedical imaging.

Photocatalysts

Quantum dots also function as photocatalysts for the light driven chemical conversion of water into hydrogen as a pathway to solar fuel. In photocatalysis, electron hole pairs formed in the dot under band gap excitation drive redox reactions in the surrounding liquid. Generally, the photocatalytic activity of the dots is related to the particle size and its degree of quantum confinement. This is because the band gap determines the chemical energy that is stored in the dot in the excited state. An obstacle for the use of quantum dots in photocatalysis is the presence of surfactants on the surface of the dots. These surfactants (or ligands) interfere with the chemical reactivity of the dots by slowing down mass transfer and electron transfer processes. Also, quantum dots made of metal chalcogenides are chemically unstable under oxidizing conditions and undergo photo corrosion reactions.

Theory

Quantum dots are theoretically described as a point like, or a zero dimensional (0D) entity. Most of their properties depend on the dimensions, shape and materials of which QDs are made. Generally QDs present different thermodynamic properties from the bulk materials of which they are made. One of these effects is the Melting-point depression. Optical properties of spherical metallic QDs are well described by the Mie scattering theory.

Quantum Confinement in Semiconductors

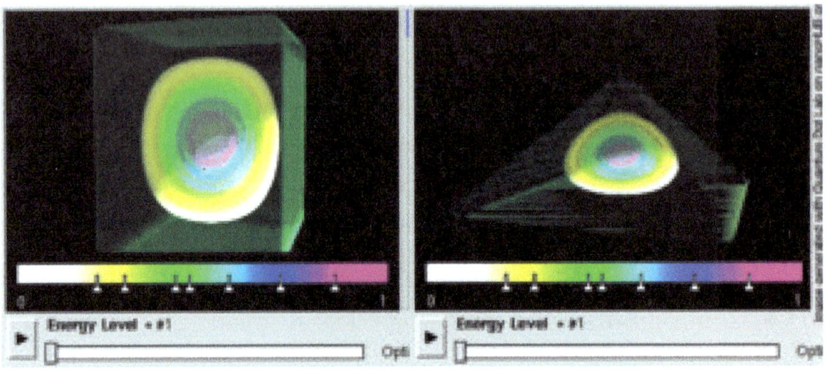

3D confined electron wave functions in a quantum dot. Here, rectangular and triangular-shaped quantum dots are shown. Energy states in rectangular dots are more *s-type* and *p-type*. However, in a triangular dot the wave functions are mixed due to confinement symmetry.

In a semiconductor crystallite whose size is smaller than twice the size of its exciton Bohr radius, the excitons are squeezed, leading to quantum confinement. The energy levels can then be predicted using the particle in a box model in which the energies of states depends on the length of the box. Comparing the quantum dots size to the Bohr radius of the electron and hole wave functions, 3 regimes can be defined. A 'strong confinement regime' is defined as the quantum dots radius being smaller than both electron and hole Bohr radius, 'weak confinement' is given when the quantum dot is larger than both. For semiconductors in which electron and hole radii are markedly different, an 'intermediate confinement regime' exists, where the quantum dot's radius is larger than the Bohr radius of one charge carrier (typically the hole), but not the other charge carrier.

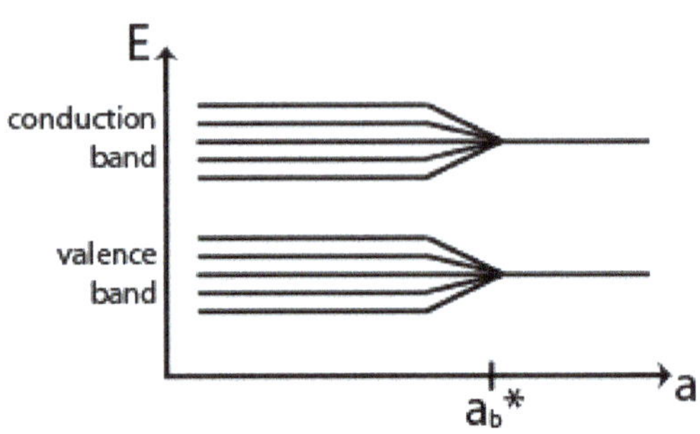

Splitting of energy levels for small quantum dots due to the quantum confinement effect. The horizontal axis is the radius, or the size, of the quantum dots and a_b* is the Exciton Bohr radius.

Band gap energy

The band gap can become smaller in the strong confinement regime as the energy levels split up. The Exciton Bohr radius can be expressed as:

$$a_b^* = \varepsilon_r \left(\frac{m}{\mu} \right) a_b$$

where a_b is the Bohr radius=0.053 nm, m is the mass, μ is the reduced mass, and ε_r is the size-dependent dielectric constant (Relative permittivity). This results in the increase in the total emission energy (the sum of the energy levels in the smaller band gaps in the strong confinement regime is larger than the energy levels in the band gaps of the original levels in the weak confinement regime) and the emission at various wavelengths. If the size distribution of QDs is not enough peaked, the convolution of multiple emission wavelengths is observed as a continuous spectra.

Confinement energy

The exciton entity can be modeled using the particle in the box. The electron and the hole can be seen as hydrogen in the Bohr model with the hydrogen nucleus replaced by the hole of positive charge and negative electron mass. Then the energy levels of the exciton can be represented as the solution to the particle in a box at the ground level (n = 1) with the mass replaced by the reduced mass. Thus by varying the size of the quantum dot, the confinement energy of the exciton can be controlled.

Bound exciton energy

There is Coulomb attraction between the negatively charged electron and the positively charged hole. The negative energy involved in the attraction is proportional to Rydberg's energy and inversely proportional to square of the size-dependent dielectric constant of the semiconductor. When the size of the semiconductor crystal is smaller than the Exciton Bohr radius, the Coulomb interaction must be modified to fit the situation.

Therefore, the sum of these energies can be represented as:

$$E_{\text{confinement}} = \frac{\hbar^2 \pi^2}{2a^2} \left(\frac{1}{m_e} + \frac{1}{m_h} \right) = \frac{\hbar^2 \pi^2}{2\mu a^2}$$

$$E_{\text{exciton}} = -\frac{1}{\epsilon_r^2} \frac{\mu}{m_e} R_y = -R_y^*$$

$$E = E_{\text{bandgap}} + E_{\text{confinement}} + E_{\text{exciton}}$$

$$= E_{\text{bandgap}} + \frac{\hbar^2 \pi^2}{2\mu a^2} - R_y^*$$

where μ is the reduced mass, a is the radius, m_e is the free electron mass, m_h is the hole mass, and ε_r is the size-dependent dielectric constant.

Although the above equations were derived using simplifying assumptions, they imply that the electronic transitions of the quantum dots will depend on their size. These quantum confinement effects are apparent only below the critical size. Larger particles do not exhibit this effect. This effect of quantum confinement on the quantum dots has been repeatedly verified experimentally and is a key feature of many emerging electronic structures.

The Coulomb interaction between confined carriers can also be studied by numerical means when results unconstrained by asymptotic approximations are pursued.

Besides confinement in all three dimensions (i.e., a quantum dot), other quantum confined semiconductors include:

- Quantum wires, which confine electrons or holes in two spatial dimensions and allow free propagation in the third.
- Quantum wells, which confine electrons or holes in one dimension and allow free propagation in two dimensions.

Models

A variety of theoretical frameworks exist to model optical, electronic, and structural properties of quantum dots. These may be broadly divided into quantum mechanical, semiclassical, and classical.

Quantum Mechanics

Quantum mechanical models and simulations of quantum dots often involve the interaction of electrons with a pseudopotential or random matrix.

Semiclassical

Semiclassical models of quantum dots frequently incorporate a chemical potential. For example, the thermodynamic chemical potential of an N-particle system is given by

$$\mu(N) = E(N) - E(N-1)$$

whose energy terms may be obtained as solutions of the Schrödinger equation. The definition of capacitance,

$$\frac{1}{C} \equiv \frac{\Delta V}{\Delta Q},$$

with the potential difference

$$\Delta V = \frac{\Delta\mu}{e} = \frac{\mu(N+\Delta N) - \mu(N)}{e}$$

may be applied to a quantum dot with the addition or removal of individual electrons,

$\Delta N = 1$ and $\Delta Q = e$.

Then

$$C(N) = \frac{e^2}{\mu(N+1) - \mu(N)} = \frac{e^2}{I(N) - A(N)}$$

is the "quantum capacitance" of a quantum dot, where we denoted by *I(N)* the ionization potential and by *A(N)* the electron affinity of the *N*-particle system.

Classical Mechanics

Classical models of electrostatic properties of electrons in quantum dots are similar in nature to the Thomson problem of optimally distributing electrons on a unit sphere.

The classical electrostatic treatment of electrons confined to spherical quantum dots is similar to their treatment in the Thomson, or plum pudding model, of the atom.

The classical treatment of both two-dimensional and three-dimensional quantum dots exhibit electron shell-filling behavior. A "periodic table of classical artificial atoms" has been described for two-dimensional quantum dots. As well, several connections have been reported between the three-dimensional Thomson problem and electron shell-filling patterns found in naturally-occurring atoms found throughout the periodic table. This latter work originated in classical electrostatic modeling of electrons in a spherical quantum dot represented by an ideal dielectric sphere.

History

The term "quantum dot" was coined in 1988. They were first discovered in a glass matrix and in colloidal solutions by Alexey Ekimov.

Bulk Semiconductors

Electrons in conduction band (and holes in the valence band) are free to move in all three dimensions of space.

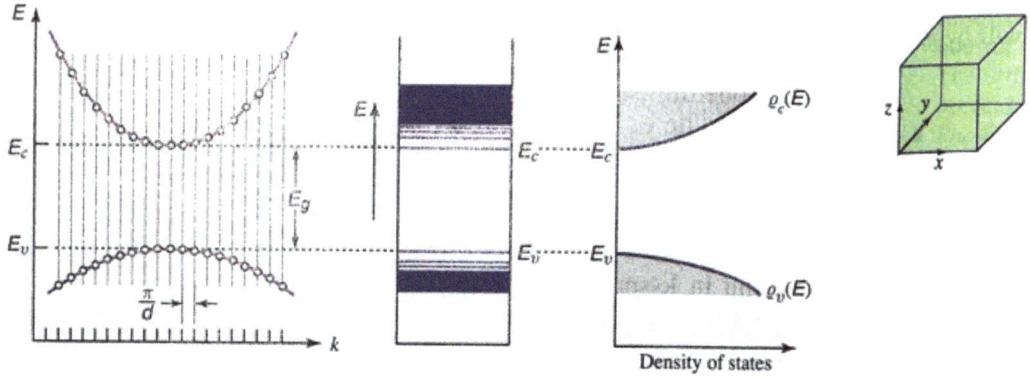

Thin Film Semiconductors

- Electrons in conduction band (and holes in the valence band) are free to move in two dimensions.
- Confined in one dimension by a potential well.
    - o Potential well created due to a larger bandgap of the semiconductors on either side of the thin film.
    - o Thinner films lead to higher energy levels.

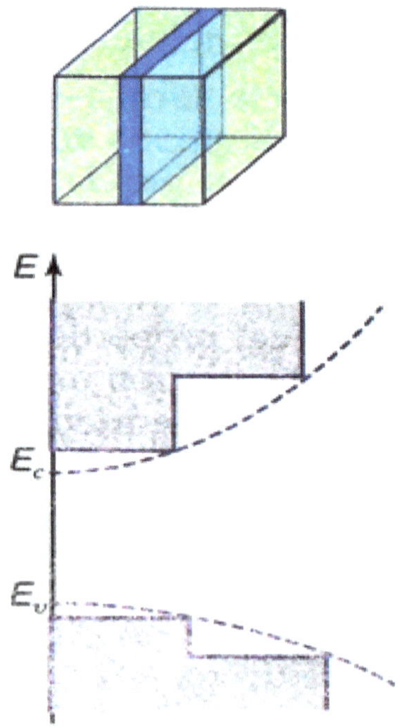

Quantum well

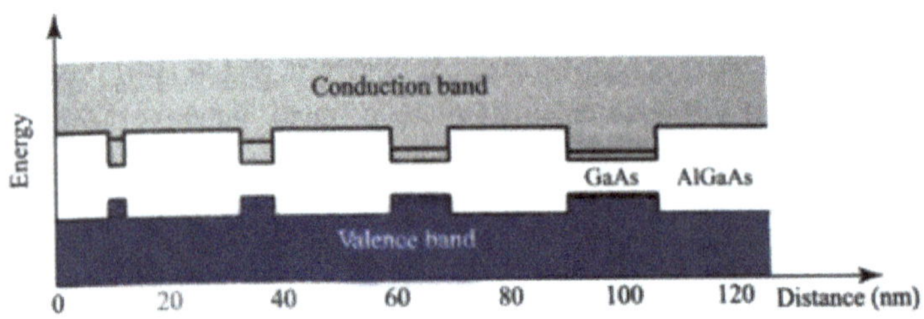

Quantum Wire

- Thin semiconductor wire surrounded by a material with a larger bandgap.
    - o Surrounding material confines electrons and holes in two dimensions (carriers can only move in one dimension) due to its larger bandgap.
        - ▪ Quantum wire acts as a potential well.

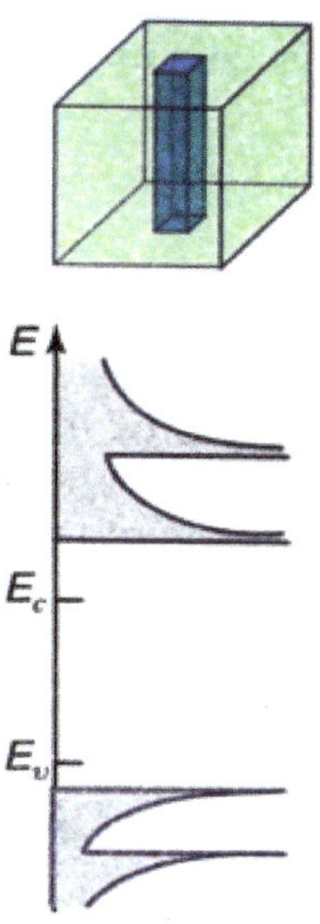

Quantum wire

Lithography

Quantum wells are covered with a polymer mask and exposed to an electron or ion beam. The surface is covered with a thin layer of metal, then cleaned and only the exposed areas keep the metal layer. Pillars are etched into the entire surface. Multiple layers are applied this way to build up the properties and size wanted. Disadvantages: slow, contamination, low density, defect formation.

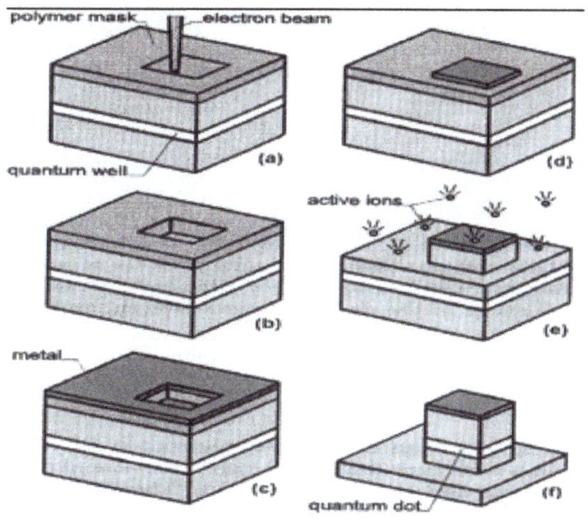

Colloidal Synthesis

- Immersion of semiconductor microcrystals in glass dielectric matrices.
- Taking a silicate glass with 1% semiconducting phase (CdS, CuCl, CdSe, or CuBr).
- Heating for several hours at high temperature.

- Formation of microcrystals of nearly equal size.
- Typically group II-VI materials (e.g. CdS, CdSe).
- Size variations ("size dispersion").

Epitaxy: Patterned Growth

- Semiconducting compounds with a smaller bandgap (GaAs) are grown on the surface of a compoundwith a larger bandgap (AlGaAs).
- Growth is restricted by coating it with a masking compound (SiO_2) and etching that mask with the shape of the required crystal cell wall shape.
- Disadvantage: density of quantum dots limited by mask pattern.

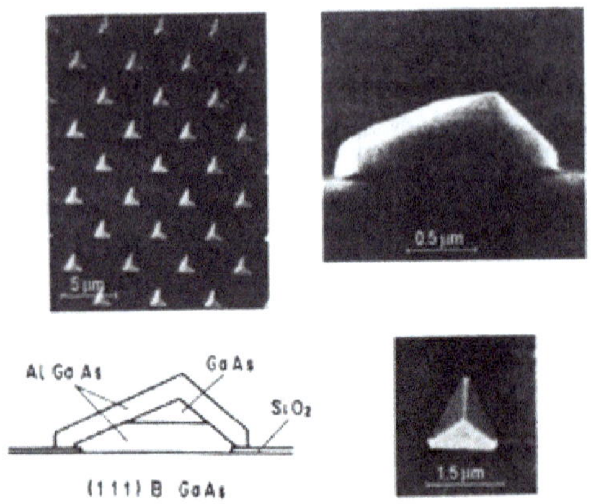

Epitaxy: Self-Organized Growth

- Uses a large difference in the lattice constants of the substrate and the crystallizing material.
- When the crystallized layer is thicker than the critical thickness, there is a strong strain on the layers.
- The breakdown results in randomly distributed islets of regular shape and size.
- Disadvantages: size and shape fluctuations, ordering.

Schematic drawing of lens-shaped self-organized quantum dot.

L. Jacak, P. Hawrylak, A. Wojs. Quantum dots

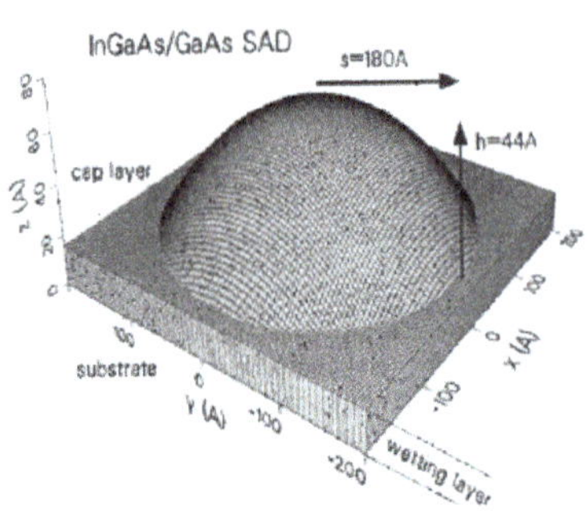

Applications

- Photovoltaic devices: solar cells
- Biology : biosensors, imaging
- Light emitting diodes: LEDs
- Quantum computation
- Flat-panel displays
- Memory elements
- Photodetectors
- Lasers

Solar Cells

- Photovoltaic effect:
    - p-n junction.
    - Sunlight excites electrons and creates electron-hole pairs.
    - Electrons concentrate on one side of the cell and holes on the other side.
    - Connecting the 2 sides creates electricity.

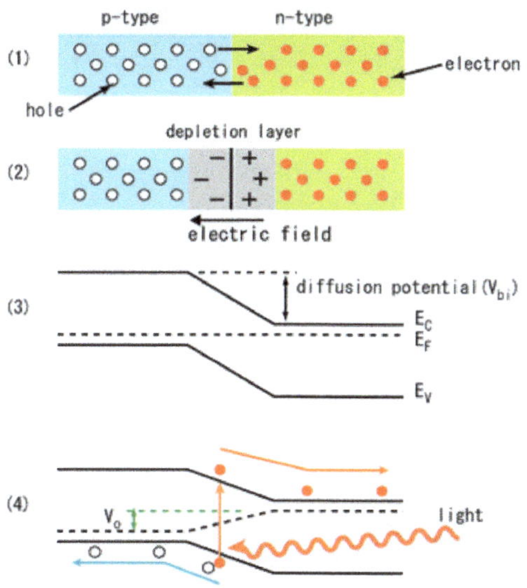

Different Generations of Solar Cells

- First generation:
    - Single crystal silicon wafer.

- o Advantages: high carrier mobility.
- o Disadvantages: most of photon energy is wasted as heat, expensive.

- Second generation:
  - o Thin-film technology.
  - o Advantages: less expensive.
  - o Disadvantages: efficiency lower compared with silicon solar cells.
- Third generation:
  - o Nanocrystal solar cells.
  - o Enhance electrical performances of the second generation while maintaining low production costs.

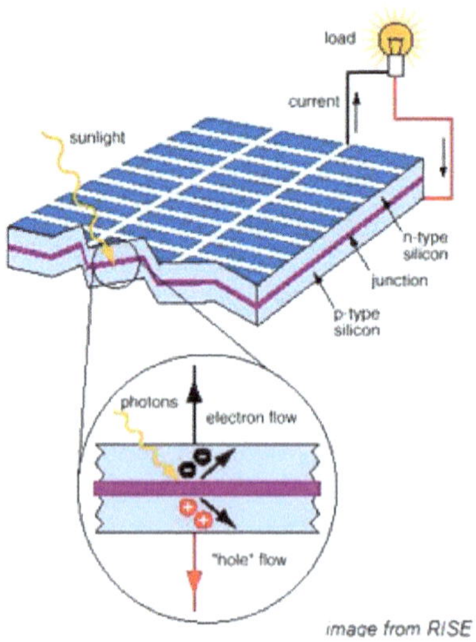

How Can Quantum Dots Improve the Efficiency?

- The quantum dot band gap is tunable and can be used to create intermediate bandgaps. The maximum theoretical efficiency of the solar cell is as high as 63.2% with this method.

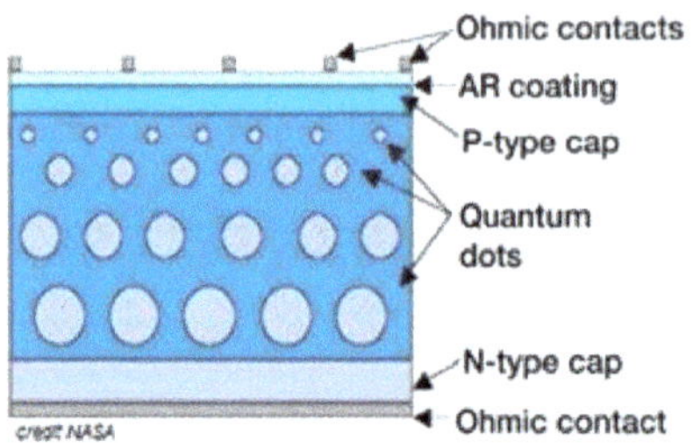

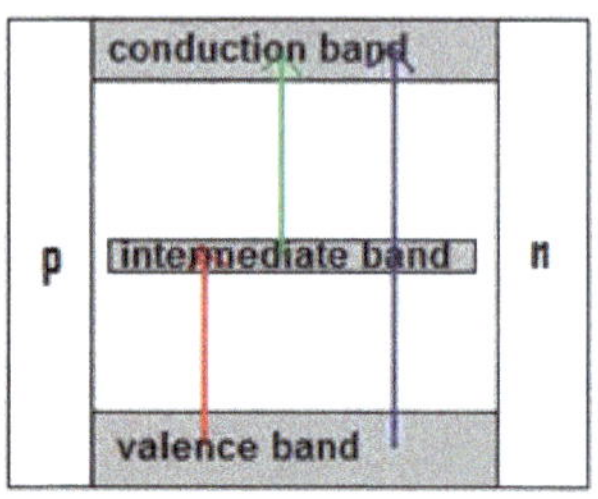

- Quantum dots can generate multiple exciton (electron-hole pairs) after collision with one photon.

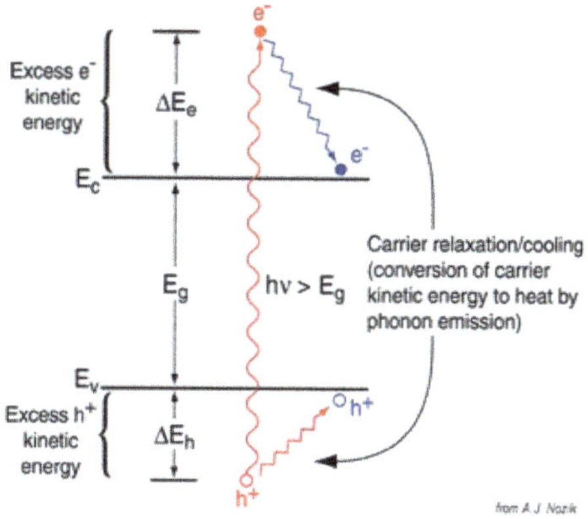

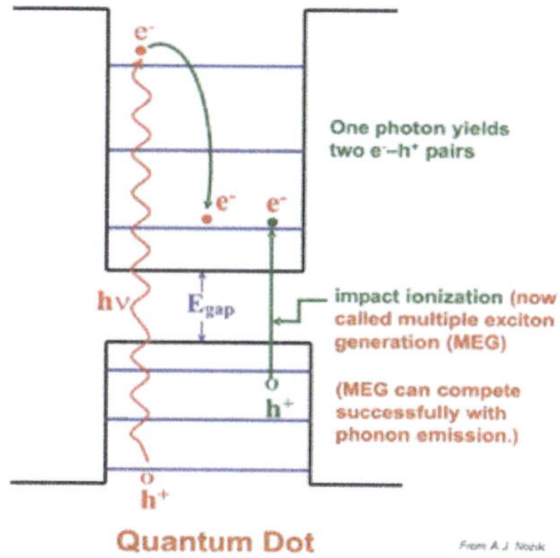

- Quantum dot:
  - o Semiconductor particle with a size in the order of the Bohr radius of the excitons.
  - o Energy levels depend on the size of the dot.
- Different methods for fabricating quantum dots.
  - o Lithography
  - o Colloidal synthesis
  - o Epitaxy
- Multiple applications.

References

- Knipping, J.; Wiggers, H.; Rellinghaus, B.; Roth, P.; Konjhodzic, D.; Meier, C. (2004). "Synthesis of high purity silicon nanoparticles in a low Pressure microwave reactor". Journal of Nanoscience and Technology. 4 (8): 1039–1044. doi:10.1166/jnn.2004.149
- Quantum Materials Corporation and the Access2Flow Consortium (2011). "Quantum materials corp achieves milestone in High Volume Production of Quantum Dots". Retrieved 7 July 2011
- Derfus, Austin M.; Chan, Warren C. W.; Bhatia, Sangeeta N. (2004-01-01). "Probing the Cytotoxicity of Semiconductor Quantum Dots". Nano Letters. 4 (1): 11–18. Bibcode:2004NanoL...4...11D. doi:10.1021/nl0347334
- Leatherdale, C. A.; Woo, W. -K.; Mikulec, F. V.; Bawendi, M. G. (2002). "On the Absorption Cross Section of CdSe Nanocrystal Quantum Dots". The Journal of Physical Chemistry B. 106 (31): 7619–7622. doi:10.1021/jp025698c
- MFTTech (24 March 2015). "LG Electronics Partners with Dow to Commercialize LGs New Ultra HD TV with Quantum Dot Technology". Retrieved 9 May 2015
- Spie (2014). "Paul Selvin Hot Topics presentation: New Small Quantum Dots for Neuroscience". SPIE Newsroom. doi:10.1117/2.3201403.17
- Marchuk, K.; Guo, Y.; Sun, W.; Vela, J.; Fang, N. (2012). "High-Precision Tracking with Non-blinking Quantum Dots Resolves Nanoscale Vertical Displacement". Journal of the American Chemical Society. 134 (14): 6108–11. doi:10.1021/ja301332t. PMID 22458433
- Lane, L. A.; Smith, A. M.; Lian, T.; Nie, S. (2014). "Compact and Blinking-Suppressed Quantum Dots for Single-Particle Tracking in Live Cells". The Journal of Physical Chemistry B. 118 (49): 14140–7. doi:10.1021/

jp5064325. PMC 4266335 . PMID 25157589

- shoori, R. C. (1996). "Electrons in artificial atoms". Nature. 379 (6564): 413–419. Bibcode:1996Natur.379..413A. doi:10.1038/379413a0
- Beane, Gary; Boldt, Klaus; Kirkwood, Nicholas; Mulvaney, Paul (7 August 2014). "Energy Transfer between Quantum Dots and Conjugated Dye Molecules". The Journal of Physical Chemistry C. 118 (31): 18079–18086. doi:10.1021/jp502033d
- Kastner, M. A. (1993). "Artificial Atoms". Physics Today. 46 (1): 24–31. Bibcode:1993PhT....46a..24K. doi:10.1063/1.881393
- Gupta, Vinay; Chaudhary, Neeraj; Srivastava, Ritu; Sharma, Gauri Datt; Bhardwaj, Ramil; Chand, Suresh (2011-07-06). "Luminscent Graphene Quantum Dots for Organic Photovoltaic Devices". Journal of the American Chemical Society. 133 (26): 9960–9963. doi:10.1021/ja2036749. PMID 21650464
- Kortshagen, U (2009). "Nonthermal plasma synthesis of semiconductor nanocrystals". J. Phys. D: Appl. Phys. 42 (11): 113001. Bibcode:2009JPhD...42k3001K. doi:10.1088/0022-3727/42/11/113001
- LaFave, T. Jr. (2013). "Correspondences between the classical electrostatic Thomson Problem and atomic electronic structure". Journal of Electrostatics. 71 (6): 1029–1035. doi:10.1016/j.elstat.2013.10.001
- Jawaid A.M.; Chattopadhyay S.; Wink D.J.; Page L.E.; Snee P.T. (2013). "A". ACS Nano. 7 (4): 3190–3197. doi:10.1021/nn305697q. PMID 23441602
- LaFave, T. Jr. (2011). "The discrete charge dielectric model of electrostatic energy". Journal of Electrostatics. 69 (5): 414–418. doi:10.1016/j.elstat.2013.10.001
- Hauser, Charlotte A. E.; Zhang, Shuguang (2010). "Peptides as biological semiconductors". Nature. 468 (7323): 516–517. Bibcode:2010Natur.468..516H. doi:10.1038/468516a. PMID 21107418
- Kolobkova, E. V.; Nikonorov, N. V.; Aseev, V. A. (2012). "Optical Technologies Silver Nanoclusters Influence on Formation of Quantum Dots in Fluorine Phosphate Glasses". Scientific and Technical Journal of Information Technologies, Mechanics and Optics. 5 (12)

6

Organic Light Emitting Diodes

A light emitting diode (LED) is a type of semiconductor that is used to produce light. A widely used type of light emitting diode is organic light emitting diodes (OLEDs). The light-creating part of an OLED is manufactured from a thin layer of organic compound. The aspects elucidated in this section are of vital importance, and provide a better understanding of organic light emitting diodes.

Light-emitting Diode

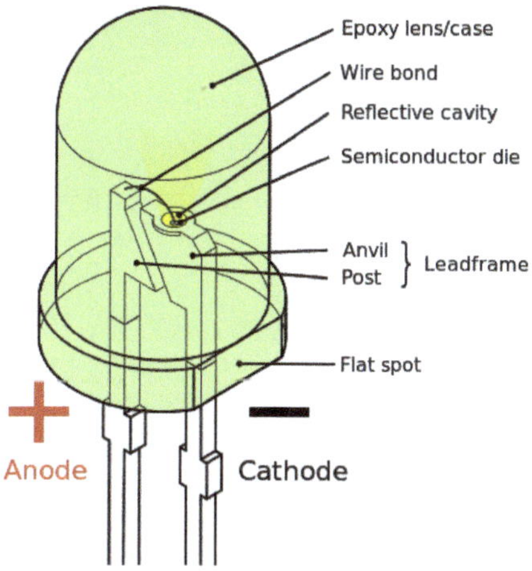

Parts of a conventional LED. The flat bottom surfaces of the anvil and post embedded inside the epoxy act as anchors, to prevent the conductors from being forcefully pulled out via mechanical strain or vibration.

A light-emitting diode (LED) is a two-lead semiconductor light source. It is a p–n junction diode that emits light when activated. When a suitable voltage is applied to the leads, electrons are able to recombine with electron holes within the device, releasing energy in the form of photons. This effect is called electroluminescence, and the color of the light (corresponding to the energy of the photon) is determined by the energy band gap of the semiconductor. LEDs are typically small (less than 1 mm^2) and integrated optical components may be used to shape the radiation pattern.

Appearing as practical electronic components in 1962, the earliest LEDs emitted low-intensity infrared light. Infrared LEDs are still frequently used as transmitting elements in remote-control circuits, such as those in remote controls for a wide variety of consumer electronics. The first visible-light LEDs were also of low intensity and limited to red. Modern LEDs are available across the visible, ultraviolet, and infrared wavelengths, with very high brightness.

Early LEDs were often used as indicator lamps for electronic devices, replacing small incandescent bulbs. They were soon packaged into numeric readouts in the form of seven-segment displays and were commonly seen in digital clocks. Recent developments have produced LEDs suitable for environmental and task lighting. LEDs have led to new displays and sensors, while their high switching rates are useful in advanced communications technology.

A bulb-shaped modern retrofit LED lamp with aluminium heat sink, a light diffusing dome and E27 screw base, using a built-in power supply working on mains voltage

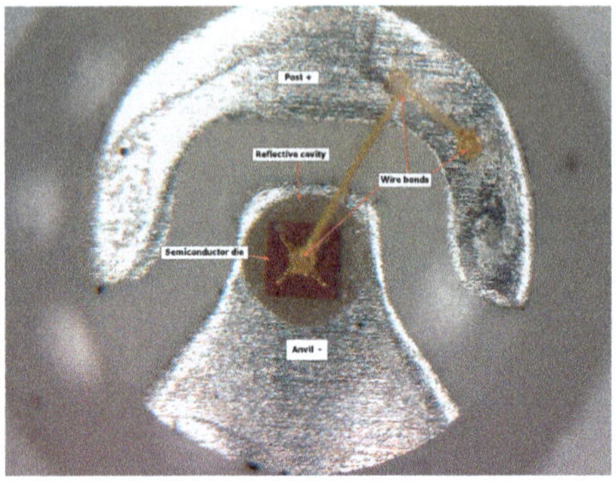

Close up image of a surface mount LED

LEDs have many advantages over incandescent light sources, including lower energy consumption, longer lifetime, improved physical robustness, smaller size, and faster switching. Light-emitting diodes are used in applications as diverse as aviation lighting, automotive headlamps, advertising, general lighting, traffic signals, camera flashes, and lighted wallpaper. As of 2017, LED lights home room lighting are as cheap or cheaper than compact fluorescent lamp sources of comparable output. They are also significantly more energy efficient and, arguably, have fewer environmental concerns linked to their disposal.

History

Discoveries and Early Devices

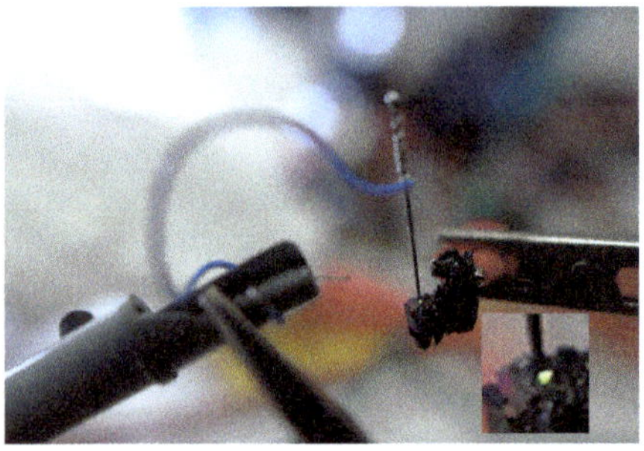

Green electroluminescence from a point contact on a crystal of SiC recreates Round's original experiment from 1907.

Electroluminescence as a phenomenon was discovered in 1907 by the British experimenter H. J. Round of Marconi Labs, using a crystal of silicon carbide and a cat's-whisker detector. Russian inventor Oleg Losev reported creation of the first LED in 1927. His research was distributed in Soviet, German and British scientific journals, but no practical use was made of the discovery for several decades. Kurt Lehovec, Carl Accardo, and Edward Jamgochian explained these first light-emitting diodes in 1951 using an apparatus employing SiC crystals with a current source of battery or pulse generator and with a comparison to a variant, pure, crystal in 1953.

Rubin Braunstein of the Radio Corporation of America reported on infrared emission from gallium arsenide (GaAs) and other semiconductor alloys in 1955. Braunstein observed infrared emission generated by simple diode structures using gallium antimonide (GaSb), GaAs, indium phosphide (InP), and silicon-germanium (SiGe) alloys at room temperature and at 77 Kelvin.

In 1957, Braunstein further demonstrated that the rudimentary devices could be used for non-radio communication across a short distance. As noted by Kroemer Braunstein "...had set up a simple optical communications link: Music emerging from a record player was used via suitable electronics to modulate the forward current of a GaAs diode. The emitted light was detected by a PbS diode some distance away. This signal was fed into an audio amplifier and played back by a loudspeaker. Intercepting the beam stopped the music. We had a great deal of fun playing with this setup." This setup presaged the use of LEDs for optical communication applications.

A Texas Instruments SNX-100 GaAs LED contained in a TO-18 transistor metal case.

In September 1961, while working at Texas Instruments in Dallas, Texas, James R. Biard and Gary Pittman discovered near-infrared (900 nm) light emission from a tunnel diode they had constructed on a GaAs substrate. By October 1961, they had demonstrated efficient light emission and signal coupling between a GaAs p-n junction light emitter and an electrically-isolated semiconductor photodetector. On August 8, 1962, Biard and Pittman filed a patent titled "Semiconductor Radiant Diode" based on their findings, which described a zinc diffused p–n junction LED with a spaced cathode contact to allow for efficient emission of infrared light under forward bias. After establishing the priority of their work based on engineering notebooks predating submissions from G.E. Labs, RCA Research Labs, IBM Research Labs, Bell Labs, and Lincoln Lab at MIT, the U.S. patent office issued the two inventors the patent for the GaAs infrared (IR) light-emitting diode, the first practical LED. Immediately after filing the patent, Texas Instruments (TI) began a project to manufacture infrared diodes. In October 1962, TI announced the first commercial LED product (the

SNX-100), which employed a pure GaAs crystal to emit a 890 nm light output. In October 1963, TI announced the first commercial hemispherical LED, the SNX-110.

The first visible-spectrum (red) LED was developed in 1962 by Nick Holonyak, Jr. while working at General Electric. Holonyak first reported his LED in the journal *Applied Physics Letters* on December 1, 1962. M. George Craford, a former graduate student of Holonyak, invented the first yellow LED and improved the brightness of red and red-orange LEDs by a factor of ten in 1972. In 1976, T. P. Pearsall created the first high-brightness, high-efficiency LEDs for optical fiber telecommunications by inventing new semiconductor materials specifically adapted to optical fiber transmission wavelengths.

Initial Commercial Development

The first commercial LEDs were commonly used as replacements for incandescent and neon indicator lamps, and in seven-segment displays, first in expensive equipment such as laboratory and electronics test equipment, then later in such appliances as TVs, radios, telephones, calculators, as well as watches. Until 1968, visible and infrared LEDs were extremely costly, in the order of US$200 per unit, and so had little practical use. The Monsanto Company was the first organization to mass-produce visible LEDs, using gallium arsenide phosphide (GaAsP) in 1968 to produce red LEDs suitable for indicators. Hewlett Packard (HP) introduced LEDs in 1968, initially using GaAsP supplied by Monsanto. These red LEDs were bright enough only for use as indicators, as the light output was not enough to illuminate an area. Readouts in calculators were so small that plastic lenses were built over each digit to make them legible. Later, other colors became widely available and appeared in appliances and equipment. In the 1970s commercially successful LED devices at less than five cents each were produced by Fairchild Optoelectronics. These devices employed compound semiconductor chips fabricated with the planar process invented by Dr. Jean Hoerni at Fairchild Semiconductor. The combination of planar processing for chip fabrication and innovative packaging methods enabled the team at Fairchild led by optoelectronics pioneer Thomas Brandt to achieve the needed cost reductions. LED producers continue to use these methods.

LED display of a TI-30 scientific calculator (ca. 1978), which uses plastic lenses to increase the visible digit size

Most LEDs were made in the very common 5 mm T1¾ and 3 mm T1 packages, but with rising power output, it has grown increasingly necessary to shed excess heat to maintain reliability, so more complex packages have been adapted for efficient heat dissipation. Packages for state-of-the-art high-power LEDs bear little resemblance to early LEDs.

Blue LED

Blue LEDs were first developed by Herbert Paul Maruska at RCA in 1972 using gallium nitride (GaN) on a sapphire substrate. SiC-types were first commercially sold in the United States by Cree in 1989. However, neither of these initial blue LEDs were very bright.

The first high-brightness blue LED was demonstrated by Shuji Nakamura of Nichia Corporation in 1994 and was based on InGaN. In parallel, Isamu Akasaki and Hiroshi Amano in Nagoya were working on developing the important GaN nucleation on sapphire substrates and the demonstration of p-type doping of GaN. Nakamura, Akasaki, and Amano were awarded the 2014 Nobel prize in physics for their work. In 1995, Alberto Barbieri at the Cardiff University Laboratory (GB) investigated the efficiency and reliability of high-brightness LEDs and demonstrated a "transparent contact" LED using indium tin oxide (ITO) on (AlGaInP/GaAs).

In 2001 and 2002, processes for growing gallium nitride (GaN) LEDs on silicon were successfully demonstrated. In January 2012, Osram demonstrated high-power InGaN LEDs grown on silicon substrates commercially.

White LEDs and the Illumination Breakthrough

The attainment of high efficiency in blue LEDs was quickly followed by the development of the first white LED. In this device a $Y_3Al_5O_{12}$:Ce (known as "YAG") phosphor coating on the emitter absorbs some of the blue emission and produces yellow light through fluorescence. The combination of that yellow with remaining blue light appears white to the eye. However, using different phosphors (fluorescent materials) it also became possible to instead produce green and red light through fluorescence. The resulting mixture of red, green and blue is not only perceived by humans as white light but is superior for illumination in terms of color rendering, whereas one cannot appreciate the color of red or green objects illuminated only by the yellow (and remaining blue) wavelengths from the YAG phosphor.

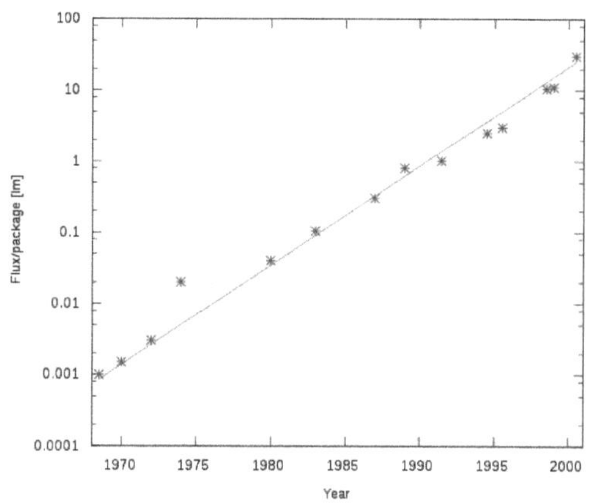

Illustration of Haitz's law, showing improvement in light output per LED over time, |with a logarithmic scale on the vertical axis

The first white LEDs were expensive and inefficient. However, the light output of LEDs has increased exponentially, with a doubling occurring approximately every 36 months since the 1960s (similar to Moore's law). This trend is generally attributed to the parallel development of other semiconductor technologies and advances in optics and materials science and has been called Haitz's law after Dr. Roland Haitz.

Light output and efficiency of blue and near-ultraviolet LEDs rose as the cost of reliable devices fell. This led to relatively high-power white-light LEDs for illumination, which are replacing incandescent and fluorescent lighting.

Experimental white LEDs have been demonstrated to produce over 300 lumens per watt of electricity; some can last up to 100,000 hours. Compared to incandescent bulbs, this is not only a huge increase in electrical efficiency but – over time – a similar or lower cost per bulb.

Working Principle

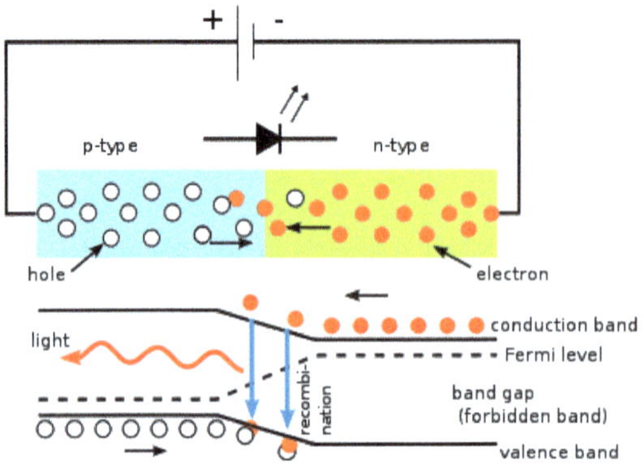

The inner workings of an LED, showing circuit (top) and band diagram (bottom)

A P-N junction can convert absorbed light energy into a proportional electric current. The same process is reversed here (i.e. the P-N junction emits light when electrical energy is applied to it). This phenomenon is generally called electroluminescence, which can be defined as the emission of light from a semi-conductor under the influence of an electric field. The charge carriers recombine in a forward-biased P-N junction as the electrons cross from the N-region and recombine with the holes existing in the P-region. Free electrons are in the conduction band of energy levels, while holes are in the valence energy band. Thus the energy level of the holes is less than the energy levels of the electrons. Some portion of the energy must be dissipated to recombine the electrons and the holes. This energy is emitted in the form of heat and light.

The electrons dissipate energy in the form of heat for silicon and germanium diodes but in gallium arsenide phosphide (GaAsP) and gallium phosphide (GaP) semiconductors, the electrons dissipate energy by emitting photons. If the semiconductor is translucent, the junction becomes the source of light as it is emitted, thus becoming a light-emitting diode. However, when the junction is reverse biased, the LED produces no light and—if the potential is great enough, the device is damaged.

Technology

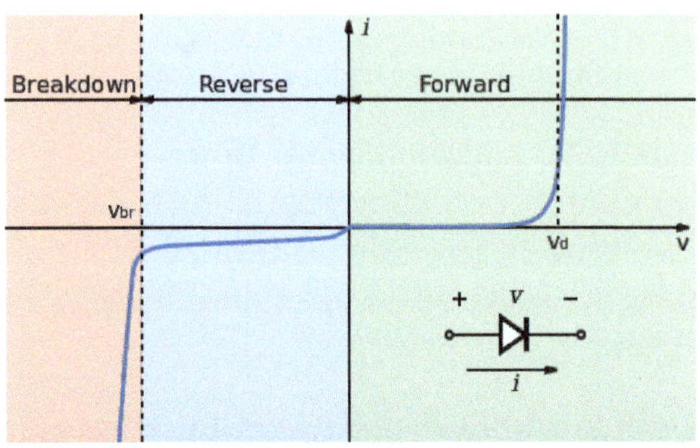

I-V diagram for a diode. An LED begins to emit light when more than 2 or 3 volts is applied. The reverse bias region uses a different vertical scale from the forward bias region to show that the leakage current is nearly constant with voltage until breakdown occurs. In forward bias, the current is small but increases exponentially with voltage.

Physics

The LED consists of a chip of semiconducting material doped with impurities to create a *p-n junction*. As in other diodes, current flows easily from the p-side, or anode, to the n-side, or cathode, but not in the reverse direction. Charge-carriers—electrons and holes—flow into the junction from electrodes with different voltages. When an electron meets a hole, it falls into a lower energy level and releases energy in the form of a photon.

The wavelength of the light emitted, and thus its color, depends on the band gap energy of the materials forming the *p-n junction*. In silicon or germanium diodes, the electrons and holes usually recombine by a *non-radiative transition*, which produces no optical emission, because these are indirect band gap materials. The materials used for the LED have a direct band gap with energies corresponding to near-infrared, visible, or near-ultraviolet light.

LED development began with infrared and red devices made with gallium arsenide. Advances in materials science have enabled making devices with ever-shorter wavelengths, emitting light in a variety of colors.

LEDs are usually built on an n-type substrate, with an electrode attached to the p-type layer deposited on its surface. P-type substrates, while less common, occur as well. Many commercial LEDs, especially GaN/InGaN, also use sapphire substrate.

Refractive Index

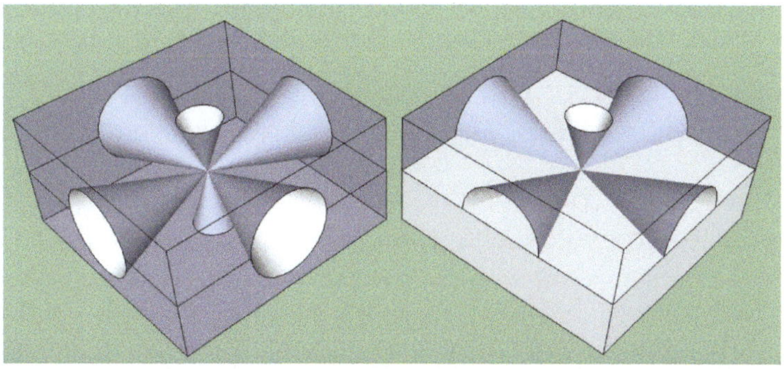

Idealized example of light emission cones in a simple square semiconductor, for a single point-source emission zone. The left illustration is for a translucent wafer, while the right illustration shows the half-cones formed when the bottom layer is opaque. The light is actually emitted equally in all directions from the point-source, but can only escape perpendicular to the semiconductor's surface and some degrees to the side, which is illustrated by the cone shapes. When the critical angle is exceeded, photons are reflected internally. The areas between the cones represent the trapped light energy wasted as heat.

Bare uncoated semiconductors such as silicon exhibit a very high refractive index relative to open air, which prevents passage of photons arriving at sharp angles relative to the air-contacting surface of the semiconductor due to total internal reflection. This property affects both the light-emission efficiency of LEDs as well as the light-absorption efficiency of photovoltaic cells. The refractive index of silicon is 3.96 (at 590 nm), while air is 1.0002926.

In general, a flat-surface uncoated LED semiconductor chip emits light only perpendicular to the semiconductor's surface, and a few degrees to the side, in a cone shape referred to as the *light cone*, *cone of light*, or the *escape cone*. The maximum angle of incidence is referred to as the critical angle. When this angle is exceeded, photons no longer escape the semiconductor but are instead reflected internally inside the semiconductor crystal as if it were a mirror.

Most materials used for LED production have very high refractive indices. This means that much of the light is reflected back into the material at the material/air surface interface. Thus, light extraction in LEDs is an important aspect of LED production, subject to much research and development. The light emission cones of a real LED wafer are far more complex than a single point-source light emission. The light emission zone is typically a two-dimensional plane between the wafers. Every atom across this plane has an individual set of emission cones. Drawing the billions of overlapping cones is impossible, so this is a simplified diagram showing the extents of all the emission cones combined. The larger side cones are clipped to show the interior features and reduce image complexity; they would extend to the opposite edges of the two-dimensional emission plane.

Internal reflections can escape through other crystalline faces if the incidence angle is low enough and the crystal is sufficiently transparent to not re-absorb the photon emission. But for a simple square LED with 90-degree angled surfaces on all sides, the faces all act as equal angle mirrors. In this case, most of the light can not escape and is lost as waste heat in the crystal.

A convoluted chip surface with angled facets similar to a jewel or fresnel lens can increase light output by distributing light perpendicular to the chip surface and far to the sides of the photon emission point.

The ideal shape of a semiconductor with maximum light output would be a microsphere with the photon emission occurring at the exact center, with electrodes penetrating to the center to contact at the emission point. All light rays emanating from the center would be perpendicular to the entire surface of the sphere, resulting in no internal reflections. A hemispherical semiconductor would also work, with the flat back-surface serving as a mirror to back-scattered photons.

Transition Coatings

After the doping of the wafer, it is cut apart into individual dies. Each die is commonly called a chip.

Many LED semiconductor chips are encapsulated or potted in clear or colored molded plastic shells. The plastic shell has three purposes:

1. Mounting the semiconductor chip in devices is easier to accomplish.
2. The tiny fragile electrical wiring is physically supported and protected from damage.

3. The plastic acts as a refractive intermediary between the relatively high-index semiconductor and low-index open air.

The third feature helps to boost the light emission from the semiconductor by acting as a diffusing lens, emitting light at a much higher angle of incidence from the light cone than the bare chip would alone.

Efficiency and Operational Parameters

Typical indicator LEDs are designed to operate with no more than 30–60 milliwatts (mW) of electrical power. Around 1999, Philips Lumileds introduced power LEDs capable of continuous use at one watt. These LEDs used much larger semiconductor die sizes to handle the large power inputs. Also, the semiconductor dies were mounted onto metal slugs to allow for heat removal from the LED die.

One of the key advantages of LED-based lighting sources is high luminous efficacy. White LEDs quickly matched and overtook the efficacy of standard incandescent lighting systems. In 2002, Lumileds made five-watt LEDs available with luminous efficacy of 18–22 lumens per watt (lm/W). For comparison, a conventional incandescent light bulb of 60–100 watts emits around 15 lm/W, and standard fluorescent lights emit up to 100 lm/W.

As of 2012, Philips had achieved the following efficacies for each color. The efficiency values show the physics – light power out per electrical power in. The lumen-per-watt efficacy value includes characteristics of the human eye and is derived using the luminosity function.

Color	Wavelength range (nm)	Typical efficiency coefficient	Typical efficacy (lm/W)
Red	$620 < \lambda < 645$	0.39	72
Red-orange	$610 < \lambda < 620$	0.29	98
Green	$520 < \lambda < 550$	0.15	93
Cyan	$490 < \lambda < 520$	0.26	75
Blue	$460 < \lambda < 490$	0.35	37

In September 2003, a new type of blue LED was demonstrated by Cree that consumes 24 mW at 20 milliamperes (mA). This produced a commercially packaged white light giving 65 lm/W at 20 mA, becoming the brightest white LED commercially available at the time, and more than four times as efficient as standard incandescents. In 2006, they demonstrated a prototype with a record white LED luminous efficacy of 131 lm/W at 20 mA. Nichia Corporation has developed a white LED with luminous efficacy of 150 lm/W at a forward current of 20 mA. Cree's XLamp XM-L LEDs, commercially available in 2011, produce 100 lm/W at their full power of 10 W, and up to 160 lm/W at around 2 W input power. In 2012, Cree announced a white LED giving 254 lm/W, and 303 lm/W in March 2014. Practical general lighting needs high-power LEDs, of one watt or more. Typical operating currents for such devices begin at 350 mA.

These efficiencies are for the light-emitting diode only, held at low temperature in a lab. Since LEDs installed in real fixtures operate at higher temperature and with driver losses, real-world efficiencies are much lower. United States Department of Energy (DOE) testing of commercial LED lamps designed to replace incandescent lamps or CFLs showed that average efficacy was still about 46 lm/W in 2009 (tested performance ranged from 17 lm/W to 79 lm/W).

Efficiency Droop

Efficiency droop is the decrease in luminous efficiency of LEDs as the electric current increases above tens of milliamperes.

This effect was initially theorized to be related to elevated temperatures. Scientists proved the opposite is true: though the life of an LED would be shortened, the efficiency droop is less severe at elevated temperatures. The mechanism causing efficiency droop was identified in 2007 as Auger recombination, which was taken with mixed reaction. In 2013, a study confirmed Auger recombination as the cause of efficiency droop.

In addition to being less efficient, operating LEDs at higher electric currents creates higher heat levels, which compromise LED lifetime. Because of this increased heat at higher currents, high-brightness LEDs have an industry standard of operating at only 350 mA, which is a compromise between light output, efficiency, and longevity.

Possible Solutions

Instead of increasing current levels, luminance is usually increased by combining multiple LEDs in one bulb. Solving the problem of efficiency droop would mean that household LED light bulbs would need fewer LEDs, which would significantly reduce costs.

Researchers at the U.S. Naval Research Laboratory have found a way to lessen the efficiency droop. They found that the droop arises from non-radiative Auger recombination of the injected carriers. They created quantum wells with a soft confinement potential to lessen the non-radiative Auger processes.

Researchers at Taiwan National Central University and Epistar Corp are developing a way to lessen the efficiency droop by using ceramic aluminium nitride (AlN) substrates, which are more thermally conductive than the commercially used sapphire. The higher thermal conductivity reduces self-heating effects.

Lifetime and Failure

Solid-state devices such as LEDs are subject to very limited wear and tear if operated at low currents and at low temperatures. Typical lifetimes quoted are 25,000 to 100,000 hours, but heat and current settings can extend or shorten this time significantly.

The most common symptom of LED (and diode laser) failure is the gradual lowering of light output and loss of efficiency. Sudden failures, although rare, can also occur. Early red LEDs were notable for their short service life. With the development of high-power LEDs, the devices are subjected to higher junction temperatures and higher current densities than traditional devices. This causes stress on the material and may cause early light-output degradation. To quantitify useful lifetime in a standardized manner, some suggest using L70 or L50, which are runtimes (typically in thousands of hours) at which a given LED reaches 70% and 50% of initial light output, respectively.

Whereas in most previous sources of light (incandescent lamps, discharge lamps, and those that burn combustible fuel, e.g. candles and oil lamps) the light results from heat, LEDs only operate if they are kept cool enough. The manufacturer commonly specifies a maximum junction temperature of 125 or 150 °C, and lower temperatures are advisable in the interests of long life. At these temperatures, relatively little heat is lost by radiation, which means that the light beam generated by an LED is cool.

The waste heat in a high-power LED (which as of 2015 can be less than half the power that it consumes) is conveyed by conduction through the substrate and package of the LED to a heat sink, which gives up the heat to the ambient air by convection. Careful thermal design is, therefore, essential, taking into account the thermal resistances of the LED's package, the heat sink and the interface between the two. Medium-power LEDs are often designed to solder directly to a printed circuit board that contains a thermally conductive metal layer. High-power LEDs are packaged in large-area ceramic packages that attach to a metal heat sink—the interface being a material with high thermal conductivity (thermal grease, phase-change material, thermally conductive pad, or thermal adhesive).

If an LED-based lamp is installed in an unventilated luminaire, or a luminaire is located in an environment that does not have free air circulation, the LED is likely to overheat, resulting in reduced life or early catastrophic failure. Thermal design is often based on an ambient temperature of 25 °C (77 °F). LEDs used in outdoor applications, such as traffic signals or in-pavement signal lights, and in climates where the temperature within the light fixture gets very high, could experience reduced output or even failure.

Since LED efficacy is higher at low temperatures, LED technology is well suited for supermarket freezer lighting. Because LEDs produce less waste heat than incandescent lamps, their use in freezers can save on refrigeration costs as well. However, they may be more susceptible to frost and snow buildup than incandescent lamps, so some LED lighting systems have been designed with an added heating circuit. Additionally, research has developed heat sink technologies that transfer heat produced within the junction to appropriate areas of the light fixture.

Colors and Materials

Conventional LEDs are made from a variety of inorganic semiconductor materials. The following table shows the available colors with wavelength range, voltage drop, and material:

Color	Wavelength [nm]	Voltage drop [ΔV]	Semiconductor material
Infrared	$\lambda > 760$	$\Delta V < 1.63$	Gallium arsenide (GaAs) Aluminium gallium arsenide (AlGaAs)
Red	$610 < \lambda < 760$	$1.63 < \Delta V < 2.03$	Aluminium gallium arsenide (AlGaAs) Gallium arsenide phosphide (GaAsP) Aluminium gallium indium phosphide (AlGaInP) Gallium(III) phosphide (GaP)
Orange	$590 < \lambda < 610$	$2.03 < \Delta V < 2.10$	Gallium arsenide phosphide (GaAsP) Aluminium gallium indium phosphide (AlGaInP) Gallium(III) phosphide (GaP)
Yellow	$570 < \lambda < 590$	$2.10 < \Delta V < 2.18$	Gallium arsenide phosphide (GaAsP) Aluminium gallium indium phosphide (AlGaInP) Gallium(III) phosphide (GaP)

Green	500 < λ < 570	1.9 < ΔV < 4.0	Traditional green: Gallium(III) phosphide (GaP) Aluminium gallium indium phosphide (AlGaInP) Aluminium gallium phosphide (AlGaP) Pure green: Indium gallium nitride (InGaN) / Gallium(III) nitride (GaN)
Blue	450 < λ < 500	2.48 < ΔV < 3.7	Zinc selenide (ZnSe) Indium gallium nitride (InGaN) Silicon carbide (SiC) as substrate Silicon (Si) as substrate—under development
Violet	400 < λ < 450	2.76 < ΔV < 4.0	Indium gallium nitride (InGaN)
Purple	Multiple types	2.48 < ΔV < 3.7	Dual blue/red LEDs, blue with red phosphor, or white with purple plastic
Ultravi-olet	λ < 400	3 < ΔV < 4.1	Indium gallium nitride (InGaN) (385-400 nm) Diamond (235 nm) Boron nitride (215 nm) Aluminium nitride (AlN) (210 nm) Aluminium gallium nitride (AlGaN) Aluminium gallium indium nitride (AlGaInN)—down to 210 nm
Pink	Multiple types	ΔV ≈ 3.3	Blue with one or two phosphor layers, yellow with red, orange or pink phosphor added afterwards, white with pink plastic, or white phosphors with pink pigment or dye over top.
White	Broad spectrum	2.8 < ΔV < 4.2	Cool / Pure White: Blue/UV diode with yellow phosphor Warm White: Blue diode with orange phosphor

Blue and Ultraviolet

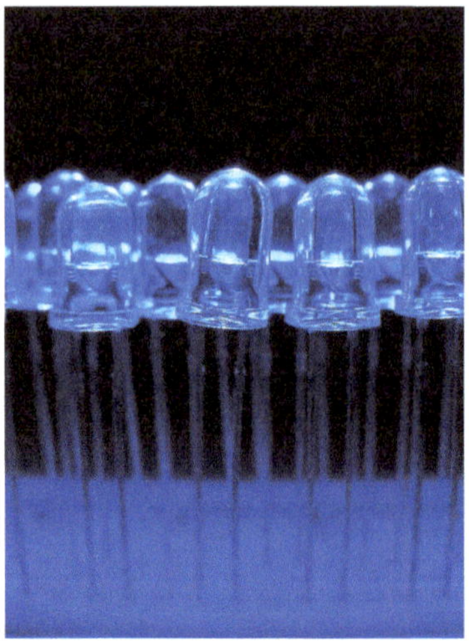

Blue LEDs

The first blue-violet LED using magnesium-doped gallium nitride was made at Stanford University in 1972 by Herb Maruska and Wally Rhines, doctoral students in materials science and engineering. At the time Maruska was on leave from RCA Laboratories, where he collaborated with Jacques Pankove on related work. In 1971, the year after Maruska left for Stanford, his RCA colleagues Pankove and Ed Miller demonstrated the first blue electroluminescence from zinc-doped gallium nitride, though the subsequent device Pankove and Miller built, the first actual gallium nitride light-emitting diode, emitted green light. In 1974 the U.S. Patent Office awarded Maruska, Rhines and Stanford professor David Stevenson a patent for their work in 1972 (U.S. Patent US3819974 A) and today, magnesium-doping of gallium nitride remains the basis for all commercial blue LEDs and laser diodes. In the early 1970s, these devices were too dim for practical use, and research into gallium nitride devices slowed. In August 1989, Cree introduced the first commercially available blue LED based on the indirect bandgap semiconductor, silicon carbide (SiC). SiC LEDs had very low efficiency, no more than about 0.03%, but did emit in the blue portion of the visible light spectrum.

In the late 1980s, key breakthroughs in GaN epitaxial growth and p-type doping ushered in the modern era of GaN-based optoelectronic devices. Building upon this foundation, Theodore Moustakas at Boston University patented a method for producing high-brightness blue LEDs using a new two-step process. Two years later, in 1993, high-brightness blue LEDs were demonstrated again by Shuji Nakamura of Nichia Corporation using a gallium nitride growth process similar to Moustakas's. Both Moustakas and Nakamura were issued separate patents, which confused the issue of who was the original inventor (partly because although Moustakas invented his first, Nakamura filed first). This new development revolutionized LED lighting, making high-power blue light sources practical, leading to the development of technologies like Blu-ray, as well as allowing the bright high-resolution screens of modern tablets and phones.

Nakamura was awarded the 2006 Millennium Technology Prize for his invention. Nakamura, Hiroshi Amano and Isamu Akasaki were awarded the Nobel Prize in Physics in 2014 for the invention of the blue LED. In 2015, a US court ruled that three companies (i.e. the litigants who had not previously settled out of court) that had licensed Nakamura's patents for production in the United States had infringed Moustakas's prior patent, and ordered them to pay licensing fees of not less than 13 million USD.

By the late 1990s, blue LEDs became widely available. They have an active region consisting of one or more InGaN quantum wells sandwiched between thicker layers of GaN, called cladding layers. By varying the relative In/Ga fraction in the InGaN quantum wells, the light emission can in theory be varied from violet to amber. Aluminium gallium nitride (AlGaN) of varying Al/Ga fraction can be used to manufacture the cladding and quantum well layers for ultraviolet LEDs, but these devices have not yet reached the level of efficiency and technological maturity of InGaN/GaN blue/green devices. If un-alloyed GaN is used in this case to form the active quantum well layers, the device emits near-ultraviolet light with a peak wavelength centred around 365 nm. Green LEDs manufactured from the InGaN/GaN system are far more efficient and brighter than green LEDs produced with non-nitride material systems, but practical devices still exhibit efficiency too low for high-brightness applications.

With nitrides containing aluminium, most often AlGaN and AlGaInN, even shorter wavelengths are achievable. Ultraviolet LEDs in a range of wavelengths are becoming available on the market. Near-UV emitters at wavelengths around 375–395 nm are already cheap and often encountered, for example, as black light lamp replacements for inspection of anti-counterfeiting UV watermarks in some documents and paper currencies. Shorter-wavelength diodes, while substantially more expensive, are commercially available for wavelengths down to 240 nm. As the photosensitivity of microorganisms approximately matches the absorption spectrum of DNA, with a peak at about 260 nm, UV LED emitting at 250–270 nm are to be expected in prospective disinfection and sterilization devices. Recent research has shown that commercially available UVA LEDs (365 nm) are already effective disinfection and sterilization devices. UV-C wavelengths were obtained in laboratories using aluminium nitride (210 nm), boron nitride (215 nm) and diamond (235 nm).

RGB

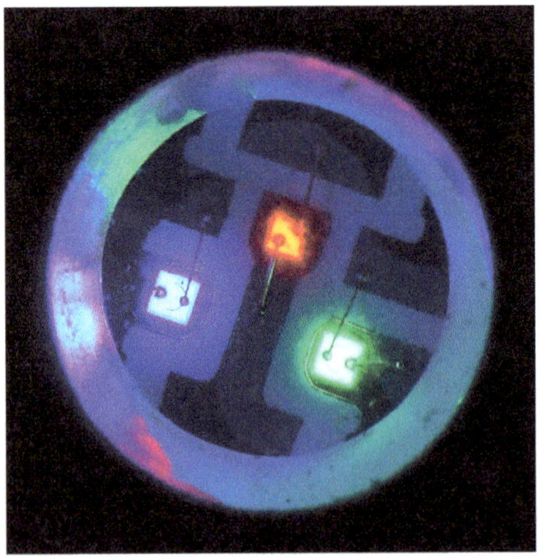

RGB-SMD-LED

RGB LEDs consist of one red, one green, and one blue LED. By independently adjusting each of the three, RGB LEDs are capable of producing a wide color gamut. Unlike dedicated-color LEDs, however, these obviously do not produce pure wavelengths. Moreover, such modules as commercially available are often not optimized for smooth color mixing.

White

There are two primary ways of producing white light-emitting diodes (WLEDs), LEDs that generate high-intensity white light. One is to use individual LEDs that emit three primary colors—red, green, and blue—and then mix all the colors to form white light. The other is to use a phosphor material to convert monochromatic light from a blue or UV LED to broad-spectrum white light, much in the same way a fluorescent light bulb works. It is important to note that the 'whiteness' of the light produced is essentially engineered to suit the human eye, and depending on the situation it may not always be appropriate to think of it as white light.

There are three main methods of mixing colors to produce white light from an LED:

- blue LED + green LED + red LED (color mixing; can be used as backlighting for displays, extremely poor for illumination due to gaps in spectrum)
- near-UV or UV LED + RGB phosphor (an LED producing light with a wavelength shorter than blue's is used to excite an RGB phosphor)
- blue LED + yellow phosphor (two complementary colors combine to form white light; more efficient than first two methods and more commonly used)

Because of metamerism, it is possible to have quite different spectra that appear white. However, the appearance of objects illuminated by that light may vary as the spectrum varies, this is the issue of Colour rendition, quite separate from Colour Temperature, where a really orange or cyan object could appear with the wrong colour and much darker as the LED or phosphor does not emit the wavelength. The best colour rendition CFL and LEDs use a mix of phosphors, resulting in less efficiency but better quality of light. Though incandescent halogen lamps have a more orange colour temperature, they are still the best easily available artificial light sources in terms of colour rendition.

RGB Systems

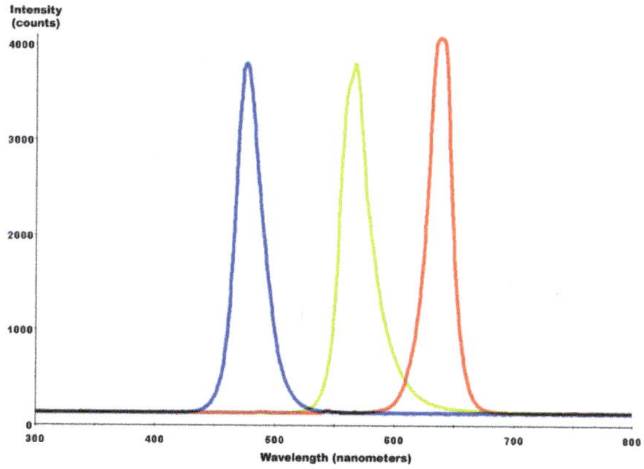

Combined spectral curves for blue, yellow-green, and high-brightness red solid-state semiconductor LEDs. FWHM spectral bandwidth is approximately 24–27 nm for all three colors.

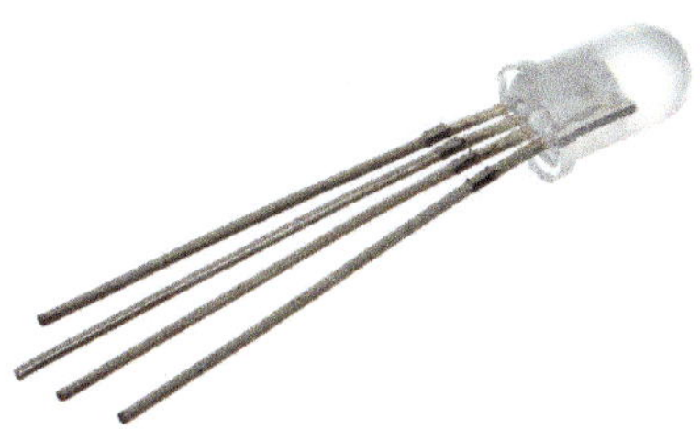

RGB LED

White light can be formed by mixing differently colored lights; the most common method is to use red, green, and blue (RGB). Hence the method is called multi-color white LEDs (sometimes referred to as RGB LEDs). Because these need electronic circuits to control the blending and diffusion of different colors, and because the individual color LEDs typically have slightly different emission patterns (leading to variation of the color depending on direction) even if they are made as a single unit, these are seldom used to produce white lighting. Nonetheless, this method has many applications because of the flexibility of mixing different colors, and in principle, this mechanism also has higher quantum efficiency in producing white light.

There are several types of multi-color white LEDs: di-, tri-, and tetrachromatic white LEDs. Several key factors that play among these different methods include color stability, color rendering capability, and luminous efficacy. Often, higher efficiency means lower color rendering, presenting a trade-off between the luminous efficacy and color rendering. For example, the dichromatic white

LEDs have the best luminous efficacy (120 lm/W), but the lowest color rendering capability. However, although tetrachromatic white LEDs have excellent color rendering capability, they often have poor luminous efficacy. Trichromatic white LEDs are in between, having both good luminous efficacy (>70 lm/W) and fair color rendering capability.

One of the challenges is the development of more efficient green LEDs. The theoretical maximum for green LEDs is 683 lumens per watt but as of 2010 few green LEDs exceed even 100 lumens per watt. The blue and red LEDs get closer to their theoretical limits.

Multi-color LEDs offer not merely another means to form white light but a new means to form light of different colors. Most perceivable colors can be formed by mixing different amounts of three primary colors. This allows precise dynamic color control. As more effort is devoted to investigating this method, multi-color LEDs should have profound influence on the fundamental method that we use to produce and control light color. However, before this type of LED can play a role on the market, several technical problems must be solved. These include that this type of LED's emission power decays exponentially with rising temperature, resulting in a substantial change in color stability. Such problems inhibit and may preclude industrial use. Thus, many new package designs aimed at solving this problem have been proposed and their results are now being reproduced by researchers and scientists. However multi-colour LEDs without phosphors can never provide good quality lighting because each LED is a narrow band source. LEDs without phosphor while a poorer solution for general lighting are the best solution for displays, either backlight of LCD, or direct LED based pixels.

Correlated color temperature (CCT) dimming for LED technology is regarded as a difficult task since binning, age and temperature drift effects of LEDs change the actual color value output. Feedback loop systems are used for example with color sensors, to actively monitor and control the color output of multiple color mixing LEDs.

Phosphor-based LEDs

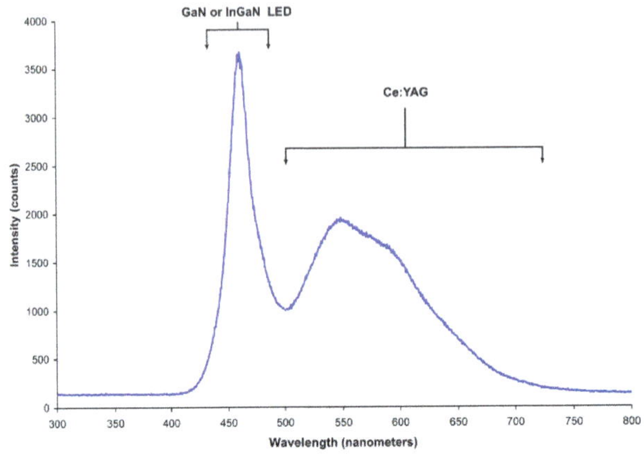

Spectrum of a white LED showing blue light directly emitted by the GaN-based LED (peak at about 465 nm) and the more broadband Stokes-shifted light emitted by the Ce^{3+}:YAG phosphor, which emits at roughly 500–700 nm

This method involves coating LEDs of one color (mostly blue LEDs made of InGaN) with phosphors of different colors to form white light; the resultant LEDs are called phosphor-based or phosphor-converted white LEDs (pcLEDs). A fraction of the blue light undergoes the Stokes shift being transformed from shorter wavelengths to longer. Depending on the color of the original LED, phosphors of different colors can be employed. If several phosphor layers of distinct colors are applied, the emitted spectrum is broadened, effectively raising the color rendering index (CRI) value of a given LED.

Phosphor-based LED efficiency losses are due to the heat loss from the Stokes shift and also other phosphor-related degradation issues. Their luminous efficacies compared to normal LEDs depend on the spectral distribution of the resultant light output and the original wavelength of the LED itself. For example, the luminous efficacy of a typical YAG yellow phosphor based white LED ranges from 3 to 5 times the luminous efficacy of the original blue LED because of the human eye's greater sensitivity to yellow than to blue (as modeled in the luminosity function). Due to the simplicity of manufacturing, the phosphor method is still the most popular method for making high-intensity white LEDs. The design and production of a light source or light fixture using a monochrome emitter with phosphor conversion is simpler and cheaper than a complex RGB system, and the majority of high-intensity white LEDs presently on the market are manufactured using phosphor light conversion.

Among the challenges being faced to improve the efficiency of LED-based white light sources is the development of more efficient phosphors. As of 2010, the most efficient yellow phosphor is still the YAG phosphor, with less than 10% Stokes shift loss. Losses attributable to internal optical losses due to re-absorption in the LED chip and in the LED packaging itself account typically for another 10% to 30% of efficiency loss. Currently, in the area of phosphor LED development, much effort is being spent on optimizing these devices to higher light output and higher operation temperatures. For instance, the efficiency can be raised by adapting better package design or by using a more suitable type of phosphor. Conformal coating process is frequently used to address the issue of varying phosphor thickness.

Some phosphor-based white LEDs encapsulate InGaN blue LEDs inside phosphor-coated epoxy. Alternatively, the LED might be paired with a remote phosphor, a preformed polycarbonate piece coated with the phosphor material. Remote phosphors provide more diffuse light, which is desirable for many applications. Remote phosphor designs are also more tolerant of variations in the LED emissions spectrum. A common yellow phosphor material is cerium-doped yttrium aluminium garnet (Ce^{3+}:YAG).

White LEDs can also be made by coating near-ultraviolet (NUV) LEDs with a mixture of high-efficiency europium-based phosphors that emit red and blue, plus copper and aluminium-doped zinc sulfide (ZnS:Cu, Al) that emits green. This is a method analogous to the way fluorescent lamps work. This method is less efficient than blue LEDs with YAG:Ce phosphor, as the Stokes shift is larger, so more energy is converted to heat, but yields light with better spectral characteristics, which render color better. Due to the higher radiative output of the ultraviolet LEDs than of the blue ones, both methods offer comparable brightness. A concern is that UV light may leak from a malfunctioning light source and cause harm to human eyes or skin.

Other White LEDs

Another method used to produce experimental white light LEDs used no phosphors at all and was based on homoepitaxially grown zinc selenide (ZnSe) on a ZnSe substrate that simultaneously emitted blue light from its active region and yellow light from the substrate.

A new style of wafers composed of gallium-nitride-on-silicon (GaN-on-Si) is being used to produce white LEDs using 200-mm silicon wafers. This avoids the typical costly sapphire substrate in relatively small 100- or 150-mm wafer sizes. The sapphire apparatus must be coupled with a mirror-like collector to reflect light that would otherwise be wasted. It is predicted that by 2020, 40% of all GaN LEDs will be made with GaN-on-Si. Manufacturing large sapphire material is difficult, while large silicon material is cheaper and more abundant. LED companies shifting from using sapphire to silicon should be a minimal investment.

Organic Light-emitting Diodes (OLEDs)

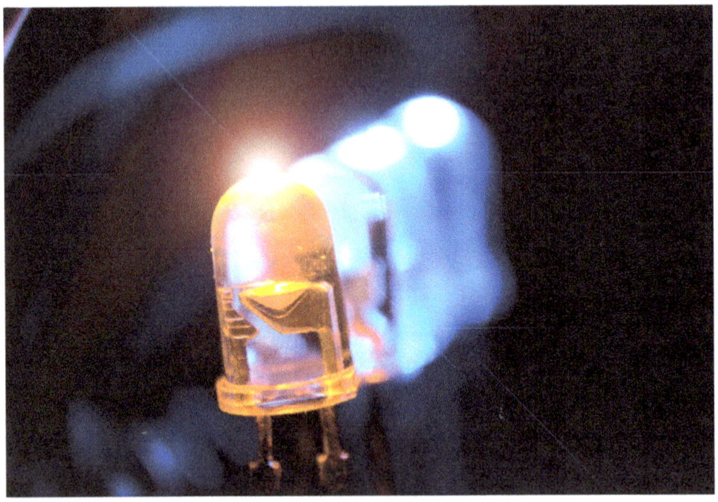

Orange light-emitting diode

In an organic light-emitting diode (OLED), the electroluminescent material comprising the emissive layer of the diode is an organic compound. The organic material is electrically conductive due to the delocalization of pi electrons caused by conjugation over all or part of the molecule, and the material therefore functions as an organic semiconductor. The organic materials can be small organic molecules in a crystalline phase, or polymers.

The potential advantages of OLEDs include thin, low-cost displays with a low driving voltage, wide viewing angle, and high contrast and color gamut. Polymer LEDs have the added benefit of printable and flexible displays. OLEDs have been used to make visual displays for portable electronic devices such as cellphones, digital cameras, and MP3 players while possible future uses include lighting and televisions.

Quantum Dot LEDs

Quantum dots (QD) are semiconductor nanocrystals with optical properties that let their emission color be tuned from the visible into the infrared spectrum. This allows quantum dot LEDs to create almost any color on the CIE diagram. This provides more color options and better color rendering than white LEDs since the emission spectrum is much narrower, characteristic of quantum confined states.

There are two types of schemes for QD excitation. One uses photo excitation with a primary light source LED (typically blue or UV LEDs are used). The other is direct electrical excitation first demonstrated by Alivisatos et al.

One example of the photo-excitation scheme is a method developed by Michael Bowers, at Vanderbilt University in Nashville, involving coating a blue LED with quantum dots that glow white in response to the blue light from the LED. This method emits a warm, yellowish-white light similar to that made by incandescent light bulbs. Quantum dots are also being considered for use in white light-emitting diodes in liquid crystal display (LCD) televisions.

In February 2011 scientists at PlasmaChem GmbH were able to synthesize quantum dots for LED applications and build a light converter on their basis, which was able to efficiently convert light from blue to any other color for many hundred hours. Such QDs can be used to emit visible or near infrared light of any wavelength being excited by light with a shorter wavelength.

The structure of QD-LEDs used for the electrical-excitation scheme is similar to basic design of OLEDs. A layer of quantum dots is sandwiched between layers of electron-transporting and hole-transporting materials. An applied electric field causes electrons and holes to move into the quantum dot layer and recombine forming an exciton that excites a QD. This scheme is commonly studied for quantum dot display. The tunability of emission wavelengths and narrow bandwidth is also beneficial as excitation sources for fluorescence imaging. Fluorescence near-field scanning optical microscopy (NSOM) utilizing an integrated QD-LED has been demonstrated.

In February 2008, a luminous efficacy of 300 lumens of visible light per watt of radiation (not per electrical watt) and warm-light emission was achieved by using nanocrystals.

Types

LEDs are produced in a variety of shapes and sizes. The color of the plastic lens is often the same as the actual color of light emitted, but not always. For instance, purple plastic is often used for infrared LEDs, and most blue devices have colorless housings. Modern high-power LEDs such as those used for lighting and backlighting are generally found in surface-mount technology (SMT) packages (not shown).

The main types of LEDs are miniature, high-power devices and custom designs such as alphanumeric or multi-color.

Miniature

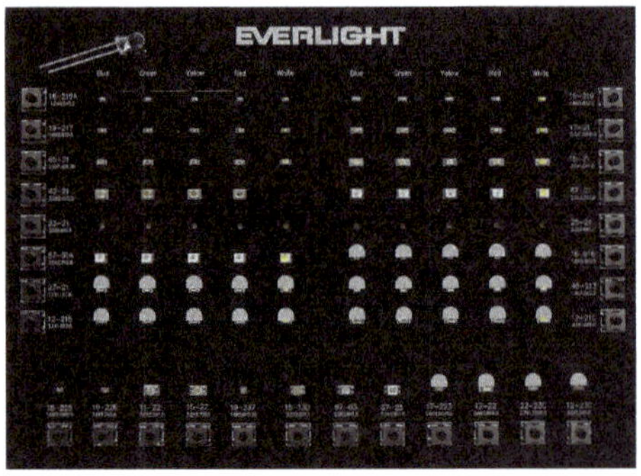

Photo of miniature surface mount LEDs in most common sizes. They can be much smaller than a traditional 5 mm lamp type LED, shown on the upper left corner.

Very small (1.6x1.6x0.35 mm) red, green, and blue surface mount miniature LED package with gold wire bonding details.

These are mostly single-die LEDs used as indicators, and they come in various sizes from 2 mm to 8 mm, through-hole and surface mount packages. They usually do not use a separate heat sink. Typical current ratings range from around 1 mA to above 20 mA. The small size sets a natural upper boundary on power consumption due to heat caused by the high current density and need for a heat sink. Often daisy chained as used in LED tapes.

Common package shapes include round, with a domed or flat top, rectangular with a flat top (as used in bar-graph displays), and triangular or square with a flat top. The encapsulation may also be clear or tinted to improve contrast and viewing angle.

Researchers at the University of Washington have invented the thinnest LED. It is made of two-dimensional (2-D) flexible materials. It is three atoms thick, which is 10 to 20 times thinner than three-dimensional (3-D) LEDs and is also 10,000 times smaller than the thickness of a human hair. These 2-D LEDs are going to make it possible to create smaller, more energy-efficient lighting, optical communication and nano lasers.

There are three main categories of miniature single die LEDs:

Low-current

Typically rated for 2 mA at around 2 V (approximately 4 mW consumption)

Standard

20 mA LEDs (ranging from approximately 40 mW to 90 mW) at around:

- 1.9 to 2.1 V for red, orange, yellow, and traditional green
- 3.0 to 3.4 V for pure green and blue
- 2.9 to 4.2 V for violet, pink, purple and white

Ultra-high-output

20 mA at approximately 2 or 4–5 V, designed for viewing in direct sunlight

5 V and 12 V LEDs are ordinary miniature LEDs that incorporate a suitable series resistor for direct connection to a 5 V or 12 V supply.

High-power

High-power light-emitting diodes attached to an LED star base (Luxeon, Lumileds)

High-power LEDs (HP-LEDs) or high-output LEDs (HO-LEDs) can be driven at currents from hundreds of mA to more than an ampere, compared with the tens of mA for other LEDs. Some can emit over a thousand lumens. LED power densities up to 300 W/cm^2 have been achieved. Since overheating is destructive, the HP-LEDs must be mounted on a heat sink to allow for heat dissipation. If the heat from an HP-LED is not removed, the device fails in seconds. One HP-LED can often replace an incandescent bulb in a flashlight, or be set in an array to form a powerful LED lamp.

Some well-known HP-LEDs in this category are the Nichia 19 series, Lumileds Rebel Led, Osram Opto Semiconductors Golden Dragon, and Cree X-lamp. As of September 2009, some HP-LEDs manufactured by Cree now exceed 105 lm/W.

Examples for Haitz's law—which predicts an exponential rise in light output and efficacy of LEDs over time—are the CREE XP-G series LED, which achieved 105 lm/W in 2009 and the Nichia 19 series with a typical efficacy of 140 lm/W, released in 2010.

AC Driven

LEDs developed by Seoul Semiconductor can operate on AC power without a DC converter. For each half-cycle, part of the LED emits light and part is dark, and this is reversed during the next half-cycle. The efficacy of this type of HP-LED is typically 40 lm/W. A large number of LED elements in series may be able to operate directly from line voltage. In 2009, Seoul Semiconductor released a high DC voltage LED, named as 'Acrich MJT', capable of being driven from AC power with a simple controlling circuit. The low-power dissipation of these LEDs affords them more flexibility than the original AC LED design.

Application-specific Variations

Flashing

Flashing LEDs are used as attention seeking indicators without requiring external electronics. Flashing LEDs resemble standard LEDs but they contain an integrated multivibrator circuit that causes the LED to flash with a typical period of one second. In diffused lens LEDs, this circuit is visible as a small black dot. Most flashing LEDs emit light of one color, but more sophisticated devices can flash between multiple colors and even fade through a color sequence using RGB color mixing.

Bi-color

Bi-color LEDs contain two different LED emitters in one case. There are two types of these. One type consists of two dies connected to the same two leads antiparallel to each other. Current flow in one direction emits one color, and current in the opposite direction emits the other color. The other type consists of two dies with separate leads for both dies and another lead for common anode or cathode so that they can be controlled independently. The most common bi-color combination is red/traditional green, however, other available combinations include amber/traditional green, red/pure green, red/blue, and blue/pure green.

Tri-color

Tri-color LEDs contain three different LED emitters in one case. Each emitter is connected to a separate lead so they can be controlled independently. A four-lead arrangement is typical with one common lead (anode or cathode) and an additional lead for each color.

RGB

RGB LEDs are tri-color LEDs with red, green, and blue emitters, in general using a four-wire connection with one common lead (anode or cathode). These LEDs can have either common positive or common negative leads. Others, however, have only two leads (positive and negative) and have a built-in tiny electronic control unit.

Decorative-multicolor

Decorative-multicolor LEDs incorporate several emitters of different colors supplied by only two lead-out wires. Colors are switched internally by varying the supply voltage.

Alphanumeric

Alphanumeric LEDs are available in seven-segment, starburst, and dot-matrix format. Seven-segment displays handle all numbers and a limited set of letters. Starburst displays can display all letters. Dot-matrix displays typically use 5x7 pixels per character. Seven-segment LED displays were in widespread use in the 1970s and 1980s, but rising use of liquid crystal displays, with their lower power needs and greater display flexibility, has reduced the popularity of numeric and alphanumeric LED displays.

Digital-RGB

Digital-RGB LEDs are RGB LEDs that contain their own "smart" control electronics. In addition to power and ground, these provide connections for data-in, data-out, and sometimes a clock or strobe signal. These are connected in a daisy chain, with the data in of the first LED sourced by a microprocessor, which can control the brightness and color of each LED independently of the others. They are used where a combination of maximum control and minimum visible electronics are needed such as strings for Christmas and LED matrices. Some even have refresh rates in the kHz range, allowing for basic video applications.

Filament

An LED filament consists of multiple LED chips connected in series on a common longitudinal substrate that forms a thin rod reminiscent of a traditional incandescent filament. These are being used as a low-cost decorative alternative for traditional light bulbs that are being phased out in many countries. The filaments require a rather high voltage to light to nominal brightness, allowing them to work efficiently and simply with mains voltages. Often a simple rectifier and capacitive current limiting are employed to create a low-cost replacement for a traditional light bulb without the complexity of the low voltage, high current converter that single die LEDs need. Usually, they are packaged in a sealed enclosure with a shape similar to lamps they were designed to replace (e.g. a bulb) and filled with inert nitrogen or carbon dioxide gas to remove heat efficiently.

Considerations for use

Power Sources

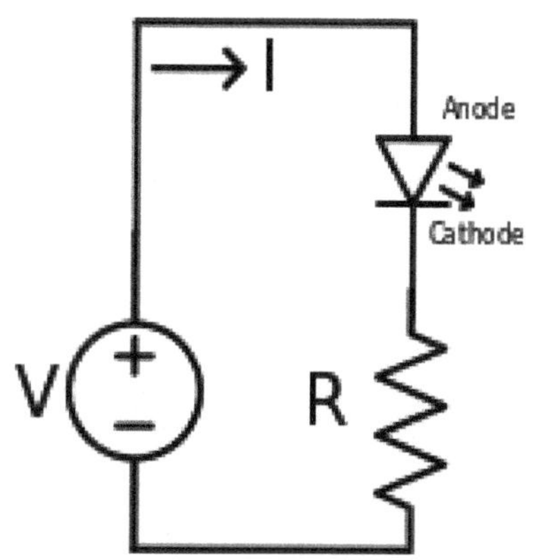

Simple LED circuit with resistor for current limiting

The current–voltage characteristic of an LED is similar to other diodes, in that the current is dependent exponentially on the voltage. This means that a small change in voltage can cause a large change in current. If the applied voltage exceeds the LED's forward voltage drop by a small amount, the current rating may be exceeded by a large amount, potentially damaging or destroying the LED. The typical solution is to use constant-current power supplies to keep the current below the LED's maximum current rating. Since most common power sources (batteries, mains) are constant-voltage sources, most LED fixtures must include a power converter, at least a current-limiting resistor. However, the high resistance of three-volt coin cells combined with the high differential resistance of nitride-based LEDs makes it possible to power such an LED from such a coin cell without an external resistor.

Electrical Polarity

As with all diodes, current flows easily from p-type to n-type material. However, no current flows and no light is emitted if a small voltage is applied in the reverse direction. If the reverse voltage grows large enough to exceed the breakdown voltage, a large current flows and the LED may be damaged. If the reverse current is sufficiently limited to avoid damage, the reverse-conducting LED is a useful noise diode.

Safety and Health

The vast majority of devices containing LEDs are "safe under all conditions of normal use", and so are classified as "Class 1 LED product"/"LED Klasse 1". At present, only a few LEDs—extremely bright LEDs that also have a tightly focused viewing angle of 8° or less—could, in theory, cause temporary blindness, and so are classified as "Class 2". The opinion of the French Agency for Food, Environmental and Occupational Health & Safety (ANSES) of 2010, on the health issues concerning LEDs, suggested banning public use of lamps in the moderate Risk Group 2, especially those with a high blue component, in places frequented by children.

In general, laser safety regulations—and the "Class 1", "Class 2", etc. system—also apply to LEDs.

While LEDs have the advantage over fluorescent lamps that they do not contain mercury, they may contain other hazardous metals such as lead and arsenic. Regarding the toxicity of LEDs when treated as waste, a study published in 2011 stated: "According to federal standards, LEDs are not hazardous except for low-intensity red LEDs, which leached Pb [lead] at levels exceeding regulatory limits (186 mg/L; regulatory limit: 5). However, according to California regulations, excessive levels of copper (up to 3892 mg/kg; limit: 2500), lead (up to 8103 mg/kg; limit: 1000), nickel (up to 4797 mg/kg; limit: 2000), or silver (up to 721 mg/kg; limit: 500) render all except low-intensity yellow LEDs hazardous."

In 2016 a statement of the American Medical Association (AMA) concerning the possible influence of blueish street lighting on the sleep-wake cycle of city-dwellers led to some controversy. So far high-pressure sodium lamps (HPS) with an orange light spectrum were the most efficient light sources commonly used in street-lighting. Now many modern street lamps are equipped with Indium gallium nitride LEDs (InGaN). These are even more efficient and mostly emit blue-rich light with a higher *correlated color temperature (CCT)*. Since light with a high CCT resembles daylight it is thought that this might have an effect on the normal circadian physiology by suppressing melatonin production in the human body. There have been no relevant studies as yet and critics claim exposure levels are not high enough to have a noticeable effect.

Advantages

- Efficiency: LEDs emit more lumens per watt than incandescent light bulbs. The efficiency of LED lighting fixtures is not affected by shape and size, unlike fluorescent light bulbs or tubes.
- Color: LEDs can emit light of an intended color without using any color filters as traditional lighting methods need. This is more efficient and can lower initial costs.
- Size: LEDs can be very small (smaller than 2 mm^2) and are easily attached to printed circuit boards.
- Warmup time: LEDs light up very quickly. A typical red indicator LED achieves full brightness in under a microsecond. LEDs used in communications devices can have even faster response times.
- Cycling: LEDs are ideal for uses subject to frequent on-off cycling, unlike incandescent and fluorescent lamps that fail faster when cycled often, or high-intensity discharge lamps (HID lamps) that require a long time before restarting.
- Dimming: LEDs can very easily be dimmed either by pulse-width modulation or lowering the forward current. This pulse-width modulation is why LED lights, particularly headlights on cars, when viewed on camera or by some people, appear to be flashing or flickering. This is a type of stroboscopic effect.
- Cool light: In contrast to most light sources, LEDs radiate very little heat in the form of IR that can cause damage to sensitive objects or fabrics. Wasted energy is dispersed as heat through the base of the LED.
- Slow failure: LEDs mostly fail by dimming over time, rather than the abrupt failure of incandescent bulbs.
- Lifetime: LEDs can have a relatively long useful life. One report estimates 35,000 to 50,000 hours of useful life, though time to complete failure may be longer. Fluorescent tubes typically are rated at about 10,000 to 15,000 hours, depending partly on the conditions of use, and incandescent light bulbs at 1,000 to 2,000 hours. Several DOE demonstrations have shown that reduced maintenance costs from this extended lifetime, rather than energy savings, is the primary factor in determining the payback period for an LED product.
- Shock resistance: LEDs, being solid-state components, are difficult to damage with external shock, unlike fluorescent and incandescent bulbs, which are fragile.
- Focus: The solid package of the LED can be designed to focus its light. Incandescent and fluorescent sources often require an external reflector to collect light and direct it in a usable manner. For larger LED packages total internal reflection (TIR) lenses are often used to the same effect. However, when large quantities of light are needed many light sources are usually deployed, which are difficult to focus or collimate towards the same target.

Disadvantages

- Initial price: LEDs are currently slightly more expensive (price per lumen) on an initial capital cost basis, than other lighting technologies. As of March 2014, at least one manufacturer claims to have reached $1 per kilolumen. The additional expense partially stems from the relatively low lumen output and the drive circuitry and power supplies needed.
- Temperature dependence: LED performance largely depends on the ambient temperature of the operating environment – or thermal management properties. Overdriving an LED in high ambient temperatures may result in overheating the LED package, eventually leading to device failure. An adequate heat sink is needed to maintain long life. This is especially important in automotive, medical, and military uses where devices must operate over a wide range of temperatures, which require low failure rates. Toshiba has produced LEDs with an operating temperature range of −40 to 100 °C, which suits the LEDs for both indoor and outdoor use in applications such as lamps, ceiling lighting, street lights, and floodlights.
- Voltage sensitivity: LEDs must be supplied with a voltage above their threshold voltage and a current below their rating. Current and lifetime change greatly with a small change in applied voltage. They thus require a current-regulated supply (usually just a series resistor for indicator LEDs).
- Color rendition: Most cool-white LEDs have spectra that differ significantly from a black body radiator like the sun or an incandescent light. The spike at 460 nm and dip at 500 nm can cause the color of objects to be perceived differently under cool-white LED illumination than sunlight or incandescent sources, due to metamerism, red surfaces being rendered particularly poorly by typical phosphor-based cool-white LEDs.
- Area light source: Single LEDs do not approximate a point source of light giving a spherical light distribution, but rather a lambertian distribution. So LEDs are difficult to apply to uses needing a spherical light field; however, different fields of light can be manipulated by the application of different optics or "lenses". LEDs cannot provide divergence below a few degrees. In contrast, lasers can emit beams with divergences of 0.2 degrees or less.
- Electrical polarity: Unlike incandescent light bulbs, which illuminate regardless of the electrical polarity, LEDs only light with correct electrical polarity. To automatically match source polarity to LED devices, rectifiers can be used.
- Blue hazard: There is a concern that blue LEDs and cool-white LEDs are now capable of exceeding safe limits of the so-called blue-light hazard as defined in eye safety specifications such as ANSI/IESNA RP-27.1–05: Recommended Practice for Photobiological Safety for Lamp and Lamp Systems.
- Light pollution: Because white LEDs, especially those with high color temperature, emit much more short wavelength light than conventional outdoor light sources such as high-pressure sodium vapor lamps, the increased blue and green sensitivity of scotopic vision means that white LEDs used in outdoor lighting cause substantially more sky glow. The American Medical Association warned on the use of high blue content white LEDs in street lighting, due to

their higher impact on human health and environment, compared to low blue content light sources (e.g. High-Pressure Sodium, PC amber LEDs, and low CCT LEDs).

- Efficiency droop: The efficiency of LEDs decreases as the electric current increases. Heating also increases with higher currents, which compromises LED lifetime. These effects put practical limits on the current through an LED in high power applications.
- Impact on insects: LEDs are much more attractive to insects than sodium-vapor lights, so much so that there has been speculative concern about the possibility of disruption to food webs.
- Use in winter conditions: Since they do not give off much heat in comparison to incandescent lights, LED lights used for traffic control can have snow obscuring them, leading to accidents.

Applications

LED uses fall into four major categories:

- Visual signals where light goes more or less directly from the source to the human eye, to convey a message or meaning
- Illumination where light is reflected from objects to give visual response of these objects
- Measuring and interacting with processes involving no human vision
- Narrow band light sensors where LEDs operate in a reverse-bias mode and respond to incident light, instead of emitting light

Indicators and Signs

The low energy consumption, low maintenance and small size of LEDs has led to uses as status indicators and displays on a variety of equipment and installations. Large-area LED displays are used as stadium displays, dynamic decorative displays, and dynamic message signs on freeways. Thin, lightweight message displays are used at airports and railway stations, and as destination displays for trains, buses, trams, and ferries.

Red and green LED traffic signals

One-color light is well suited for traffic lights and signals, exit signs, emergency vehicle lighting, ships' navigation lights or lanterns (chromacity and luminance standards being set under the Convention on the International Regulations for Preventing Collisions at Sea 1972, Annex I and the CIE) and LED-based Christmas lights. In cold climates, LED traffic lights may remain snow-covered. Red or yellow LEDs are used in indicator and alphanumeric displays in environments where night vision must be retained: aircraft cockpits, submarine and ship bridges, astronomy observatories, and in the field, e.g. night time animal watching and military field use.

Automotive applications for LEDs continue to grow.

Because of their long life, fast switching times, and visibility in broad daylight due to their high output and focus, LEDs have been used in brake lights for cars' high-mounted brake lights, trucks, and buses, and in turn signals for some time. However, many vehicles now use LEDs for their rear light clusters. The use in brakes improves safety, due to a great reduction in the time needed to light fully, or faster rise time, up to 0.5 second faster than an incandescent bulb. This gives drivers behind more time to react. In a dual intensity circuit (rear markers and brakes) if the LEDs are not pulsed at a fast enough frequency, they can create a phantom array, where ghost images of the LED appear if the eyes quickly scan across the array. White LED headlamps are beginning to appear. Using LEDs has styling advantages because LEDs can form much thinner lights than incandescent lamps with parabolic reflectors.

Due to the relative cheapness of low output LEDs, they are also used in many temporary uses such as glowsticks, throwies, and the photonic textile Lumalive. Artists have also used LEDs for LED art.

Weather and all-hazards radio receivers with Specific Area Message Encoding (SAME) have three LEDs: red for warnings, orange for watches, and yellow for advisories and statements whenever issued.

Lighting

With the development of high-efficiency and high-power LEDs, it has become possible to use LEDs in lighting and illumination. To encourage the shift to LED lamps and other high-efficiency lighting, the US Department of Energy has created the L Prize competition. The Philips Lighting North America LED bulb won the first competition on August 3, 2011, after successfully completing 18 months of intensive field, lab, and product testing.

LEDs are used as street lights and in other architectural lighting. The mechanical robustness and long lifetime are used in automotive lighting on cars, motorcycles, and bicycle lights. LED light emission may be efficiently controlled by using nonimaging optics principles.

LED street lights are employed on poles and in parking garages. In 2007, the Italian village of Torraca was the first place to convert its entire illumination system to LEDs.

LEDs are used in aviation lighting. Airbus has used LED lighting in its Airbus A320 Enhanced since 2007, and Boeing uses LED lighting in the 787. LEDs are also being used now in airport and heliport lighting. LED airport fixtures currently include medium-intensity runway lights, runway centerline lights, taxiway centerline and edge lights, guidance signs, and obstruction lighting.

LEDs are also used as a light source for DLP projectors, and to backlight LCD televisions (referred to as LED TVs) and laptop displays. RGB LEDs raise the color gamut by as much as 45%. Screens for TV and computer displays can be made thinner using LEDs for backlighting.

The lack of IR or heat radiation makes LEDs ideal for stage lights using banks of RGB LEDs that can easily change color and decrease heating from traditional stage lighting, as well as medical lighting where IR-radiation can be harmful. In energy conservation, the lower heat output of LEDs also means air conditioning (cooling) systems have less heat in need of disposal.

LEDs are small, durable and need little power, so they are used in handheld devices such as flashlights. LED strobe lights or camera flashes operate at a safe, low voltage, instead of the 250+ volts commonly found in xenon flashlamp-based lighting. This is especially useful in cameras on mobile phones, where space is at a premium and bulky voltage-raising circuitry is undesirable.

LEDs are used for infrared illumination in night vision uses including security cameras. A ring of LEDs around a video camera, aimed forward into a retroreflective background, allows chroma keying in video productions.

LED for miners, to increase visibility inside mines

LEDs are used in mining operations, as cap lamps to provide light for miners. Research has been done to improve LEDs for mining, to reduce glare and to increase illumination, reducing risk of injury to the miners.

LEDs are now used commonly in all market areas from commercial to home use: standard lighting, AV, stage, theatrical, architectural, and public installations, and wherever artificial light is used.

LEDs are increasingly finding uses in medical and educational applications, for example as mood enhancement, and new technologies such as AmBX, exploiting LED versatility. NASA has even sponsored research for the use of LEDs to promote health for astronauts.

Data Communication and Other Signalling

Light can be used to transmit data and analog signals. For example, lighting white LEDs can be used in systems assisting people to navigate in closed spaces while searching necessary rooms or objects.

Assistive listening devices in many theaters and similar spaces use arrays of infrared LEDs to send sound to listeners' receivers. Light-emitting diodes (as well as semiconductor lasers) are used to send data over many types of fiber optic cable, from digital audio over TOSLINK cables to the very high bandwidth fiber links that form the Internet backbone. For some time, computers were commonly equipped with IrDA interfaces, which allowed them to send and receive data to nearby machines via infrared.

Because LEDs can cycle on and off millions of times per second, very high data bandwidth can be achieved.

Sustainable Lighting

Efficient lighting is needed for sustainable architecture. In 2009, US Department of Energy testing results on LED lamps showed an average efficacy of 35 lm/W, below that of typical CFLs, and as low as 9 lm/W, worse than standard incandescent bulbs. A typical 13-watt LED lamp emitted 450 to 650 lumens, which is equivalent to a standard 40-watt incandescent bulb.

However, as of 2011, there are LED bulbs available as efficient as 150 lm/W and even inexpensive low-end models typically exceed 50 lm/W, so that a 6-watt LED could achieve the same results as a standard 40-watt incandescent bulb. The latter has an expected lifespan of 1,000 hours, whereas an LED can continue to operate with reduced efficiency for more than 50,000 hours.

See the chart below for a comparison of common light types:

	LED	**CFL**	**Incandescent**
Lightbulb Projected Lifespan	50,000 hours	10,000 hours	1,200 hours
Watts Per Bulb (equiv. 60 watts)	10	14	60
Cost Per Bulb	$2.00	$7.00	$1.25
kWh of Electricity Used Over 50,000 Hours	500	700	3000
Cost of Electricity (@ 0.10 per kWh)	$50	$70	$300
Bulbs Needed for 50,000 Hours of Use	1	5	42
Equivalent 50,000 Hours Bulb Expense	$2.00	$35.00	$52.50
TOTAL Cost for 50,000 Hours	$52.00	$105.00	$352.50

Energy Consumption

In the US, one kilowatt-hour (3.6 MJ) of electricity currently causes an average 1.34 pounds (610 g) of CO_2 emission. Assuming the average light bulb is on for 10 hours a day, a 40-watt bulb causes 196 pounds (89 kg) of CO_2 emission per year. The 6-watt LED equivalent only causes 30 pounds (14 kg) of CO_2 over the same time span. A building's carbon footprint from lighting can, therefore, be reduced by 85% by exchanging all incandescent bulbs for new LEDs—if a building previously used only incandescent bulbs.

In practice, most buildings that use a lot of lighting use fluorescent lighting, which has 22% luminous efficiency compared with 5% for filaments, so changing to LED lighting would still give a 34% reduction in electrical power use and carbon emissions.

The reduction in carbon emissions depends on the source of electricity. Nuclear power in the United States produced 19.2% of electricity in 2011, so reducing electricity consumption in the U.S. reduces carbon emissions more than in France (75% nuclear electricity) or Norway (almost entirely hydroelectric).

Replacing lights that spend the most time lit results in the most savings, so LED lights in infrequently used locations bring a smaller return on investment.

Light Sources for Machine Vision Systems

Machine vision systems often require bright and homogeneous illumination, so features of interest are easier to process. LEDs are often used for this purpose, and this is likely to remain one of their major uses until the price drops low enough to make signaling and illumination uses more widespread. Barcode scanners are the most common example of machine vision, and many low-cost products use red LEDs instead of lasers. Optical computer mice are an example of LEDs in machine vision, as it is used to provide an even light source on the surface for the miniature camera within the mouse. LEDs constitute a nearly ideal light source for machine vision systems for several reasons:

- The size of the illuminated field is usually comparatively small and machine vision systems are often quite expensive, so the cost of the light source is usually a minor concern. However, it might not be easy to replace a broken light source placed within complex machinery, and here the long service life of LEDs is a benefit.
- LED elements tend to be small and can be placed with high density over flat or even-shaped substrates (PCBs etc.) so that bright and homogeneous sources that direct light from tightly controlled directions on inspected parts can be designed. This can often be obtained with small, low-cost lenses and diffusers, helping to achieve high light densities with control over lighting levels and homogeneity. LED sources can be shaped in several configurations (spot lights for reflective illumination; ring lights for coaxial illumination; backlights for contour illumination; linear assemblies; flat, large format panels; dome sources for diffused, omnidirectional illumination).
- LEDs can be easily strobed (in the microsecond range and below) and synchronized with imaging. High-power LEDs are available allowing well-lit images even with very short light pulses. This is often used to obtain crisp and sharp "still" images of quickly moving parts.

- LEDs come in several different colors and wavelengths, allowing easy use of the best color for each need, where different color may provide better visibility of features of interest. Having a precisely known spectrum lets tightly matched filters be used to separate informative bandwidth or to reduce disturbing effects of ambient light. LEDs usually operate at comparatively low working temperatures, simplifying heat management, and dissipation. This allows using plastic lenses, filters, and diffusers. Waterproof units can also easily be designed, allowing use in harsh or wet environments (food, beverage, oil industries).

 LED daytime running lights of Audi A4	 Traffic light using LED
 LED digital display that can display four digits and points	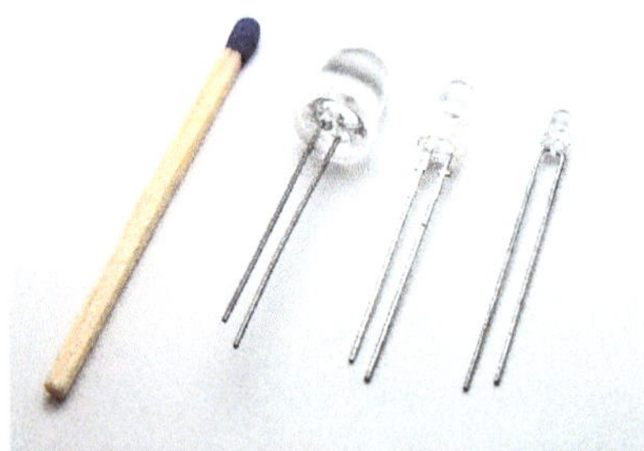 Different sized LEDs. 8 mm, 5 mm and 3 mm, with a wooden match-stick for scale.

Other Applications

LED wallpaper by Meystyle

The light from LEDs can be modulated very quickly so they are used extensively in optical fiber and free space optics communications. This includes remote controls, such as for TVs, VCRs, and LED Computers, where infrared LEDs are often used. Opto-isolators use an LED combined with a photodiode or phototransistor to provide a signal path with electrical isolation between two circuits.

This is especially useful in medical equipment where the signals from a low-voltage sensor circuit (usually battery-powered) in contact with a living organism must be electrically isolated from any possible electrical failure in a recording or monitoring device operating at potentially dangerous voltages. An optoisolator also lets information be transferred between circuits that don't share a common ground potential.

Many sensor systems rely on light as the signal source. LEDs are often ideal as a light source due to the requirements of the sensors. LEDs are used as motion sensors, for example in optical computer mice. The Nintendo Wii's sensor bar uses infrared LEDs. Pulse oximeters use them for measuring oxygen saturation. Some flatbed scanners use arrays of RGB LEDs rather than the typical cold-cathode fluorescent lamp as the light source. Having independent control of three illuminated colors allows the scanner to calibrate itself for more accurate color balance, and there is no need for warm-up. Further, its sensors only need be monochromatic, since at any one time the page being scanned is only lit by one color of light. Since LEDs can also be used as photodiodes, they can be used for both photo emission and detection. This could be used, for example, in a touchscreen that registers reflected light from a finger or stylus. Many materials and biological systems are sensitive to, or dependent on, light. Grow lights use LEDs to increase photosynthesis in plants, and bacteria and viruses can be removed from water and other substances using UV LEDs for sterilization.

LEDs have also been used as a medium-quality voltage reference in electronic circuits. The forward voltage drop (e.g. about 1.7 V for a normal red LED) can be used instead of a Zener diode in low-voltage regulators. Red LEDs have the flattest I/V curve above the knee. Nitride-based LEDs have a fairly steep I/V curve and are useless for this purpose. Although LED forward voltage is far more current-dependent than a Zener diode, Zener diodes with breakdown voltages below 3 V are not widely available.

The progressive miniaturization of low-voltage lighting technology, such as LEDs and OLEDs, suitable to incorporate into low-thickness materials has fostered experimentation in combining light sources and wall covering surfaces for interior walls. The new possibilities offered by these developments have prompted some designers and companies, such as Meystyle, Ingo Maurer, Lomox and Philips, to research and develop proprietary LED wallpaper technologies, some of which are currently available for commercial purchase. Other solutions mainly exist as prototypes or are in the process of being further refined.

OLED

An organic light-emitting diode (OLED) is a light-emitting diode (LED) in which the emissive electroluminescent layer is a film of organic compound that emits light in response to an electric current. This layer of organic semiconductor is situated between two electrodes; typically, at least one of these electrodes is transparent. OLEDs are used to create digital displays in devices such as television screens, computer monitors, portable systems such as mobile phones, handheld game consoles and PDAs. A major area of research is the development of white OLED devices for use in solid-state lighting applications.

Prototype OLED lighting panels

There are two main families of OLED: those based on small molecules and those employing polymers. Adding mobile ions to an OLED creates a light-emitting electrochemical cell (LEC) which has a slightly different mode of operation. An OLED display can be driven with a passive-matrix (PMOLED) or active-matrix (AMOLED) control scheme. In the PMOLED scheme, each row (and line) in the display is controlled sequentially, one by one, whereas AMOLED control uses a thin-film transistor backplane to directly access and switch each individual pixel on or off, allowing for higher resolution and larger display sizes.

An OLED display works without a backlight; thus, it can display deep black levels and can be thinner and lighter than a liquid crystal display (LCD). In low ambient light conditions (such as a dark room), an OLED screen can achieve a higher contrast ratio than an LCD, regardless of whether the LCD uses cold cathode fluorescent lamps or an LED backlight.

History

André Bernanose and co-workers at the Nancy-Université in France made the first observations of electroluminescence in organic materials in the early 1950s. They applied high alternating voltages in air to materials such as acridine orange, either deposited on or dissolved in cellulose or cellophane thin films. The proposed mechanism was either direct excitation of the dye molecules or excitation of electrons.

In 1960 Martin Pope and some of his co-workers at New York University developed ohmic dark-injecting electrode contacts to organic crystals. They further described the necessary energetic requirements (work functions) for hole and electron injecting electrode contacts. These contacts are the basis of charge injection in all modern OLED devices. Pope's group also first observed direct current (DC) electroluminescence under vacuum on a single pure crystal of anthracene and on anthracene crystals doped with tetracene in 1963 using a small area silver electrode at 400 volts. The proposed mechanism was field-accelerated electron excitation of molecular fluorescence.

Pope's group reported in 1965 that in the absence of an external electric field, the electroluminescence in anthracene crystals is caused by the recombination of a thermalized electron and hole, and that the conducting level of anthracene is higher in energy than the exciton energy level. Also in 1965, W. Helfrich and W. G. Schneider of the National Research Council in Canada produced double injection recombination electroluminescence for the first time in an anthracene single crystal using hole and electron injecting electrodes, the forerunner of modern double-injection devices. In the same year, Dow Chemical researchers patented a method of preparing electroluminescent cells using high-voltage (500–1500 V) AC-driven (100–3000 Hz) electrically insulated one millimetre thin layers of a melted phosphor consisting of ground anthracene powder, tetracene, and graphite powder. Their proposed mechanism involved electronic excitation at the contacts between the graphite particles and the anthracene molecules.

Roger Partridge made the first observation of electroluminescence from polymer films at the National Physical Laboratory in the United Kingdom. The device consisted of a film of poly(N-vinylcarbazole) up to 2.2 micrometers thick located between two charge injecting electrodes. The results of the project were patented in 1975 and published in 1983.

The First Practical OLEDs

Hong Kong-born American physical chemist Ching W. Tang and his co-worker Steven Van Slyke at Eastman Kodak built the first practical OLED device in 1987. This was a revolution for the technology. This device used a novel two-layer structure with separate hole transporting and electron transporting layers such that recombination and light emission occurred in the middle of the organic layer; this resulted in a reduction in operating voltage and improvements in efficiency.

Research into polymer electroluminescence culminated in 1990 with J. H. Burroughes *et al.* at the Cavendish Laboratory in Cambridge reporting a high efficiency green light-emitting polymer based device using 100 nm thick films of poly(p-phenylene vinylene).

Universal Display Corporation holds the majority of patents concerning the commercialization of OLEDs.

Working Principle

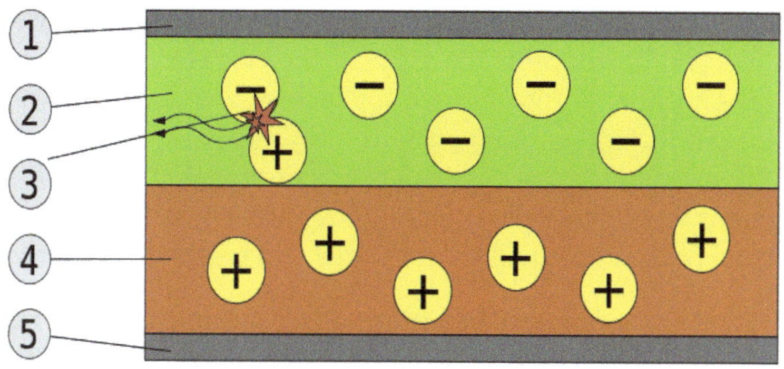

Schematic of a bilayer OLED: 1. Cathode (−), 2. Emissive Layer, 3. Emission of radiation, 4. Conductive Layer, 5. Anode (+)

A typical OLED is composed of a layer of organic materials situated between two electrodes, the anode and cathode, all deposited on a substrate. The organic molecules are electrically conductive as a result of delocalization of pi electrons caused by conjugation over part or all of the molecule. These materials have conductivity levels ranging from insulators to conductors, and are therefore considered organic semiconductors. The highest occupied and lowest unoccupied molecular orbitals (HOMO and LUMO) of organic semiconductors are analogous to the valence and conduction bands of inorganic semiconductors.

Originally, the most basic polymer OLEDs consisted of a single organic layer. One example was the first light-emitting device synthesised by J. H. Burroughes *et al.*, which involved a single layer of poly(p-phenylene vinylene). However multilayer OLEDs can be fabricated with two or more layers in order to improve device efficiency. As well as conductive properties, different materials may be chosen to aid charge injection at electrodes by providing a more gradual electronic profile, or block a charge from reaching the opposite electrode and being wasted. Many modern OLEDs incorporate a simple bilayer structure, consisting of a conductive layer and an emissive layer. More recent developments in OLED architecture improves quantum efficiency (up to 19%) by using a graded heterojunction. In the graded heterojunction architecture, the composition of hole and electron-transport materials varies continuously within the emissive layer with a dopant emitter. The graded heterojunction architecture combines the benefits of both conventional architectures by improving charge injection while simultaneously balancing charge transport within the emissive region.

During operation, a voltage is applied across the OLED such that the anode is positive with respect to the cathode. Anodes are picked based upon the quality of their optical transparency, electrical conductivity, and chemical stability. A current of electrons flows through the device from cathode to anode, as electrons are injected into the LUMO of the organic layer at the cathode and withdrawn from the HOMO at the anode. This latter process may also be described as the injection of electron holes into the HOMO. Electrostatic forces bring the electrons and the holes towards each other and they recombine forming an exciton, a bound state of the electron and hole. This happens closer to the emissive layer, because in organic semiconductors holes are generally more mobile than electrons. The decay of this excited state results in a relaxation of the energy levels of the electron, accompanied by emission of radiation whose frequency is in the visible region. The frequency of this radiation depends on the band gap of the material, in this case the difference in energy between the HOMO and LUMO.

As electrons and holes are fermions with half integer spin, an exciton may either be in a singlet state or a triplet state depending on how the spins of the electron and hole have been combined. Statistically three triplet excitons will be formed for each singlet exciton. Decay from triplet states (phosphorescence) is spin forbidden, increasing the timescale of the transition and limiting the internal efficiency of fluorescent devices. Phosphorescent organic light-emitting diodes make use of spin–orbit interactions to facilitate intersystem crossing between singlet and triplet states, thus obtaining emission from both singlet and triplet states and improving the internal efficiency.

Indium tin oxide (ITO) is commonly used as the anode material. It is transparent to visible light and has a high work function which promotes injection of holes into the HOMO level of the organic layer. A typical conductive layer may consist of PEDOT:PSS as the HOMO level of this material generally lies between the work function of ITO and the HOMO of other commonly used polymers, reducing the energy barriers for hole injection. Metals such as barium and calcium are often used for the cathode as they have low work functions which promote injection of electrons into the LUMO of the organic layer. Such metals are reactive, so they require a capping layer of aluminium to avoid degradation.

Experimental research has proven that the properties of the anode, specifically the anode/hole transport layer (HTL) interface topography plays a major role in the efficiency, performance, and lifetime of organic light emitting diodes. Imperfections in the surface of the anode decrease anode-organic film interface adhesion, increase electrical resistance, and allow for more frequent formation of non-emissive dark spots in the OLED material adversely affecting lifetime. Mechanisms to decrease anode roughness for ITO/glass substrates include the use of thin films and self-assembled monolayers. Also, alternative substrates and anode materials are being considered to increase OLED performance and lifetime. Possible examples include single crystal sapphire substrates treated with gold (Au) film anodes yielding lower work functions, operating voltages, electrical resistance values, and increasing lifetime of OLEDs.

Single carrier devices are typically used to study the kinetics and charge transport mechanisms of an organic material and can be useful when trying to study energy transfer processes. As current through the device is composed of only one type of charge carrier, either electrons or holes, recombination does not occur and no light is emitted. For example, electron only devices can be obtained by replacing ITO with a lower work function metal which increases the energy barrier of hole injection. Similarly, hole only devices can be made by using a cathode made solely of aluminium, resulting in an energy barrier too large for efficient electron injection.

Carrier Balance

Balanced charge injection and transfer are required to get high internal efficiency, pure emission of luminance layer without contaminated emission from charge transporting layers, and high stability. A common way to balance charge is optimizing the thickness of the charge transporting layers but is hard to control. Another way is using the exciplex. Exciplex formed between hole-transporting (p-type) and electron-transporting (n-type) side chains to localize electron-hole pairs. Energy is then transferred to luminophore and provide high efficiency. An example of using exciplex is grafting Oxadiazole and carbazole side units in red diketopyrrolopyrrole-doped Copolymer main chain shows improved external quantum efficiency and color purity in no optimized OLED.

Material Technologies

Small Molecules

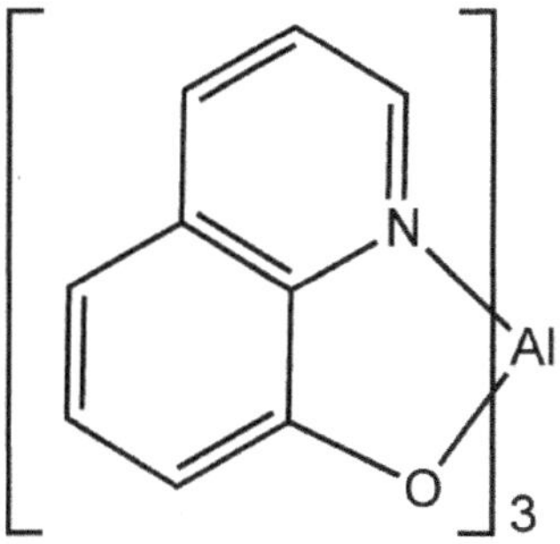

Alq_3, commonly used in small molecule OLEDs

Efficient OLEDs using small molecules were first developed by Dr. Ching W. Tang *et al.* at Eastman Kodak. The term OLED traditionally refers specifically to this type of device, though the term SM-OLED is also in use.

Molecules commonly used in OLEDs include organometallic chelates (for example Alq_3, used in the organic light-emitting device reported by Tang *et al.*), fluorescent and phosphorescent dyes and conjugated dendrimers. A number of materials are used for their charge transport properties, for example triphenylamine and derivatives are commonly used as materials for hole transport layers. Fluorescent dyes can be chosen to obtain light emission at different wavelengths, and compounds such as perylene, rubrene and quinacridone derivatives are often used. Alq_3 has been used as a green emitter, electron transport material and as a host for yellow and red emitting dyes.

The production of small molecule devices and displays usually involves thermal evaporation in a vacuum. This makes the production process more expensive and of limited use for large-area devices, than other processing techniques. However, contrary to polymer-based devices, the vacuum deposition process enables the formation of well controlled, homogeneous films, and the construction of very complex multi-layer structures. This high flexibility in layer design, enabling distinct charge transport and charge blocking layers to be formed, is the main reason for the high efficiencies of the small molecule OLEDs.

Coherent emission from a laser dye-doped tandem SM-OLED device, excited in the pulsed regime, has been demonstrated. The emission is nearly diffraction limited with a spectral width similar to that of broadband dye lasers.

Researchers report luminescence from a single polymer molecule, representing the smallest possible organic light-emitting diode (OLED) device. Scientists will be able to optimize substances to produce more powerful light emissions. Finally, this work is a first step towards making molecule-sized components that combine electronic and optical properties. Similar components could form the basis of a molecular computer.

Polymer light-emitting Diodes

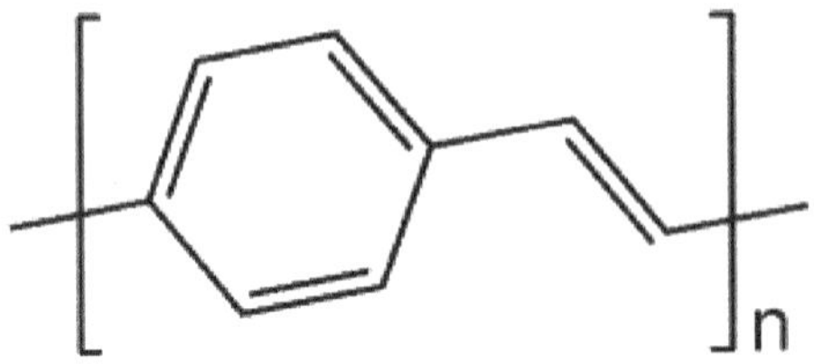

poly(*p*-phenylene vinylene), used in the first PLED

Polymer light-emitting diodes (PLED), also light-emitting polymers (LEP), involve an electroluminescent conductive polymer that emits light when connected to an external voltage. They are used as a thin film for full-spectrum colour displays. Polymer OLEDs are quite efficient and require a relatively small amount of power for the amount of light produced.

Vacuum deposition is not a suitable method for forming thin films of polymers. However, polymers can be processed in solution, and spin coating is a common method of depositing thin polymer films. This method is more suited to forming large-area films than thermal evaporation. No vacuum is required, and the emissive materials can also be applied on the substrate by a technique derived from commercial inkjet printing. However, as the application of subsequent layers tends to dissolve those already present, formation of multilayer structures is difficult with these methods. The metal cathode may still need to be deposited by thermal evaporation in vacuum. An alternative method to vacuum deposition is to deposit a Langmuir-Blodgett film.

Typical polymers used in pleaded displays include derivatives of poly(*p*-phenylene vinylene) and polyfluorene. Substitution of side chains onto the polymer backbone may determine the colour of emitted light or the stability and solubility of the polymer for performance and ease of processing. While unsubstituted poly(p-phenylene vinylene) (PPV) is typically insoluble, a number of PPVs and related poly(naphthalene vinylene)s (PNVs) that are soluble in organic solvents or water have been prepared via ring opening metathesis polymerization. These water-soluble polymers or conjugated poly electrolytes (CPEs) also can be used as hole injection layers alone or in combination with nanoparticles like graphene.

Phosphorescent Materials

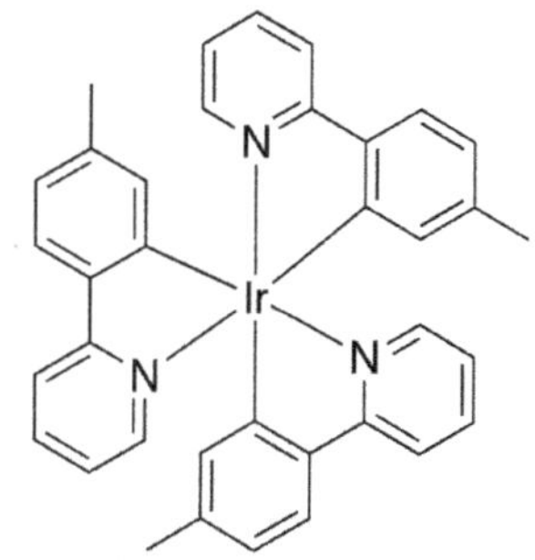

Ir(mppy)$_3$, a phosphorescent dopant which emits green light.

Phosphorescent organic light emitting diodes use the principle of electrophosphorescence to convert electrical energy in an OLED into light in a highly efficient manner, with the internal quantum efficiencies of such devices approaching 100%.

Typically, a polymer such as poly(N-vinylcarbazole) is used as a host material to which an organometallic complex is added as a dopant. Iridium complexes such as Ir(mppy)$_3$ are currently the focus of research, although complexes based on other heavy metals such as platinum have also been used.

The heavy metal atom at the centre of these complexes exhibits strong spin-orbit coupling, facilitating intersystem crossing between singlet and triplet states. By using these phosphorescent materials, both singlet and triplet excitons will be able to decay radiatively, hence improving the internal quantum efficiency of the device compared to a standard pleaded where only the singlet states will contribute to emission of light.

Applications of OLEDs in solid state lighting require the achievement of high brightness with good CIE coordinates (for white emission). The use of macromolecular species like polyhedral oligomeric silsesquioxanes (POSS) in conjunction with the use of phosphorescent species such as Ir for printed OLEDs have exhibited brightnesses as high as 10,000 cd/m^2.

Device Architectures

Structure

Bottom or top emission

Bottom or top distinction refers not to orientation of the OLED display, but to the direction that emitted light exits the device. OLED devices are classified as bottom emission devices if light emitted passes through the transparent or semi-transparent bottom electrode and substrate on which the panel was manufactured. Top emission devices are classified based on whether or not the light emitted from the OLED device exits through the lid that is added following fabrication of the device. Top-emitting OLEDs are better suited for active-matrix applications as they can be more easily integrated with a non-transparent transistor backplane. The TFT array attached to the bottom substrate on which AMOLEDs are manufactured are typically non-transparent, resulting in considerable blockage of transmitted light if the device followed a bottom emitting scheme.

Transparent OLEDs

Transparent OLEDs use transparent or semi-transparent contacts on both sides of the device to create displays that can be made to be both top and bottom emitting (transparent). TOLEDs can greatly improve contrast, making it much easier to view displays in bright sunlight. This technology can be used in Head-up displays, smart windows or augmented reality applications.

Graded heterojunction

Graded heterojunction OLEDs gradually decrease the ratio of electron holes to electron transporting chemicals. This results in almost double the quantum efficiency of existing OLEDs.

Stacked OLEDs

Stacked OLEDs use a pixel architecture that stacks the red, green, and blue subpixels on top of one another instead of next to one another, leading to substantial increase in gamut and color depth, and greatly reducing pixel gap. Currently, other display technologies have the RGB (and RGBW) pixels mapped next to each other decreasing potential resolution.

Inverted OLED

In contrast to a conventional OLED, in which the anode is placed on the substrate, an Inverted OLED uses a bottom cathode that can be connected to the drain end of an n-channel TFT especially for the low cost amorphous silicon TFT backplane useful in the manufacturing of AMOLED displays.

Patterning Technologies

Patternable organic light-emitting devices use a light or heat activated electroactive layer. A latent material (PEDOT-TMA) is included in this layer that, upon activation, becomes highly efficient as a hole injection layer. Using this process, light-emitting devices with arbitrary patterns can be prepared.

Colour patterning can be accomplished by means of laser, such as radiation-induced sublimation transfer (RIST).

Organic vapour jet printing (OVJP) uses an inert carrier gas, such as argon or nitrogen, to transport evaporated organic molecules (as in organic vapour phase deposition). The gas is expelled through a micrometre-sized nozzle or nozzle array close to the substrate as it is being translated. This allows printing arbitrary multilayer patterns without the use of solvents.

Conventional OLED displays are formed by vapor thermal evaporation (VTE) and are patterned by shadow-mask. A mechanical mask has openings allowing the vapor to pass only on the desired location.

Like ink jet material depositioning, inkjet etching (IJE) deposits precise amounts of solvent onto a substrate designed to selectively dissolve the substrate material and induce a structure or pattern. Inkjet etching of polymer layers in OLED's can be used to increase the overall out-coupling efficiency. In OLEDs, light produced from the emissive layers of the OLED is partially transmitted out of the device and partially trapped inside the device by total internal reflection (TIR). This trapped light is wave-guided along the interior of the device until it reaches an edge where it is dissipated by either absorption or emission. Inkjet etching can be used to selectively alter the polymeric layers of OLED structures to decrease overall TIR and increase out-coupling efficiency of the OLED. Compared to a non-etched polymer layer, the structured polymer layer in the OLED structure from the IJE process helps to decrease the TIR of the OLED device. IJE solvents are commonly organic instead of water based due to their non-acidic nature and ability to effectively dissolve materials at temperatures under the boiling point of water.

Backplane Technologies

For a high resolution display like a TV, a TFT backplane is necessary to drive the pixels correctly. Currently, low temperature polycrystalline silicon (LTPS) – thin-film transistor (TFT) is used for commercial AMOLED displays. LTPS-TFT has variation of the performance in a display, so various compensation circuits have been reported. Due to the size limitation of the excimer laser used for LTPS, the AMOLED size was limited. To cope with the hurdle related to the panel size, amorphous-silicon/microcrystalline-silicon backplanes have been reported with large display prototype demonstrations.

Fabrication

Transfer-printing is an emerging technology to assemble large numbers of parallel OLED and AMOLED devices efficiently. It takes advantage of standard metal deposition, photolithography, and etching to create alignment marks commonly on glass or other device substrates. Thin polymer adhesive layers are applied to enhance resistance to particles and surface defects. Microscale ICs are transfer-printed onto the adhesive surface and then baked to fully cure adhesive layers. An additional photosensitive polymer layer is applied to the substrate to account for the topography caused by the printed ICs, reintroducing a flat surface. Photolithography and etching removes some polymer layers to uncover conductive pads on the ICs. Afterwards, the anode layer is applied to the device backplane to form bottom electrode. OLED layers are applied to the anode layer with conventional vapor deposition, and covered with a conductive metal electrode layer. As of 2011 transfer-printing was capable to print onto target substrates up to 500mm X 400mm. This size limit needs to expand for transfer-printing to become a common process for the fabrication of large OLED/AMOLED displays.

Advantages

Demonstration of a 4.1" prototype flexible display from Sony

The different manufacturing process of OLEDs lends itself to several advantages over flat panel displays made with LCD technology.

Lower cost in the future

OLEDs can be printed onto any suitable substrate by an inkjet printer or even by screen printing, theoretically making them cheaper to produce than LCD or plasma displays. However, fabrication of the OLED substrate is currently more costly than that of a TFT LCD, until mass production methods lower costs through scalability. Roll-to-roll vapor-deposition methods for organic devices do allow mass production of thousands of devices per minute for minimal cost; however, this technique also induces problems: devices with multiple layers can be challenging to make because of registration—lining up the different printed layers to the required degree of accuracy.

Lightweight and flexible plastic substrates

OLED displays can be fabricated on flexible plastic substrates, leading to the possible fabrication of flexible organic light-emitting diodes for other new applications, such as roll-up displays embedded in fabrics or clothing. If a substrate like polyethylene terephthalate (PET) can be used, the displays may be produced inexpensively. Furthermore, plastic substrates are shatter-resistant, unlike the glass displays used in LCD devices.

Better picture quality

OLEDs enable a greater contrast ratio and wider viewing angle compared to LCDs, because OLED pixels emit light directly. Furthermore, OLED pixel colors appear correct and unshifted, even as the viewing angle approaches 90° from the normal.

Better power efficiency and thickness

LCDs filter the light emitted from a backlight, allowing a small fraction of light through. Thus, they cannot show true black. However, an inactive OLED element does not produce light or consume power, allowing true blacks. Removing the backlight also makes OLEDs lighter because some substrates are not needed. When looking at top-emitting OLEDs, thickness also plays a role when talking about index match layers (IMLs). Emission intensity is enhanced when the IML thickness is 1.3–2.5 nm. The refractive value and the matching of the optical IMLs property, including the device structure parameters, also enhance the emission intensity at these thicknesses.

Response time

OLEDs also have a much faster response time than an LCD. Using response time compensation technologies, the fastest modern LCDs can reach response times as low as 1 ms for their fastest color transition, and are capable of refresh frequencies as high as 240 Hz. According to LG, OLED response times are up to 1,000 times faster than LCD, putting conservative estimates at under 10 μs (0.01 ms), which could theoretically accommodate refresh frequencies approaching 100 kHz (100,000 Hz). Due to their extremely fast response time, OLED displays can also be easily designed to be strobed, creating an effect similar to CRT flicker in order to avoid the sample-and-hold behavior seen on both LCDs and some OLED displays, which creates the perception of motion blur.

Disadvantages

LEP (light emitting polymer) display showing partial failure

An old OLED display showing wear

Lifespan

The biggest technical problem for OLEDs was the limited lifetime of the organic materials. One 2008 technical report on an OLED TV panel found that "After 1,000 hours the blue luminance degraded by 12%, the red by 7% and the green by 8%." In particular, blue OLEDs historically have had a lifetime of around 14,000 hours to half original brightness (five years at 8 hours a day) when used for flat-panel displays. This is lower than the typical lifetime of LCD, LED or PDP technology. Each currently is rated for about 25,000–40,000 hours to half brightness, depending on manufacturer and model. Degradation occurs because of the accumulation of nonradiative recombination centers and luminescence quenchers in the emissive zone. It is said that the chemical breakdown in the semiconductors occurs in four steps: 1) recombination of charge carriers through the absorption of UV light, 2) homolytic dissociation, 3) subsequent radical addition reactions that form π radicals, and 4) disproportionation between two radicals resulting in hydrogen-atom transfer reactions. However, some manufacturers' displays aim to increase the lifespan of OLED displays, pushing their expected life past that of LCD displays by improving light outcoupling, thus achieving the same brightness at a lower drive current. In 2007, experimental OLEDs were created which can sustain 400 cd/m^2 of luminance for over 198,000 hours for green OLEDs and 62,000 hours for blue OLEDs.

Color balance

Additionally, as the OLED material used to produce blue light degrades significantly more rapidly than the materials that produce other colors, blue light output will decrease relative to the other colors of light. This variation in the differential color output will change the color balance of the display and is much more noticeable than a decrease in overall luminance. This can be avoided partially by adjusting color balance, but this may require advanced control circuits and interaction with the user, which is unacceptable for users. More commonly, though, manufacturers optimize the size of the R, G and B subpixels to reduce the current density through the subpixel in order to equalize lifetime at full luminance. For example, a blue subpixel may be 100% larger than the green subpixel. The red subpixel may be 10% smaller than the green.

Efficiency of blue OLEDs

Improvements to the efficiency and lifetime of blue OLEDs is vital to the success of OLEDs as replacements for LCD technology. Considerable research has been invested in developing blue OLEDs with high external quantum efficiency as well as a deeper blue color. External quantum efficiency values of 20% and 19% have been reported for red (625 nm) and green (530 nm) diodes, respectively. However, blue diodes (430 nm) have only been able to achieve maximum external quantum efficiencies in the range of 4% to 6%.

Water damage

Water can instantly damage the organic materials of the displays. Therefore, improved sealing processes are important for practical manufacturing. Water damage especially may limit the longevity of more flexible displays.

Outdoor performance

As an emissive display technology, OLEDs rely completely upon converting electricity to light, unlike most LCDs which are to some extent reflective. e-paper leads the way in efficiency with ~ 33% ambient light reflectivity, enabling the display to be used without any internal light source. The metallic cathode in an OLED acts as a mirror, with reflectance approaching 80%, leading to poor readability in bright ambient light such as outdoors. However, with the proper application of a circular polarizer and antireflective coatings, the diffuse reflectance can be reduced to less than 0.1%. With 10,000 fc incident illumination (typical test condition for simulating outdoor illumination), that yields an approximate photopic contrast of 5:1. Recent advances in OLED technologies, however, enable OLEDs to become actually better than LCDs in bright sunlight. The Super AMOLED display in the Galaxy S5, for example, was found to outperform all LCD displays on the market in terms of brightness and reflectance.

Power consumption

While an OLED will consume around 40% of the power of an LCD displaying an image that is primarily black, for the majority of images it will consume 60–80% of the power of an LCD. However, an OLED can use more than three times as much power to display an image with a white background, such as a document or web site. This can lead to reduced battery life in mobile devices, when white backgrounds are used.

Manufacturers and Commercial uses

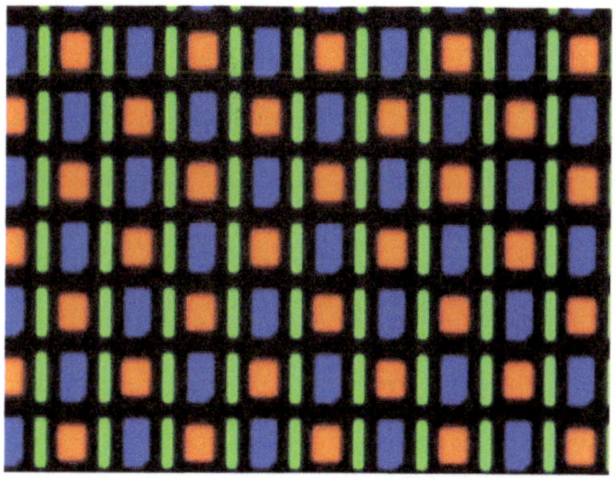

Magnified image of the AMOLED screen on the Google Nexus One smartphone using the RGBG system of the PenTile Matrix Family.

A 3.8 cm (1.5 in) OLED display from a Creative ZEN V media player

OLED technology is used in commercial applications such as displays for mobile phones and portable digital media players, car radios and digital cameras among others. Such portable applications favor the high light output of OLEDs for readability in sunlight and their low power drain. Portable displays are also used intermittently, so the lower lifespan of organic displays is less of an issue. Prototypes have been made of flexible and rollable displays which use OLEDs' unique characteristics. Applications in flexible signs and lighting are also being developed. Philips Lighting have made OLED lighting samples under the brand name "Lumiblade" available online and Novaled AG based in Dresden, Germany, introduced a line of OLED desk lamps called "Victory" in September, 2011.

OLED lighting in a shopping mall in Aachen, Germany

OLEDs have been used in most Motorola and Samsung color cell phones, as well as some HTC, LG and Sony Ericsson models. Nokia has also introduced some OLED products including the N85 and the N86 8MP, both of which feature an AMOLED display. OLED technology can also be found in digital media players such as the Creative ZEN V, the iriver clix, the Zune HD and the Sony Walkman X Series.

The Google and HTC Nexus One smartphone includes an AMOLED screen, as does HTC's own Desire and Legend phones. However, due to supply shortages of the Samsung-produced displays, certain HTC models will use Sony's SLCD displays in the future, while the Google and Samsung Nexus S smartphone will use "Super Clear LCD" instead in some countries.

OLED displays were used in watches made by Fossil (JR-9465) and Diesel (DZ-7086).

Other manufacturers of OLED panels include Anwell Technologies Limited (Hong Kong), AU Optronics (Taiwan), Chimei Innolux Corporation (Taiwan), LG (Korea), and others.

In 2009, Shearwater Research introduced the Predator as the first color OLED diving computer available with a user replaceable battery.

DuPont stated in a press release in May 2010 that they can produce a 50-inch OLED TV in two minutes with a new printing technology. If this can be scaled up in terms of manufacturing, then the total cost of OLED TVs would be greatly reduced. DuPont also states that OLED TVs made with this less expensive technology can last up to 15 years if left on for a normal eight-hour day.

The use of OLEDs may be subject to patents held by Universal Display Corporation, Eastman Kodak, DuPont, General Electric, Royal Philips Electronics, numerous universities and others. There are by now thousands of patents associated with OLEDs, both from larger corporations and smaller technology companies.

RIM, the maker of BlackBerry smartphones, uses OLED displays in their BlackBerry 10 devices.

Flexible OLED displays are already being produced and these are used by manufacturers to create curved displays such as the Galaxy S7 Edge but so far there they are not in devices that can be flexed by the consumer. Apart from the screen itself the circuit boards and batteries would need to be flexible. Samsung demonstrated a roll-out display in 2016.

Fashion

Textiles incorporating OLEDs are an innovation in the fashion world and pose for a way to integrate lighting to bring inert objects to a whole new level of fashion. The hope is to combine the comfort and low cost properties of textile with the OLEDs properties of illumination and low energy consumption. Although this scenario of illuminated clothing is highly plausible, challenges are still a road block. Some issues include: the lifetime of the OLED, rigidness of flexible foil substrates, and the lack of research in making more fabric like photonic textiles.

Samsung Applications

By 2004 Samsung, South Korea's largest conglomerate, was the world's largest OLED manufacturer, producing 40% of the OLED displays made in the world, and as of 2010 has a 98% share of the global AMOLED market. The company is leading the world of OLED industry, generating $100.2 million out of the total $475 million revenues in the global OLED market in 2006. As of 2006, it held more than 600 American patents and more than 2800 international patents, making it the largest owner of AMOLED technology patents.

Samsung SDI announced in 2005 the world's largest OLED TV at the time, at 21 inches (53 cm). This OLED featured the highest resolution at the time, of 6.22 million pixels. In addition, the company adopted active matrix based technology for its low power consumption and high-resolution qualities. This was exceeded in January 2008, when Samsung showcased the world's largest and thinnest OLED TV at the time, at 31 inches (78 cm) and 4.3 mm.

In May 2008, Samsung unveiled an ultra-thin 12.1 inch (30 cm) laptop OLED display concept, with a 1,280×768 resolution with infinite contrast ratio. According to Woo Jong Lee, Vice President of the Mobile Display Marketing Team at Samsung SDI, the company expected OLED displays to be used in notebook PCs as soon as 2010.

In October 2008, Samsung showcased the world's thinnest OLED display, also the first to be "flappable" and bendable. It measures just 0.05 mm (thinner than paper), yet a Samsung staff member said that it is "technically possible to make the panel thinner". To achieve this thickness, Samsung etched an OLED panel that uses a normal glass substrate. The drive circuit was formed by low-temperature polysilicon TFTs. Also, low-molecular organic EL materials were employed. The pixel count of the display is 480 × 272. The contrast ratio is 100,000:1, and the luminance is 200 cd/m^2. The colour reproduction range is 100% of the NTSC standard.

In the same month, Samsung unveiled what was then the world's largest OLED Television at 40-inch with a Full HD resolution of 1920 × 1080 pixels. In the FPD International, Samsung stated that its 40-inch OLED Panel is the largest size currently possible. The panel has a contrast ratio of 1,000,000:1, a colour gamut of 107% NTSC, and a luminance of 200 cd/m^2 (peak luminance of 600 cd/m^2).

At the Consumer Electronics Show (CES) in January 2010, Samsung demonstrated a laptop computer with a large, transparent OLED display featuring up to 40% transparency and an animated OLED display in a photo ID card.

Samsung's latest AMOLED smartphones use their Super AMOLED trademark, with the Samsung Wave S8500 and Samsung i9000 Galaxy S being launched in June 2010. In January 2011 Samsung announced their Super AMOLED Plus displays, which offer several advances over the older Super AMOLED displays: real stripe matrix (50% more sub pixels), thinner form factor, brighter image and an 18% reduction in energy consumption.

At CES 2012, Samsung introduced the first 55" TV screen that uses Super OLED technology.

On January 8, 2013, at CES Samsung unveiled a unique curved 4K Ultra S9 OLED television, which they state provides an "IMAX-like experience" for viewers.

On August 13, 2013, Samsung announced availability of a 55-inch curved OLED TV (model KN-55S9C) in the US at a price point of $8999.99.

On September 6, 2013, Samsung launched its 55-inch curved OLED TV (model KE55S9C) in the United Kingdom with John Lewis.

Samsung introduced the *Galaxy Round* smartphone in the Korean market in October 2013. The device features a 1080p screen, measuring 5.7 inches (14 cm), that curves on the vertical axis in a rounded case. The corporation has promoted the following advantages: A new feature called "Round Interaction" that allows users to look at information by tilting the handset on a flat surface with the screen off, and the feel of one continuous transition when the user switches between home screens.

Sony Applications

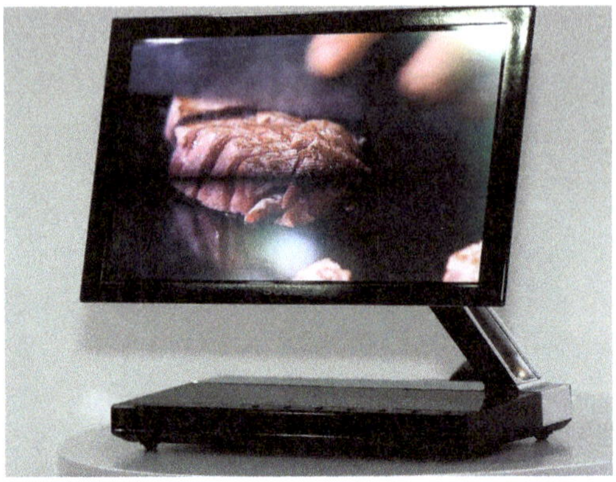

Sony XEL-1, the world's first OLED TV. (front)

The Sony CLIÉ PEG-VZ90 was released in 2004, being the first PDA to feature an OLED screen. Other Sony products to feature OLED screens include the MZ-RH1 portable minidisc recorder, released in 2006 and the Walkman X Series.

At the 2007 Las Vegas Consumer Electronics Show (CES), Sony showcased 11-inch (28 cm, resolution 960×540) and 27-inch (68.5 cm), full HD resolution at 1920 × 1080 OLED TV models.

Both claimed 1,000,000:1 contrast ratios and total thicknesses (including bezels) of 5 mm. In April 2007, Sony announced it would manufacture 1000 11-inch (28 cm) OLED TVs per month for market testing purposes. On October 1, 2007, Sony announced that the 11-inch (28 cm) model, now called the XEL-1, would be released commercially; the XEL-1 was first released in Japan in December 2007.

In May 2007, Sony publicly unveiled a video of a 2.5-inch flexible OLED screen which is only 0.3 millimeters thick. At the Display 2008 exhibition, Sony demonstrated a 0.2 mm thick 3.5 inch (9 cm) display with a resolution of 320×200 pixels and a 0.3 mm thick 11 inch (28 cm) display with 960×540 pixels resolution, one-tenth the thickness of the XEL-1.

In July 2008, a Japanese government body said it would fund a joint project of leading firms, which is to develop a key technology to produce large, energy-saving organic displays. The project involves one laboratory and 10 companies including Sony Corp. NEDO said the project was aimed at developing a core technology to mass-produce 40 inch or larger OLED displays in the late 2010s.

In October 2008, Sony published results of research it carried out with the Max Planck Institute over the possibility of mass-market bending displays, which could replace rigid LCDs and plasma screens. Eventually, bendable, see-through displays could be stacked to produce 3D images with much greater contrast ratios and viewing angles than existing products.

Sony exhibited a 24.5" (62 cm) prototype OLED 3D television during the Consumer Electronics Show in January 2010.

In January 2011, Sony announced the PlayStation Vita handheld game console (the successor to the PSP) will feature a 5-inch OLED screen.

On February 17, 2011, Sony announced its 25" (63.5 cm) OLED Professional Reference Monitor aimed at the Cinema and high end Drama Post Production market.

On June 25, 2012, Sony and Panasonic announced a joint venture for creating low cost mass production OLED televisions by 2013.

LG Applications

As of 2010, LG Electronics produced one model of OLED television, the 15 inch 15EL9500 and had announced a 31" (78 cm) OLED 3D television for March 2011. On December 26, 2011, LG officially announced the "world's largest 55" OLED panel" and featured it at CES 2012. In late 2012, LG announces the launch of the 55EM9600 OLED television in Australia.

In January 2015, LG Display signed a long term agreement with Universal Display Corporation for the supply of OLED materials and the right to use their patented OLED emitters.

Mitsubishi Applications

Lumiotec is the first company in the world developing and selling, since January 2011, mass-produced OLED lighting panels with such brightness and long lifetime. Lumiotec is a joint venture of

Mitsubishi Heavy Industries, ROHM, Toppan Printing, and Mitsui & Co. On June 1, 2011, Mitsubishi installed a 6-meter OLED 'sphere' in Tokyo's Science Museum.

Recom Group/Video Name Tag Applications

On January 6, 2011, Los Angeles based technology company Recom Group introduced the first small screen consumer application of the OLED at the Consumer Electronics Show in Las Vegas. This was a 2.8" (7 cm) OLED display being used as a wearable video name tag. At the Consumer Electronics Show in 2012, Recom Group introduced the world's first video mic flag incorporating three 2.8" (7 cm) OLED displays on a standard broadcaster's mic flag. The video mic flag allowed video content and advertising to be shown on a broadcasters standard mic flag.

BMW

BMW plans to use OLEDs in tail lights and interior lights in their future cars; however, OLEDs are currently too dim to be used for brake lights, headlights and indicators.

Research

In 2014, Mitsubishi Chemical Corporation (MCC), a subsidiary of the Mitsubishi Chemical Holdings developed an organic light-emitting diode (OLED) panel with a life of 30,000 hours, twice that of conventional OLED panels.

The search for efficient OLED materials has been extensively supported by simulation methods. By now it is possible to calculate important properties completely computationally, independent of experimental input. This allows cost-efficient pre-screening of materials, prior to expensive synthesis and experimental characterisation.

Function of OLED

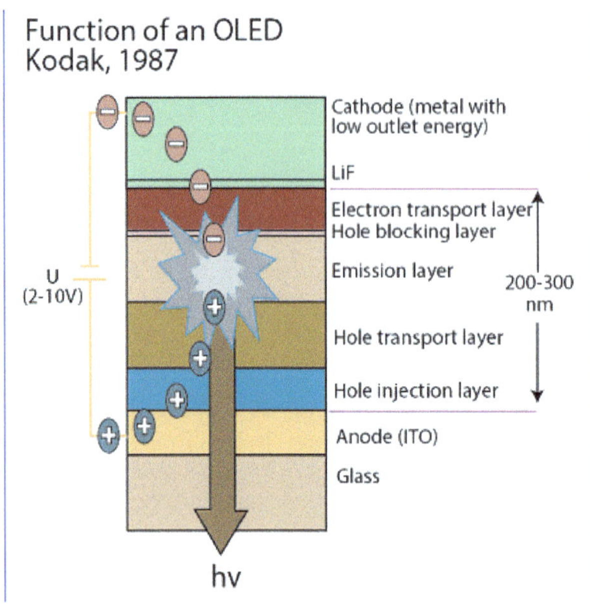

OLED is a display device that sandwiches carbon based films between the two electrodes and when voltage is applied creates light.

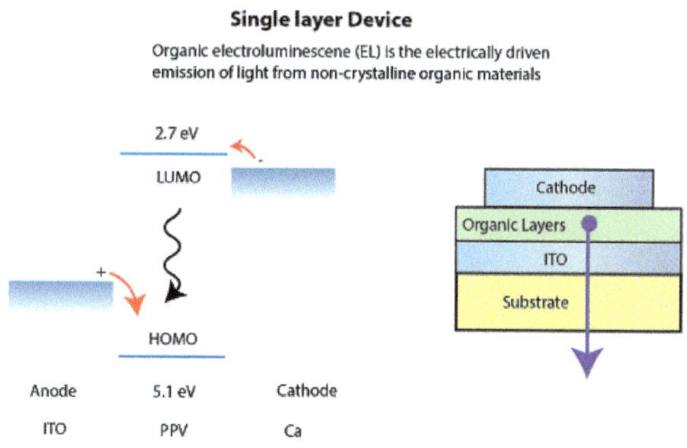

Energy level diagram of a two-layer OLED

- HOMO, LUMO of the HTL is slightly above that of the ETL

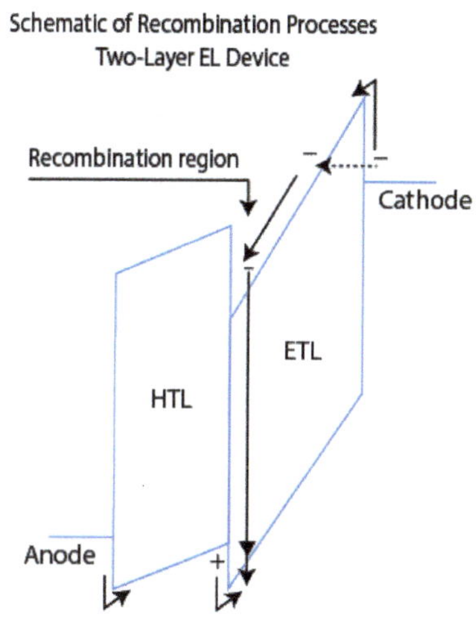

Chemistry Behind Emission

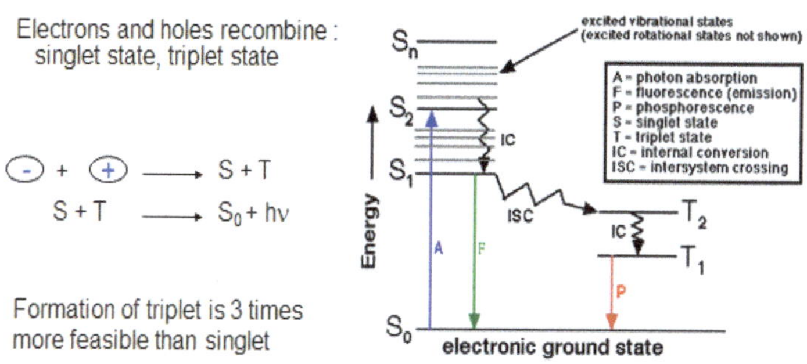

Factors Influencing Efficiency

1. Efficiency of electrons and holes recombination.
2. Efficiency of excited state formation upon annihilation.
3. Quantum yield of emission of excited state.

OLED's Based on Type of Materials

1. Polymer Light Emitting Diode (PLEDs)

eg :Conjugated molecules

2. Small Molecule Organic Light Emitting Diodes (SMOLEDs).

eg : Metal chelates

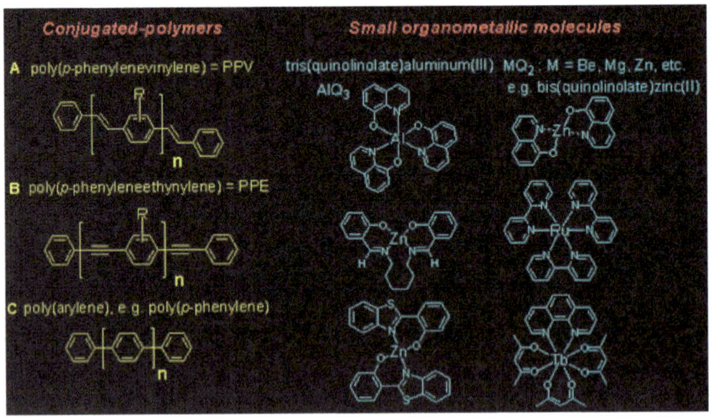

SMOLEDs

Criteria Metal chelates must satisfy

- Thermally stable,
- Highly luminescent in the solid state,
- Thin-film forming upon vacuum deposition
- Capable of transporting electrons.

First Report on Thin Film Electroluminescence

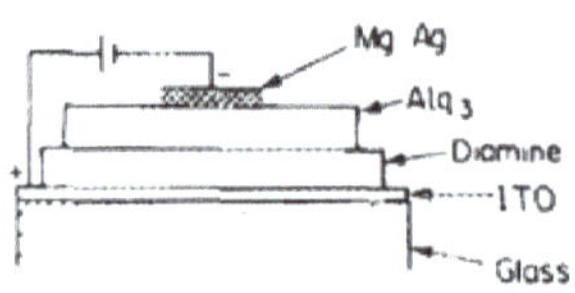

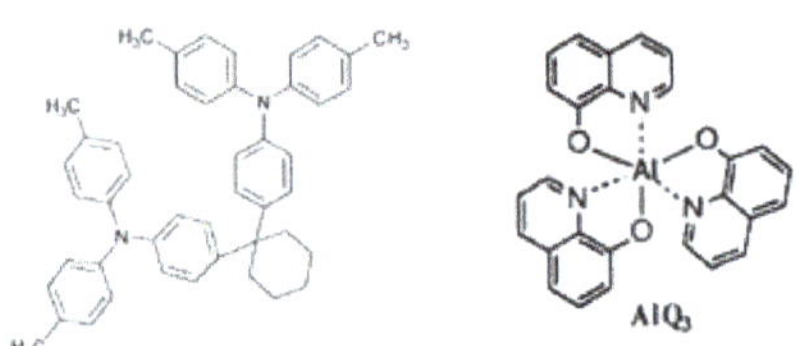

Mg:Ag – 10:1

Luminescent film - 600 A°

Diamine – 750A°

- Relatively High voltage (80-100 V)
  - ◇ Inject charge into organic crystals
- Low work function alloy-cathode
- Organic layers, cathode were vacuum deposited.

Key Factors

- Morphological properties of organic layers are critical.
- Thin films must be smooth and continuous.
- Mg is susceptible to atmospheric oxidation and corrosion.
- Ag improves the sticking coefficient of the metal to the organic layer.
- A dc voltage of less than 10V drives the diode.

Rules Governing the Fluorescence of Metal Chelates

(1) Paramagnetic metal ions: Essentially non-fluorescent

(2) Increasing atomic number: Fluorescence reduced

$$InQ_3 < GaQ_3 < AlQ_3$$

(3) Covalent nature of the metal-ligand bonding increased: Emission shifts to longer wavelength.

Light Emitting Polymers

1. Dendrimers: They are highly branched structures built up from monomer units with precisely controlled architectures.

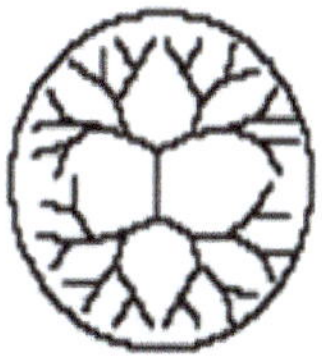

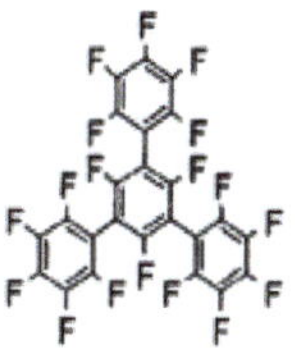

2. Long chain conjugated molecules:

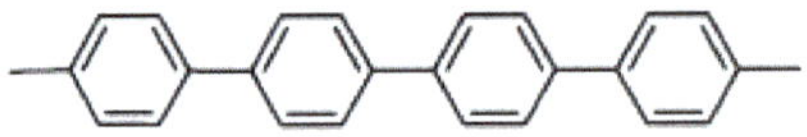

polyphenylene

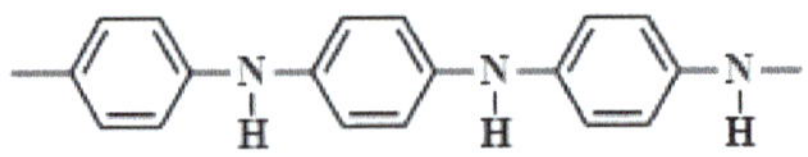

polyaniline

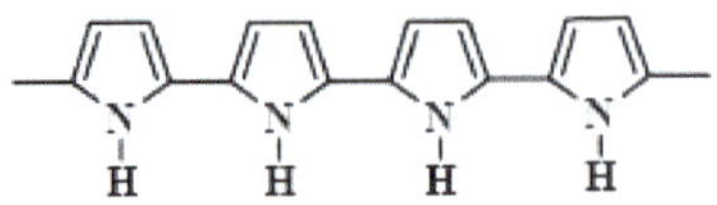

polypyrrole

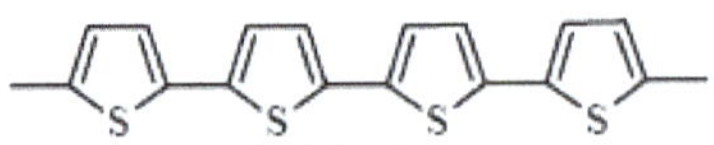

polythiophene

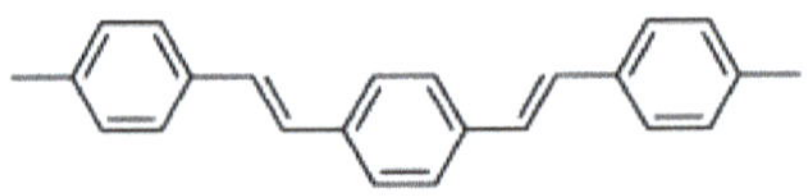

poly(phenylenevinylene)

Semiconducting Property

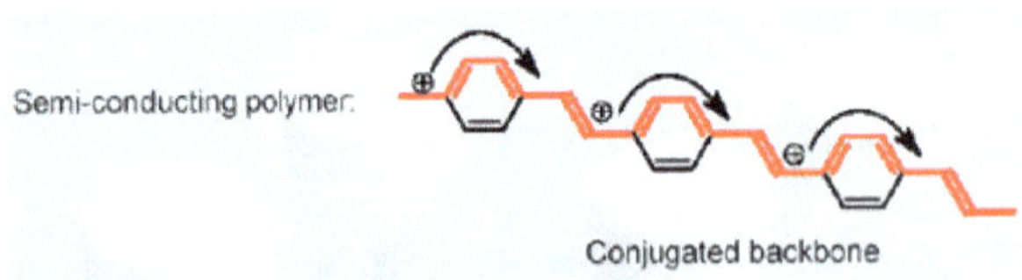

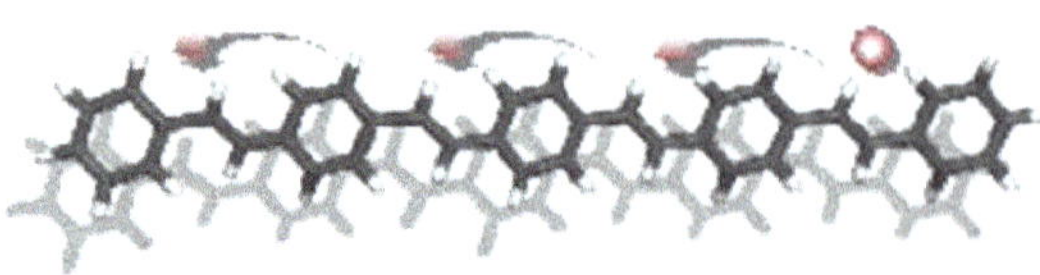

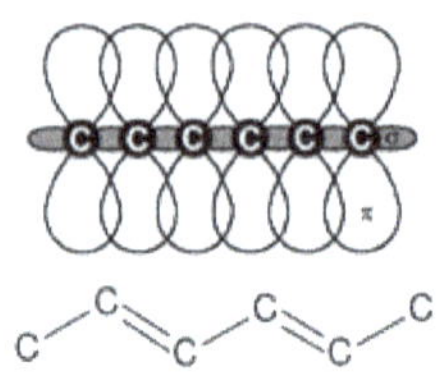

Electroluminescent behaviour

- Semiconducting properties: delocalised π-electron bonding
- π and π* orbitals form delocalised valence and conduction wavefunctions, which support mobile charge carriers
- Electrons and holes capture: polymer film
- Form neutral bound excited state: Exciton
- Due to confinement, energy difference between singlet and triplet may be large
- Inorganic semiconductors, organic dyes: deposited sublimation or vapor deposition

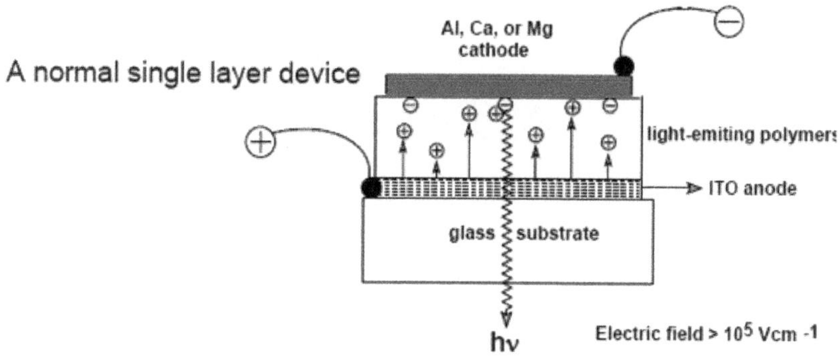

Fluorescent conjugated polymers: deposited from solution by spin-coating or Langmuir Blodgett technique.

Multilayer Devices

Increase efficiency of devices - electron injection has to be significantly boosted.

Electron-conducting/holeblocking (ECHB) layer.

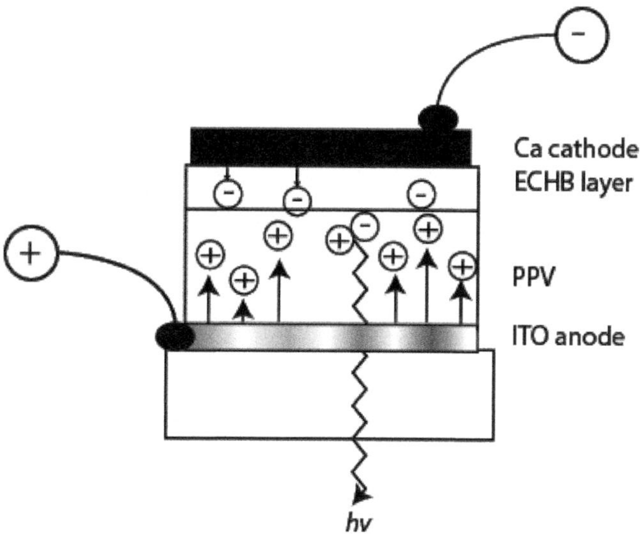

Other Applications

- FOLED: Flexible OLED
- PHOLED: Phosphorescent OLED
- TOLED: Transparent OLED
- SOLED: Stacked OLED
- PMOLED: Passive Matrix OLED
- AMOLED: Active Matrix OLED

Future Research

Solutions for the following:

- Susceptibility towards oxidative degradation
- Lifetimes remains lower
- Photooxidation produces carbonyl defects that quench fluorescence
- OLED is a display device that sandwiches carbon based films between the two electrodes and when voltage is applied creates light
- SMOLED's & LEP's are its technology branches.
- Chemical modifications to the structure can tune the emission over the entire visible region.
- Multilayer devices and dopants also play a role in tuning emission.

Phosphorescent organic light-emitting diode

Phosphorescent organic light-emitting diodes (PHOLED) are a type of organic light-emitting diode (OLED) that use the principle of phosphorescence to obtain higher internal efficiencies than fluorescent OLEDs. This technology is currently under development by many industrial and academic research groups.

Method of Operation

Like all types of OLED, phosphorescent OLEDs emit light due to the electroluminescence of an organic semiconductor layer in an electric current. Electrons and holes are injected into the organic layer at the electrodes and form excitons, a bound state of the electron and hole.

Electrons and holes are both fermions with half integer spin. An exciton formed by the recombination of two such particles may either be in a singlet state or a triplet state, depending on how the spins have been combined. Statistically, there is a 25% probability of forming a singlet state and

75% probability of forming a triplet state. Decay of the excitons results in the production of light through spontaneous emission.

In OLEDs using fluorescent organic molecules only, the decay of triplet excitons is quantum mechanically forbidden by selection rules, meaning that the lifetime of triplet excitons is long and phosphorescence is not readily observed. Hence it would be expected that in fluorescent OLEDs only the formation of singlet excitons results in the emission of useful radiation, placing a theoretical limit on the internal quantum efficiency (the percentage of excitons formed that result in emission of a photon) of 25%.

However, phosphorescent OLEDs generate light from both triplet and singlet excitons, allowing the internal quantum efficiency of such devices to reach nearly 100%.

This is commonly achieved by doping a host molecule with an organometallic complex. These contain a heavy metal atom at the centre of the molecule, for example platinum or iridium, of which the green emitting complex $Ir(mppy)_3$ is just one of many examples. The large spin-orbit interaction experienced by the molecule due to this heavy metal atom facilitates intersystem crossing, a process which mixes the singlet and triplet character of excited states. This reduces the lifetime of the triplet state, therefore phosphorescence is readily observed.

Applications

Due to their potentially high level of energy efficiency, even when compared to other OLEDs, PHOLEDs are being studied for potential use in large-screen displays such as computer monitors or television screens, as well as general lighting needs. One potential use of PHOLEDs as lighting devices is to cover walls with large area PHOLED light panels. This would allow entire rooms to glow uniformly, rather than require the use of light bulbs which distribute light unequally throughout a room. The United States Department of Energy has recognized the potential for massive energy savings via the use of this technology and therefore has awarded $200,000 USD in contracts to develop PHOLED products for general lighting applications.

Challenges

One problem that currently hampers the widespread adoption of this highly energy efficient technology is that the average lifetimes of red and green PHOLEDs are often tens of thousands of hours longer than those of blue PHOLEDs. This may cause displays to become visually distorted much sooner than would be acceptable for a commercially viable device.

References

- Narendran, N.; Y. Gu (2005). "Life of LED-based white light sources". IEEE/OSA Journal of Display Technology. 1 (1): 167–171. Bibcode:2005JDisT...1..167N. doi:10.1109/JDT.2005.852510
- Lee, Thomas H. (2004). The design of CMOS radio-frequency integrated circuits. Cambridge University Press. p. 20. ISBN 0-521-83539-9
- Wolinsky, Howard (February 5, 2005). "U. of I.'s Holonyak out to take some of Edison's luster". Chicago Sun-Times. Retrieved July 29, 2007
- Moreno, I.; Sun, C. C. (2008). "Modeling the radiation pattern of LEDs". Optics Express. 16 (3): 1808–1819. doi:10.1364/OE.16.001808. PMID 18542260

- Bardsley, J. N. (2004). "International OLED Technology Roadmap". IEEE Journal of Selected Topics in Quantum Electronics. 10: 3–4. doi:10.1109/JSTQE.2004.824077
- Mueller, Gerd (2000) Electroluminescence I, Academic Press, ISBN 0-12-752173-9, p. 67, "escape cone of light" from semiconductor, illustrations of light cones on p. 69
- Zheludev, N. (2007). "The life and times of the LED: a 100-year history" (free-download PDF). Nature Photonics. 1 (4): 189–192. Bibcode:2007NaPho...1..189Z. doi:10.1038/nphoton.2007.34
- Luginbuhl, C. (2014). "The impact of light source spectral power distribution on sky glow". Journal of Quantitative Spectroscopy and Radiative Transfer. 139: 21–26. doi:10.1016/j.jqsrt.2013.12.004
- Lipták, Bela G. (2005) Instrument Engineers' Handbook: Process control and optimization, CRC Press, ISBN 0-8493-1081-4 p. 537, "cone of light" in context of optical fibers
- Taniyasu, Y.; Kasu, M.; Makimoto, T. (2006). "An aluminium nitride light-emitting diode with a wavelength of 210 nanometres". Nature. 441 (7091): 325–328. doi:10.1038/nature04760. PMID 16710416
- Chang, Yi-Lu; Lu, Zheng-Hong (2013). "White Organic Light-Emitting Diodes for Solid-State Lighting". Journal of Display Technology. PP (99): 1. Bibcode:2013JDisT...9..459C. doi:10.1109/JDT.2013.2248698
- Elektrotechnik Gesamtband Technische Mathematik Kommunikationselektronik (in German) (1st ed.). Westermann. 1997. p. 171. ISBN 3142212515
- Bessho, M; Shimizu, K (2012). "Latest trends in LED lighting". Electronics and Communications in Japan. 95 (1): 1–7. doi:10.1002/ecj.10394
- Kallmann, H.; Pope, M. (1960). "Positive Hole Injection into Organic Crystals". The Journal of Chemical Physics. 32: 300. Bibcode:1960JChPh..32..300K. doi:10.1063/1.1700925
- Heeger, A. J. (1993) in W. R. Salaneck, I. Lundstrom, B. Ranby, Conjugated Polymers and Related Materials, Oxford, 27–62. ISBN 0-19-855729-9
- "GaN-based blue light emitting device development by Akasaki and Amano" (PDF). Takeda Award 2002 Achievement Facts Sheet. The Takeda Foundation. April 5, 2002. Retrieved November 28, 2007
- K. Lehovec; C. A. Accardo; AND E. Jamgochian (1951). "Injected Light Emission of Silicon Carbide Crystals" (free-download HTML). Physical Review. 83 (3): 603–607. doi:10.1103/PhysRev.83.603
- Mark, Peter; Helfrich, Wolfgang (1962). "Space-Charge-Limited Currents in Organic Crystals". Journal of Applied Physics. 33: 205. Bibcode:1962JAP....33..205M. doi:10.1063/1.1728487
- "Luminus Products CST-90 Series Datasheet" (PDF). Luminus Devices. Archived from the original (PDF) on 2010-03-31. Retrieved October 25, 2009.

Permissions

All chapters in this book are published with permission under the Creative Commons Attribution Share Alike License or equivalent. Every chapter published in this book has been scrutinized by our experts. Their significance has been extensively debated. The topics covered herein carry significant information for a comprehensive understanding. They may even be implemented as practical applications or may be referred to as a beginning point for further studies.

We would like to thank the editorial team for lending their expertise to make the book truly unique. They have played a crucial role in the development of this book. Without their invaluable contributions this book wouldn't have been possible. They have made vital efforts to compile up to date information on the varied aspects of this subject to make this book a valuable addition to the collection of many professionals and students.

This book was conceptualized with the vision of imparting up-to-date and integrated information in this field. To ensure the same, a matchless editorial board was set up. Every individual on the board went through rigorous rounds of assessment to prove their worth. After which they invested a large part of their time researching and compiling the most relevant data for our readers.

The editorial board has been involved in producing this book since its inception. They have spent rigorous hours researching and exploring the diverse topics which have resulted in the successful publishing of this book. They have passed on their knowledge of decades through this book. To expedite this challenging task, the publisher supported the team at every step. A small team of assistant editors was also appointed to further simplify the editing procedure and attain best results for the readers.

Apart from the editorial board, the designing team has also invested a significant amount of their time in understanding the subject and creating the most relevant covers. They scrutinized every image to scout for the most suitable representation of the subject and create an appropriate cover for the book.

The publishing team has been an ardent support to the editorial, designing and production team. Their endless efforts to recruit the best for this project, has resulted in the accomplishment of this book. They are a veteran in the field of academics and their pool of knowledge is as vast as their experience in printing. Their expertise and guidance has proved useful at every step. Their uncompromising quality standards have made this book an exceptional effort. Their encouragement from time to time has been an inspiration for everyone.

The publisher and the editorial board hope that this book will prove to be a valuable piece of knowledge for students, practitioners and scholars across the globe.

Index